T0190150

Developments in Mathematics

Volume 53

Series editors

Krishnaswami Alladi, Gainesville, USA
Hershel M. Farkas, Jerusalem, Israel

More information about this series at http://www.springer.com/series/5834

Guorong Wang · Yimin Wei
Sanzheng Qiao

Generalized Inverses: Theory and Computations

Science Press
Beijing

Springer

Guorong Wang
Department of Mathematics
Shanghai Normal University
Shanghai
China

Sanzheng Qiao
Department of Computing and Software
McMaster University
Hamilton, ON
Canada

Yimin Wei
Department of Mathematics
Fudan University
Shanghai
China

ISSN 1389-2177 ISSN 2197-795X (electronic)
Developments in Mathematics
ISBN 978-981-13-4340-7 ISBN 978-981-13-0146-9 (eBook)
https://doi.org/10.1007/978-981-13-0146-9

Jointly Published with Science Press, Beijing, China

The print edition is not for sale in China Mainland. Customers from China Mainland please order the print book from: Science Press, Beijing, China.

Mathematics Subject Classification (2010): 15A09, 65Fxx, 47A05

Printed on acid-free paper

This Springer imprint is published by the registered company Springer Nature Singapore Pte Ltd. part of Springer Nature
The registered company address is: 152 Beach Road, #21-01/04 Gateway East, Singapore 189721, Singapore

Preface

The concept of the generalized inverses was first introduced by I. Fredholm [81] in 1903. He proposed a generalized inverse of an integral operator, called pseudoinverse. The generalized inverses of differential operators were brought up in D. Hilbert's [107] discussion of the generalized Green's functions in 1904. For a history of the generalized inverses of differential operators, the reader is referred to W. Reid's paper [189] in 1931.

The generalized inverse of a matrix was first introduced by E. H. Moore [166] in 1920, where a unique generalized inverse by means of projectors of matrices is defined. Little was done in the next 30 years until the mid-1950s, when the discoveries of the least-squares properties of certain generalized inverses and the relationship of the generalized inverses to solutions of linear systems brought new interests in the subject. In particular, R. Penrose [174] showed in 1955 that Moore's inverse is the unique matrix satisfying four matrix equations. This important discovery revived the study of the generalized inverses. In honor of Moore and Penrose's contribution, this unique generalized inverse is called the Moore–Penrose inverse.

The theory, applications, and computational methods for the generalized inverses have been developing rapidly during the last 50 years. One milestone is the publication of several books and monographs [9, 19, 92, 187] on the subject in 1970s. Particularly, the excellent volume by Ben-Israel and Greville [9] has made a long-lasting impact on the subject. The other milestone is the publications of the two volumes of proceedings. The first, edited by M. Z. Nashed, is the volume of the proceedings [167] of the Advanced Seminar on the Generalized Inverses and Applications held at the University of Wisconsin–Madison in 1973. It is an excellent and extensive survey book. It contains 14 survey papers on the theory, computations and applications of the generalized inverses, and a comprehensive bibliography that includes all related references up to 1975. The other, edited by S. L. Campbell, is the volume of the proceedings [18] of the AMS Regional Conference held in Columbia, South Carolina, in 1976. It is a new survey book consisting of 12 papers on the latest applications of the generalized inverses. The volume describes the developments in the research directions and the types of the generalized inverses since the mid-1970s. Prior to this period, due to the applications in statistics, research often centered in the generalized inverses for solving linear systems and the generalized inverses with the least-squares properties. Recent studies focus on such topics as: infinite-dimensional theory, numerical computation, matrices of special types (Boolean, integral), matrices over algebraic structures other than real or complex field, systems theory, and non-equation solving generalized inverses.

I have been teaching and conducting research in the generalized inverses of matrices since 1976. I gave a course "Generalized Inverses of Matrices" and held many seminars for graduate students majoring in Computational Mathematics in our department. Since 1979, my colleagues, graduated students, and I have obtained a number of results on the generalized inverses in the areas of perturbation theory, condition numbers, recursive algorithms, finite algorithms, embedding algorithms, parallel algorithms, the generalized inverses of rank-r modified matrices and Hessenberg matrices, extensions of the Cramer's rule, and the representation and approximation of the generalized inverses of linear operators. Dozens of papers have been published in refereed journals in China and other countries. They have drawn attention from researchers around the world. I have received letters from more than ten universities in eight countries, USA, Germany, Sweden, etc., requesting papers or seeking academic contacts. Colleagues in China show strong interests and support in our work and request a systematic presentation of our work. With the support of the Academia Sinica Publishing Foundation and the National Natural Science Foundation of China, Science Press published my book "Generalized Inverses of Matrices and Operators" [241] in Chinese in 1994. That book is noticed and well received by researchers and colleagues in China. It has been adopted by several universities as a textbook or reference book for graduate courses. The book was reprinted in 1998.

In order to improve graduate teaching and international academic exchange, I was encouraged to write this English version based on the Chinese version. This English version is not a direct translation of the Chinese version. In addition to the contents in the Chinese version, this book includes the contents from more than 100 papers since 1994. The final product is an entirely new book, while the spirit of the Chinese version still lives. For example, Sects. 2, 3, and 5 of Chap. 3; Sect. 1 of Chap. 6; Sects. 4 and 5 of Chap. 7; Sects. 1, 4, and 5 of Chap. 8, Chaps. 4, 10, and 11 are all new.

Yimin Wei of Fudan University in China and Qiao Sanzheng of McMaster University in Canada were two of my former excellent students. They have made many achievements in the area of the generalized inverses and are recognized internationally. I would not be able to finish this book without their cooperation.

We would like to thank A. Ben-Israel, Jianmin Miao of Rutgers University; R. E. Hartwig, S. L. Campbell, and C. D. Meyer, Jr. of North Carolina State University; and C. W. Groetsch of University of Cincinnati. The texts [9], [19], and [92] undoubtedly have had an influence on this book. We also thank Erxiong Jiang of Shanghai University, Zihao Cao of Fudan University, Musheng Wei of East China Normal University and Yonglin Chen of Nanjing Normal

University for their help and advice in the subject for many years, and my doctoral student Yaomin Yu for typing this book.

I would appreciate any comments and corrections from the readers.

Finally, I am indebted to the support by the Graduate Textbook Publishing Foundation of Shanghai Education Committee and Shanghai Normal University.

June 2003   Guorong Wang
                                                                      Shanghai Normal University

Preface to the Second Edition

Since the first publication of the book more than one decade ago, we have witnessed exciting developments in the study of the generalized inverses. We are encouraged by colleagues, Science Press, and Springer to update our book. This edition is the result of their encouragement. To include recent developments, this edition has two new chapters on the generalized inverses of special matrices and an updated bibliography. The new chapter six is about the generalized inverses of structured matrices, such as Toeplitz matrix and more general matrices of low displacement rank. It discusses the structure of the generalized inverses of structured matrices and presents efficient algorithms for computing the generalized inverses by exploiting the structure. The new chapter ten studies the generalized inverses of polynomial matrices, that is, matrices whose entries are polynomials. Remarks and references are updated to include recent publications. More than seventy publications are added to the bibliography.

To Science Press and Springer, we are grateful for their encouragement of the publication of this new edition. We would like to thank the reviewers for their constructive comments, which helped us improve the presentation and readability of the book.

Also, we would like to thank National Natural Science Foundation of China under grant 11171222 for supporting Wang Guorong, International Cooperation Project of Shanghai Municipal Science and Technology Commission under grant 16510711200 for supporting Yimin Wei and Sanzheng Qiao, National Natural Science Foundation of China under grant 11771099 and Key Laboratory of Mathematics for Nonlinear Science of Fudan University for supporting Yimin Wei, and Natural Science and Engineering Council of Canada under grant RGPIN-2014-04252 for supporting Sanzheng Qiao.

Shanghai, China	Guorong Wang
Shanghai, China	Yimin Wei
Hamilton, Canada	Sanzheng Qiao
November 2017	Shanghai Normal University

Contents

Notations

Matrices: For the matrices A and B, and the indices α and β

I	Identity matrix
A^T	Transpose of A
A^*	Complex conjugate and transpose of A
$A^{\#}$	Weighted conjugate transpose of A
A^{-1}	Inverse of A
$A^{(1)}$	$\{1\}$-inverses of A
$A^{(1,3)}, A^{(1,3M)}$	$\{1,3\}$-, $\{1,3M\}$-inverses of A
$A^{(1,4)}, A^{(1,4N)}$	$\{1,4\}$-, $\{1,4N\}$-inverses of A
$A_{T,S}^{(1,2)}$	$\{1,2\}$-inverse of A with prescribed range T and null space S
$A_{T,S}^{(2)}$	$\{2\}$-inverse of A with prescribed range T and null space S
$A^{\dagger}$	Moore–Penrose inverse of A
$A_{MN}^{\dagger}$	Weighted Moore–Penrose inverse of A
A_d	Drazin inverse of A
A_g	Group inverse of A
$A_{d,W}$	W-weighted Drazin inverse of A
$A_{(L)}^{(-1)}$	Bott–Duffin inverse of A
$A_{(L)}^{(\dagger)}$	Generalized Bott–Duffin inverse of A
$A[\alpha, \beta]$ or $A_{\alpha\beta}$	Submatrix of A having row indices α and column indices β
$A[\alpha]$ or A_α	Submatrix $A_{\alpha\alpha}$ of A
$A[\alpha, *]$ or $A_{\alpha *}$	Submatrix of A consisting of the rows indexed by α
$A[*, \beta]$ or $A_{*\beta}$	Submatrix of A consisting of the columns indexed by β
$A[\alpha', \beta']$	Submatrix obtained from A by deleting the rows indexed by α and the columns indexed by β
$A[\alpha']$	Submatrix $A[\alpha', \alpha']$ of A
$\mathrm{adj}(A)$	Adjoint matrix of A

$C_k(A)$ kth compound matrix of A

$A(j \to \mathbf{b})$ Matrix obtained from A by replacing the jth column with the vector $\mathbf{b}$

$A(\mathbf{d}^{\mathrm{T}} \leftarrow i)$ Matrix obtained from A by replacing the ith row with the row vector $\mathbf{d}^{\mathrm{T}}$

$A \otimes B$ Kronecker product of A and B

Sets and Spaces: For the matrices A and B

$\mathcal{N}(A)$ Null space of A

$\mathcal{N}_c(A)$ Subspace complementary to $\mathcal{N}(A)$

$\mathcal{N}(A,B)$ Null space of (A,B)

$\mathcal{R}(A)$ Range of A

$\mathcal{R}_c(A)$ Subspace complementary to $R(A)$

$\mathcal{R}(A,B)$ Range of (A,B)

$\mathbb{R}, \mathbb{C}$ Fields of real, complex numbers

$\mathbb{R}^n, \mathbb{C}^n$ n-dimensional real, complex vector spaces

$\mathbb{R}^{m \times n}, \mathbb{C}^{m \times n}$ Sets of $m \times n$ matrices over $\mathbb{R}, \mathbb{C}$

$\mathbb{R}_r^{m \times n}, \mathbb{C}_r^{m \times n}$ Sets of $m \times n$ matrices of rank r over $\mathbb{R}, \mathbb{C}$

Index sets: For $A \in \mathbb{R}_r^{m \times n}$, and the indices α and β

$Q_{k,n}$ $Q_{k,n} = \{\alpha : \alpha = (\alpha_1, \cdots, \alpha_k), \ 1 \leq \alpha_1 < \cdots < \alpha_k \leq n\}$

$\mathcal{I}(A)$ $\mathcal{I}(A) = \{I \in Q_{r,m} : \mathrm{rank}(A_{I*}) = r\}$

$\mathcal{J}(A)$ $\mathcal{J}(A) = \{J \in Q_{r,n} : \mathrm{rank}(A_{*J}) = r\}$

$\mathcal{B}(A)$ $\mathcal{B}(A) = \mathcal{I}(A) \times \mathcal{J}(A) = \{(I,J) \in Q_{r,m} \times Q_{r,n} : \mathrm{rank}(A_{IJ}) = r\}$

$\mathcal{I}(\alpha)$ $\mathcal{I}(\alpha) = \{I \in \mathcal{I}(A) : \alpha \subset J\}$

$\mathcal{J}(\beta)$ $\mathcal{J}(\beta) = \{J \in \mathcal{J}(A) : \beta \subset J\}$

$\mathcal{B}(\alpha, \beta)$ $\mathcal{B}(\alpha, \beta) = \mathcal{I}(\alpha) \times \mathcal{J}(\beta)$

Miscellaneous: For the matrix A

$\det(A)$ Determinant of A

$\frac{\partial}{\partial |A_{\alpha\beta}|}|A|$ The coefficient of $\det(A_{\alpha\beta})$ in the Laplace expansion of $\det(A)$

$\frac{\partial}{\partial a_{ij}}|A|$ Cofactor of a_{ij}

$\mathrm{Vol}(A)$ Volume of A, $\mathrm{Vol}(A) = \sqrt{\sum_{(I,J) \in \mathcal{N}(A)} det^2(A_{IJ})}$

$\mathrm{rank}(A)$ Rank of A

$\mathrm{null}(A)$ Nullity of A

$\mathrm{Ind}(A)$ Index of A

$\mathrm{tr}(A)$ Trace of A

$\lambda(A)$ Spectrum of A

$\sigma(A)$ Set of singular values of A

$\mu_{MN}(A)$ Set of weighted (M, N) singular values of A, where M and N are Hermitian positive definite matrices

$\rho(A)$ Spectral radius of A

$\kappa(A)$ Condition number with respect to the inverse of A

$\kappa_{MN}(A)$ Condition number with respect to the weighted Moore–Penrose inverse of A

$\kappa_2(A)$ Condition number with respect to the Moore–Penrose inverse of A

$\kappa_d(A)$ Condition number with respect to the Drazin inverse of A

$\dim(L)$ Dimension of a space L

$P_{L,M}$ Projector on a space L along a space M

P_L Orthogonal projector on L along $L^\perp$

p.d.   Positive definite

L-p.d. L-positive definite

p.s.d.   Positive semi-definite

L-p.s.d. L-positive semi-definite

$\|\cdot\|_p$ ℓ_p-norm

Chapter 1
Equation Solving Generalized Inverses

There are various ways to introduce the generalized inverses. We introduce them by considering the problem of solving systems of linear equations. Various generalized inverses are introduced in terms of solving various systems of linear equations inconsistent and consistent systems. We also show how the generalized inverses can be used for expressing solution of a matrix equation, common solution of two systems of linear equations, and common solution of two matrix equations.

1.1 Moore-Penrose Inverse

In this section, the Moore-Penrose inverse is introduced. Its definition and some basic properties are given in Sect. 1.1.1. Before establishing a relation between the Moore-Penrose inverse and the full-rank factorization in Sect. 1.1.3, we briefly review the concept of the range and null space of a matrix and some properties of the matrix rank in Sect. 1.1.2. Finally, Sect. 1.1.4 shows how the Moore-Penrose inverse plays a role in finding the minimum-norm least-squares solution of an inconsistent system of linear equations.

Let $\mathbb{C}$ ($\mathbb{R}$) be the field of complex (real) numbers, $\mathbb{C}^n$ ($\mathbb{R}^n$) the vector space of n-tuples of complex (real) numbers over $\mathbb{C}$ ($\mathbb{R}$), $\mathbb{C}^{m \times n}$ ($\mathbb{R}^{m \times n}$) the class of $m \times n$ complex (real) matrices, $\mathbb{C}_r^{m \times n}$ ($\mathbb{R}_r^{m \times n}$) the class of $m \times n$ complex (real) matrices of rank r, and $\mathcal{R}(A) = \{\mathbf{y} \in \mathbb{C}^m : \mathbf{y} = A\mathbf{x}, \mathbf{x} \in \mathbb{C}^n\}$ the range of $A \in \mathbb{C}^{m \times n}$. It is well known that for every nonsingular matrix $A \in \mathbb{C}_n^{n \times n}$ there exists a unique matrix $X \in \mathbb{C}_n^{n \times n}$ satisfying

$$AX = I_n \quad \text{and} \quad XA = I_n,$$

where I_n is the identity matrix of order n. This X is called the inverse of A, denoted by $X = A^{-1}$. The nonsingular system of linear equations

$$A\mathbf{x} = \mathbf{b} \quad (A \in \mathbb{C}_n^{n \times n}, \mathbf{b} \in \mathbb{C}^n)$$

© Springer Nature Singapore Pte Ltd. and Science Press 2018
G. Wang et al., *Generalized Inverses: Theory and Computations*,
Developments in Mathematics 53, https://doi.org/10.1007/978-981-13-0146-9_1

has a unique solution

$$\mathbf{x} = A^{-1}\mathbf{b}.$$

In general, A may be singular or rectangular, the system may have no solution or multiple solutions. Specifically, the consistent system of linear equations

$$A\mathbf{x} = \mathbf{b} \quad (A \in \mathbb{C}^{m \times n}, \ m < n, \ \mathbf{b} \in \mathcal{R}(A)) \tag{1.1.1}$$

has many solutions. Whereas, the inconsistent system of linear equations

$$A\mathbf{x} = \mathbf{b} \quad (A \in \mathbb{C}^{m \times n}, \ \mathbf{b} \notin \mathcal{R}(A)) \tag{1.1.2}$$

has no solution, however, it has a least-squares solution.

Can we find a suitable matrix X, such that $\mathbf{x} = X\mathbf{b}$ is some kind of solution of the general system $A\mathbf{x} = \mathbf{b}$? This X is called the equation solving generalized inverse. A generalized inverse should reduce to the regular inverse A^{-1} when A is nonsingular. The Moore-Penrose inverse and the $\{i, j, k\}$ inverses, which will be discussed in Sect. 1.2, are the classes of the generalized inverses.

1.1.1 Definition and Basic Properties of $A^{\dagger}$

Let A^* denote the complex conjugate and transpose of A. In the case when $A \in \mathbb{C}_n^{m \times n}$ is of full column rank, A^*A is a nonsingular matrix of order n and the least-squares solution $\mathbf{x}$ of the overdetermined system of linear equation (1.1.2) can be obtained by solving the following normal equations

$$A^*A\mathbf{x} = A^*\mathbf{b}, \tag{1.1.3}$$

specifically, $\mathbf{x} = (A^*A)^{-1}A^*\mathbf{b}$. Define

$$X = (A^*A)^{-1}A^*.$$

It can be verified that the above defined X is the unique matrix satisfying the following four conditions, known as the Penrose conditions,

(1) $AXA = A$,
(2) $XAX = X$,
(3) $(AX)^* = AX$,
(4) $(XA)^* = XA$.

The matrix X satisfying the above four conditions is called the Moore-Penrose generalized inverse of A, denoted by $A^{\dagger}$. Thus the least-squares solution of (1.1.3) is $\mathbf{x} = A^{\dagger}\mathbf{b}$.

Especially, if $m = n = \text{rank}(A)$, we have

$$A^\dagger = (A^*A)^{-1}A^* = A^{-1}(A^*)^{-1}A^* = A^{-1},$$

showing that the Moore-Penrose inverse $A^\dagger$ reduces to the usual inverse A^{-1} when A is nonsingular.

For a general m-by-n matrix A, we have the following definition.

Definition 1.1.1 Let $A \in \mathbb{C}^{m \times n}$, then the matrix $X \in \mathbb{C}^{n \times m}$ satisfying the Penrose conditions (1)–(4) is called the Moore-Penrose inverse of A, abbreviated as the M-P inverse and denoted by $A^\dagger$.

The following theorem shows that the above defined generalized inverse uniquely exists for any $A \in \mathbb{C}^{m \times n}$.

Theorem 1.1.1 *The generalized inverse X satisfying the Penrose conditions (1)–(4) is existent and unique.*

Proof Let $A \in \mathbb{C}_r^{m \times n}$, then A can be decomposed as $A = Q^*RP$ (see for example [1]), where Q and P are unitary matrices of orders m and n respectively and

$$R = \begin{bmatrix} R_{11} & O \\ O & O \end{bmatrix} \in \mathbb{C}^{m \times n},$$

where R_{11} is a nonsingular upper triangular matrix of order r. Denote

$$R^\dagger = \begin{bmatrix} R_{11}^{-1} & O \\ O & O \end{bmatrix} \in \mathbb{C}^{n \times m},$$

then $X = P^*R^\dagger Q$ satisfies the Penrose conditions (1)–(4). Indeed,

$$AXA = Q^*RPP^*R^\dagger QQ^*RP = Q^*RP = A,$$
$$XAX = P^*R^\dagger QQ^*RPP^*R^\dagger Q = P^*R^\dagger Q = X,$$
$$(AX)^* = (Q^*RPP^*R^\dagger Q)^* = \left(Q^* \begin{bmatrix} I_r & O \\ O & O \end{bmatrix} Q \right)^* = AX,$$
$$(XA)^* = (P^*R^\dagger QQ^*RP)^* = \left(P^* \begin{bmatrix} I_r & O \\ O & O \end{bmatrix} P \right)^* = XA.$$

Therefore, for any $A \in \mathbb{C}_r^{m \times n}$, $X = A^\dagger$ always exists.

The uniqueness of X is proved as follows. If X_1 and X_2 both satisfy the Penrose conditions (1)–(4), then

$$\begin{aligned}
X_1 &= X_1 A X_1 = X_1 A X_2 A X_1 \\
&= X_1 (A X_2)^* (A X_1)^* = X_1 (A X_1 A X_2)^* \\
&= X_1 (A X_2)^* = X_1 A X_2 \\
&= X_1 A X_2 A X_2 = (X_1 A)^* (X_2 A)^* X_2 \\
&= (X_2 A X_1 A)^* X_2 = (X_2 A)^* X_2 \\
&= X_2 A X_2 = X_2.
\end{aligned}$$

This completes the proof.   □

Now that we have shown the existence and uniqueness of the M-P inverse, we list some of its properties.

Theorem 1.1.2 *Let* $A \in \mathbb{C}^{m \times n}$, *then*

(1) $(A^\dagger)^\dagger = A$;

(2) $(\lambda A)^\dagger = \lambda^\dagger A^\dagger$, *where* $\lambda \in \mathbb{C}$, $\lambda^\dagger = \begin{cases} \lambda^{-1}, & \lambda \neq 0, \\ 0, & \lambda = 0; \end{cases}$

(3) $(A^*)^\dagger = (A^\dagger)^*$;

(4) $(AA^*)^\dagger = (A^*)^\dagger A^\dagger$; $(A^*A)^\dagger = A^\dagger (A^*)^\dagger$;

(5) $A^\dagger = (A^*A)^\dagger A^* = A^* (AA^*)^\dagger$;

(6) $A^* = A^* A A^\dagger = A^\dagger A A^*$;

(7) *If* $\mathrm{rank}(A) = n$, *then* $A^\dagger A = I_n$; *If* $\mathrm{rank}(A) = m$, *then* $AA^\dagger = I_m$;

(8) $(UAV)^\dagger = V^* A^\dagger U^*$, *when* U *and* V *are unitary matrices.*

The above properties can be verified by using Definition 1.1.1. The proof is left as an exercise.

1.1.2   Range and Null Space of a Matrix

Definition 1.1.2 For $A \in \mathbb{C}^{m \times n}$, we denote the range of A as

$$\mathcal{R}(A) = \{\mathbf{y} \in \mathbb{C}^m : \mathbf{y} = A\mathbf{x}, \ \mathbf{x} \in \mathbb{C}^n\}$$

and the null space of A as

$$\mathcal{N}(A) = \{\mathbf{x} \in \mathbb{C}^n : A\mathbf{x} = \mathbf{0}\}.$$

Using the above definitions, we can prove that

$$\mathcal{R}(A)^{\perp} = \mathcal{N}(A^*),$$

where $\mathcal{R}(A)^{\perp}$ is the orthogonal complementary subspace of $\mathcal{R}(A)$, i.e., the set of all the vectors in $\mathbb{C}^m$ which are orthogonal to every vector in $\mathcal{R}(A)$. Every vector $\mathbf{x} \in \mathbb{C}^m$ can be expressed uniquely as the sum:

$$\mathbf{x} = \mathbf{y} + \mathbf{z}, \quad \mathbf{y} \in \mathcal{R}(A), \ \mathbf{z} \in \mathcal{R}(A)^{\perp}.$$

Theorem 1.1.3 *The basic properties of the range and null space:*

(1) $\mathcal{R}(A) = \mathcal{R}(AA^{\dagger}) = \mathcal{R}(AA^*)$;
(2) $\mathcal{R}(A^{\dagger}) = \mathcal{R}(A^*) = \mathcal{R}(A^{\dagger}A) = \mathcal{R}(A^*A)$;
(3) $\mathcal{R}(I - A^{\dagger}A) = \mathcal{N}(A^{\dagger}A) = \mathcal{N}(A) = \mathcal{R}(A^*)^{\perp}$;
(4) $\mathcal{R}(I - AA^{\dagger}) = \mathcal{N}(AA^{\dagger}) = \mathcal{N}(A^{\dagger}) = \mathcal{N}(A^*) = \mathcal{R}(A)^{\perp}$;
(5) $\mathcal{R}(AB) = \mathcal{R}(A) \Leftrightarrow \operatorname{rank}(AB) = \operatorname{rank}(A)$;
(6) $\mathcal{N}(AB) = \mathcal{N}(B) \Leftrightarrow \operatorname{rank}(AB) = \operatorname{rank}(B)$.

The proof is left as an exercise.
    The following properties of rank are used in this book.

Lemma 1.1.1 *Let $A \in \mathbb{C}^{m \times n}$, $E_A = I_m - AA^{\dagger}$, and $F_A = I_n - A^{\dagger}A$, then*

(1) $\operatorname{rank}(A) = \operatorname{rank}(A^{\dagger}) = \operatorname{rank}(A^{\dagger}A) = \operatorname{rank}(AA^{\dagger})$;
(2) $\operatorname{rank}(A) = m - \operatorname{rank}(E_A)$, $\operatorname{rank}(A) = n - \operatorname{rank}(F_A)$;
(3) $\operatorname{rank}(AA^*) = \operatorname{rank}(A) = \operatorname{rank}(A^*A)$.

The proof is left as an exercise.

1.1.3 Full-Rank Factorization

The columns of a full column rank matrix form a basis for the range of the matrix. Likewise, the rows of a full row rank matrix form a basis for the space spanned by the rows of the matrix. In this subsection, we show that a non-null matrix that is of neither full column rank nor full row rank can be expressed as a product of a matrix of full column rank and a matrix of full row rank. We call a factorization with the above property the full-rank factorization of a non-null matrix. This factorization turns out to be a powerful tool in the study of the generalized inverses.

Theorem 1.1.4 *Let $A \in \mathbb{C}_r^{m \times n}$, $r > 0$, then there exist matrices $F \in \mathbb{C}_r^{m \times r}$ and $G \in \mathbb{C}_r^{r \times n}$ such that*

$$A = FG.$$

Proof Let $A = [\mathbf{a}_1, \mathbf{a}_2, \cdots, \mathbf{a}_n]$ and F be any matrix whose columns form a basis for $\mathcal{R}(A)$, then $F = [\mathbf{f}_1, \mathbf{f}_2, \cdots, \mathbf{f}_r] \in \mathbb{C}_r^{m \times r}$ and every column $\mathbf{a}_i$ of A is uniquely representable as a linear combination of the columns of F:

$$\mathbf{a}_i = g_{1i}\mathbf{f}_1 + g_{2i}\mathbf{f}_2 + \cdots + g_{ri}\mathbf{f}_r, \quad i = 1, 2, \ldots, n.$$

Hence

$$A = [\mathbf{a}_1, \mathbf{a}_2, \cdots, \mathbf{a}_n]$$

$$= [\mathbf{f}_1, \mathbf{f}_2, \cdots, \mathbf{f}_r]
\begin{bmatrix}
g_{11} & g_{12} & \cdots & g_{1n} \\
g_{21} & g_{22} & \cdots & g_{2n} \\
\vdots & \vdots & \cdots & \vdots \\
g_{r1} & g_{r2} & \cdots & g_{rn}
\end{bmatrix}$$

$$\equiv FG.$$

The matrix $G \in \mathbb{C}_r^{r \times n}$ is uniquely determined. Obviously, $\text{rank}(G) \leq r$. On the other hand,

$$\text{rank}(G) \geq \text{rank}(FG) = r.$$

Thus $\text{rank}(G) = r$.   □

Let $A = FG$ be a full-rank factorization of A and $C \in \mathbb{C}_r^{r \times r}$, then

$$A = (FC)(C^{-1}G) \equiv F_1 G_1,$$

which is also a full-rank factorization of A. This shows that the full-rank factorization of A is not unique. A practical algorithm for the full-rank factorization is given in Chap. 5. MacDuffe [2] pointed out that a full-rank factorization of A leads to an explicit formula for the M-P inverse $A^\dagger$ of A.

Theorem 1.1.5 *Let $A \in \mathbb{C}_r^{m \times n}$, $r > 0$, and its full-rank factorization $A = FG$, then*

$$A^\dagger = G^*(F^*AG^*)^{-1}F^* = G^*(GG^*)^{-1}(F^*F)^{-1}F^*. \tag{1.1.4}$$

Proof First we show that F^*AG^* is nonsingular. From $A = FG$,

$$F^*AG^* = (F^*F)(GG^*),$$

and F^*F and GG^* are r-by-r matrices. Also by Lemma 1.1.1, both are of rank r. Thus F^*AG^* is the product of two nonsingular matrices, therefore F^*AG^* is nonsingular and

$$(F^*AG^*)^{-1} = (GG^*)^{-1}(F^*F)^{-1}.$$

Let

$$X = G^*(GG^*)^{-1}(F^*F)^{-1}F^*,$$

the right side of (1.1.4). Using the above expression for X, it is easy to verify that X satisfies the Penrose conditions (1)–(4). By the uniqueness of the M-P inverse $A^\dagger$, (1.1.4) is therefore established. ☐

1.1.4 Minimum-Norm Least-Squares Solution

In Sect. 1.1.1, we introduced the definition of the Moore-Penrose inverse by considering the least-squares solution of an inconsistent system of linear equations, where the coefficient matrix is of full column rank. In this subsection, we consider the problem of solving a general inconsistent system of linear equations and its relation with the Moore-Penrose inverse.

Let $\mathbf{x} = [x_1, x_2, \cdots, x_p]^* \in \mathbb{C}^p$, then

$$\|\mathbf{x}\|_2 = \left(\sum_{i=1}^p |x_i|^2 \right)^{1/2} = (\mathbf{x}^*\mathbf{x})^{1/2}$$

is the 2-norm of $\mathbf{x}$. For simplicity, we set $\|\mathbf{x}\| = \|\mathbf{x}\|_2$.

If $\mathbf{u}, \mathbf{v} \in \mathbb{C}^p$ and $(\mathbf{u}, \mathbf{v}) = 0$, i.e., $\mathbf{u}$ and $\mathbf{v}$ are orthogonal, then

$$\|\mathbf{u} + \mathbf{v}\|^2 = (\mathbf{u} + \mathbf{v}, \mathbf{u} + \mathbf{v}) = (\mathbf{u}, \mathbf{u}) + (\mathbf{v}, \mathbf{u}) + (\mathbf{u}, \mathbf{v}) + (\mathbf{v}, \mathbf{v})$$
$$= \|\mathbf{u}\|^2 + \|\mathbf{v}\|^2,$$

which is the Pythagorean theorem.

Considering the problem of finding a solution $\mathbf{x}$ for the general system of linear equations (1.1.2):

$$A\mathbf{x} = \mathbf{b} \quad (A \in \mathbb{C}^{m \times n}, \mathbf{b} \notin \mathcal{R}(A)),$$

we look for an $\mathbf{x}$ minimizing the residual $\|A\mathbf{x} - \mathbf{b}\|$.

Definition 1.1.3 Let $A \in \mathbb{C}^{m \times n}$ and $\mathbf{b} \in \mathbb{C}^m$, then a vector $\mathbf{u} \in \mathbb{C}^n$ is called a least-squares solution of $A\mathbf{x} = \mathbf{b}$ if $\|A\mathbf{u} - \mathbf{b}\| \leq \|A\mathbf{v} - \mathbf{b}\|$ for all $\mathbf{v} \in \mathbb{C}^n$.

A system of linear equations may have many least-squares solutions. In many applications, the least-squares solution with minimum norm is of interest.

Definition 1.1.4 Let $A \in \mathbb{C}^{m \times n}$ and $\mathbf{b} \in \mathbb{C}^m$, then a vector $\mathbf{u} \in \mathbb{C}^n$ is called the minimum-norm least-squares solution of $A\mathbf{x} = \mathbf{b}$ if $\mathbf{u}$ is a least-squares solution of $A\mathbf{x} = \mathbf{b}$ and $\|\mathbf{u}\| < \|\mathbf{w}\|$ for any other least-squares solution $\mathbf{w}$.

When $\mathbf{b} \in \mathcal{R}(A)$, the system of linear equations $A\mathbf{x} = \mathbf{b}$ is consistent, then the solution and the least-squares solution of $A\mathbf{x} = \mathbf{b}$ obviously coincide.

The next theorem shows the relation between the minimum-norm least-squares solution of (1.1.2) and the M-P inverse $A^\dagger$.

Theorem 1.1.6 *Let $A \in \mathbb{C}^{m \times n}$ and $\mathbf{b} \in \mathbb{C}^m$, then $A^\dagger \mathbf{b}$ is the minimum-norm least-squares solution of* (1.1.2).

Proof Let $\mathbf{b} = \mathbf{b}_1 + \mathbf{b}_2$, where

$$\mathbf{b}_1 = AA^\dagger \mathbf{b} \in \mathcal{R}(A) \quad \text{and} \quad \mathbf{b}_2 = (I - AA^\dagger)\mathbf{b} \in \mathcal{R}(A)^\perp,$$

then $A\mathbf{x} - \mathbf{b}_1 \in \mathcal{R}(A)$ and

$$\|A\mathbf{x} - \mathbf{b}\|^2 = \|A\mathbf{x} - \mathbf{b}_1 + (-\mathbf{b}_2)\|^2 = \|A\mathbf{x} - \mathbf{b}_1\|^2 + \|\mathbf{b}_2\|^2.$$

Thus $\mathbf{x}$ is a least-squares solution if and only if $\mathbf{x}$ is a solution of the consistent system $A\mathbf{x} = AA^\dagger \mathbf{b}$. It is obvious that $A^\dagger \mathbf{b}$ is a particular solution. From Theorem 1.1.3,

$$\mathcal{N}(A) = \{(I - A^\dagger A)\mathbf{h} : \mathbf{h} \in \mathbb{C}^n\},$$

thus the general solution of the consistent system $A\mathbf{x} = AA^\dagger \mathbf{b}$ is given by

$$\mathbf{x} = A^\dagger \mathbf{b} + (I - A^\dagger A)\mathbf{h}, \quad \mathbf{h} \in \mathbb{C}^n.$$

Since

$$\|A^\dagger \mathbf{b}\|^2 < \|A^\dagger \mathbf{b}\|^2 + \|(I - A^\dagger A)\mathbf{h}\|^2$$
$$= \|A^\dagger \mathbf{b} + (I - A^\dagger A)\mathbf{h}\|^2, \quad (I - A^\dagger A)\mathbf{h} \neq \mathbf{0},$$

$\mathbf{x} = A^\dagger \mathbf{b}$ is the minimum-norm least-squares solution of (1.1.2). $\square$

In some applications, the minimality of a least-squares solution is important, in others it is not important. If the minimality is not important, then the next theorem can be very useful.

Theorem 1.1.7 *Let $A \in \mathbb{C}^{m \times n}$ and $\mathbf{b} \in \mathbb{C}^m$, then the following statements are equivalent:*

(1) $\mathbf{u}$ *is a least-squares solution of* $A\mathbf{x} = \mathbf{b}$;
(2) $\mathbf{u}$ *is a solution of* $A\mathbf{x} = AA^\dagger \mathbf{b}$;
(3) $\mathbf{u}$ *is a solution of* $A^* A\mathbf{x} = A^* \mathbf{b}$;
(4) $\mathbf{u}$ *is of the form* $A^\dagger \mathbf{b} + \mathbf{h}$, *where* $\mathbf{h} \in \mathcal{N}(A)$.

Proof From the proof of Theorem 1.1.6, we know that (1), (2) and (4) are equivalent. If (1) holds, then premultiplying $A\mathbf{u} = \mathbf{b}$ with A^* gives (3). On the other hand, premultiplying $A^* A\mathbf{u} = A^* \mathbf{b}$ with $A^{*\dagger}$ gives

$$A\mathbf{u} = AA^\dagger \mathbf{b},$$

consequently,
$$\mathbf{u} = A^\dagger(AA^\dagger\mathbf{b}) + \mathbf{h} = A^\dagger\mathbf{b} + \mathbf{h}, \quad \mathbf{h} \in \mathcal{N}(A).$$

Thus (4) holds.   □

Notice that the equations in statement (3) of Theorem 1.1.7 do not involve $A^\dagger$ and are consistent. They are called the normal equations and play an important role in certain areas of statistics.

Exercises 1.1

1. Prove Theorem 1.1.2.
2. Prove that $\mathcal{R}(A) = \mathcal{N}(A^*)^\perp$.
3. Prove that $\operatorname{rank}(AA^*) = \operatorname{rank}(A) = \operatorname{rank}(A^*A)$.
4. Prove that $\mathcal{R}(AA^*) = \mathcal{R}(A)$, $\mathcal{N}(A^*A) = \mathcal{N}(A)$.
5. Prove that

$$\mathcal{R}(AB) = \mathcal{R}(A) \Leftrightarrow \operatorname{rank}(AB) = \operatorname{rank}(A);$$
$$\mathcal{N}(AB) = \mathcal{N}(B) \Leftrightarrow \operatorname{rank}(AB) = \operatorname{rank}(B).$$

6. Prove Theorem 1.1.3.
7. Show that if $A = FG$ is a full-rank factorization, then

$$A^\dagger = G^\dagger F^\dagger.$$

8. If $\mathbf{a}$ and $\mathbf{b}$ are column vectors, then
   (1) $\mathbf{a}^\dagger = (\mathbf{a}^*\mathbf{a})^\dagger\mathbf{a}^*$;
   (2) $(\mathbf{ab}^*)^\dagger = (\mathbf{a}^*\mathbf{a})^\dagger(\mathbf{b}^*\mathbf{b})^\dagger\mathbf{ba}^*$.
9. Show that $H^\dagger = H$ if and only if $H^* = H$ and $H^2 = H$.
10. If U and V are unitary matrices, show that

$$(UAV)^\dagger = V^*A^\dagger U^*$$

    for any matrix A for which the product UAV is defined.
11. Show that if $A \in \mathbb{C}^{m \times n}$ and $\operatorname{rank}(A) = 1$, then $A^\dagger = \alpha^{-1}A^*$, where $\alpha = \operatorname{tr}(A^*A) = \sum_{i,j} |a_{ij}|^2$.
12. Show that if $X \in \mathbb{C}^{m \times n}$,

$$\mathbf{x}_0 = \begin{bmatrix} 1 \\ 1 \\ \vdots \\ 1 \end{bmatrix} \in \mathbb{C}^m, \quad X_1 = [\mathbf{x}_0 \ X] \in \mathbb{C}^{m \times (n+1)},$$

and

$$\mathbf{b} \in \mathbb{C}^n, \quad \beta_0 \in \mathbb{C}, \quad \mathbf{b}_1 = \begin{bmatrix} \beta_0 \\ \mathbf{b} \end{bmatrix} \in \mathbb{C}^{n+1},$$

then $\mathbf{b}_1$ is a least-squares solution of $X_1\mathbf{b}_1 = \mathbf{y}$ if and only if $\beta_0 = m^{-1}\mathbf{x}_0^*(\mathbf{y} - X\mathbf{b})$ and $\mathbf{b}$ is a least-squares solution of

$$(I - m^{-1}\mathbf{x}_0\mathbf{x}_0^*)X\mathbf{b} = (I - m^{-1}\mathbf{x}_0\mathbf{x}_0^*)\mathbf{y}.$$

1.2   The $\{i, j, k\}$ Inverses

We discussed the relations between the minimum-norm least-squares solution of an inconsistent system of linear equations (1.1.2) and the M-P inverse in Sect. 1.1. In this section, we introduce the $\{i, j, k\}$ inverses and their relations with the solution of other linear equations and some matrix equations.

1.2.1   The $\{1\}$ Inverse and the Solution of a Consistent System of Linear Equations

If $A \in \mathbb{C}_n^{n \times n}$, then one of the characteristics of A^{-1} is that for every $\mathbf{b}$, $A^{-1}\mathbf{b}$ is the solution of $A\mathbf{x} = \mathbf{b}$. One might ask: for a general $A \in \mathbb{C}^{m \times n}$, what are the characteristics of a matrix $X \in \mathbb{C}^{n \times m}$ such that $X\mathbf{b}$ is a solution of the consistent system of linear equations (1.1.1)?

If $AX\mathbf{b} = \mathbf{b}$ is true for every $\mathbf{b} \in \mathcal{R}(A)$, it is clear that

$$AXA = A,$$

i.e., the Penrose condition (1) holds. Conversely, suppose X satisfies $AXA = A$. For every $\mathbf{b} \in \mathcal{R}(A)$ there exists an $\mathbf{x}_\mathbf{b} \in \mathbb{C}^n$ such that $A\mathbf{x}_\mathbf{b} = \mathbf{b}$. Therefore $AX\mathbf{b} = AXA\mathbf{x}_\mathbf{b} = A\mathbf{x}_\mathbf{b} = \mathbf{b}$ for every $\mathbf{b} \in \mathcal{R}(A)$. The following theorem is a formal statement of the above argument.

Theorem 1.2.1 *For $A \in \mathbb{C}^{m \times n}$, $X \in \mathbb{C}^{n \times m}$ is a matrix such that $X\mathbf{b}$ is a solution of $A\mathbf{x} = \mathbf{b}$ for every $\mathbf{b} \in \mathcal{R}(A)$ if and only if X satisfies*

$$AXA = A.$$

Definition 1.2.1 A matrix X satisfying the Penrose condition (1) $AXA = A$ is called the equation solving generalized inverse for $AXA = A$ or $\{1\}$ inverse of A and is denoted by $X = A^{(1)}$ or $X \in A\{1\}$, where $A\{1\}$ denotes the set of all $\{1\}$ inverses of A.

1.2.2 The $\{1, 4\}$ Inverse and the Minimum-Norm Solution of a Consistent System

Similar to the $\{1\}$ inverse, we define the $\{1, 4\}$ inverse and show its relation with minimum-norm solution of a consistent system. A consistent system of equations may have many solutions. Suppose we seek $X \in \mathbb{C}^{n \times m}$ such that, in addition to being an equation solving inverse for consistent linear equations (1.1.1), for each $\mathbf{b} \in \mathcal{R}(A)$, it also satisfies $\|X\mathbf{b}\| < \|\mathbf{z}\|$ for all $\mathbf{z} \neq X\mathbf{b}$ and $\mathbf{z} \in \{\mathbf{x} : A\mathbf{x} = \mathbf{b}\}$. That is, for each $\mathbf{b} \in \mathcal{R}(A)$, $X\mathbf{b}$ is the solution with minimal-norm.

Let $A^{(1,4)}$ denote a matrix satisfying the Penrose conditions (1) and (4). If $\mathbf{b} \in \mathcal{R}(A)$, then $AA^{(1,4)}\mathbf{b} = \mathbf{b}$ and the solutions and the least-squares solutions of the consistent system of linear equation (1.1.1) coincide. Therefore the least-squares solutions satisfy

$$A\mathbf{x} = \mathbf{b} = AA^{(1,4)}\mathbf{b}.$$

It is clear that $A^{(1,4)}\mathbf{b}$ is a least-squares solution. Note that $(I - A^{(1,4)}A)\mathbf{h} \in \mathcal{N}(A)$. So a general least-squares solution can be presented by

$$A^{(1,4)}\mathbf{b} + (I - A^{(1,4)}A)\mathbf{h}, \quad \mathbf{h} \in \mathbb{C}^n.$$

Since

$$A^{(1,4)}\mathbf{b} = A^{(1,4)}A\mathbf{y} = (A^{(1,4)}A)^*\mathbf{y} = A^*A^{(1,4)*}\mathbf{y} \in \mathcal{R}(A^*),$$

we have

$$(A^{(1,4)}\mathbf{b}, \; (I - A^{(1,4)}A)\mathbf{h}) = 0.$$

By the Pythagorean theorem, we get

$$\|A^{(1,4)}\mathbf{b}\|^2 \leq \|A^{(1,4)}\mathbf{b}\|^2 + \|(I - A^{(1,4)}A)\mathbf{h}\|^2$$
$$= \|A^{(1,4)}\mathbf{b} + (I - A^{(1,4)}A)\mathbf{h}\|^2.$$

It then follows that $A^{(1,4)}\mathbf{b}$ is the minimum-norm least-squares solution of (1.1.1) and is also the minimum-norm solution of (1.1.1). Therefore, for each $\mathbf{b} \in \mathcal{R}(A)$, we require that $X\mathbf{b} = A^{(1,4)}\mathbf{b}$, that is,

$$XA = A^{(1,4)}A,$$

which is equivalent to the Penrose conditions (1) and (4):

$$AXA = A \quad \text{and} \quad (XA)^* = XA.$$

The next theorem is a formal statement of what we have just shown.

Theorem 1.2.2 *Let $A \in \mathbb{C}^{m \times n}$ and $\mathbf{b} \in \mathcal{R}(A)$, then X is a matrix such that $AX\mathbf{b} = \mathbf{b}$ and $\|X\mathbf{b}\| < \|\mathbf{z}\|$ for all $\mathbf{z} \neq X\mathbf{b}$ and $\mathbf{z} \in \{\mathbf{x} : A\mathbf{x} = \mathbf{b}\}$ if and only if X satisfies*

$$AXA = A, \quad and \quad (XA)^* = XA.$$

Definition 1.2.2 A matrix X satisfying the Penrose conditions (1) and (4) is called the generalized inverse for the minimum-norm solution of the consistent system of linear equations (1.1.1) or the $\{1, 4\}$ inverse of A, and is denoted by $X = A^{(1,4)}$ or $X \in A\{1, 4\}$, where $A\{1, 4\}$ denotes the set of all $\{1, 4\}$ inverses of A.

1.2.3   The $\{1, 3\}$ Inverse and the Least-Squares Solution of An Inconsistent System

For $A \in \mathbb{C}^{m \times n}$, $\mathbf{b} \in \mathbb{C}^m$ and $\mathbf{b} \notin \mathcal{R}(A)$, the vector $A^\dagger \mathbf{b}$ is the minimum-norm least-squares solution of the inconsistent system of linear equations (1.1.2). In some applications, the minimality of the norm of a least-squares solution is not important, one might settle for any least-squares solution without considering the size of its norm.

Suppose $A \in \mathbb{C}^{m \times n}$, let us try to determine the characteristics of a matrix X such that $X\mathbf{b}$ is a least-squares solution of (1.1.2).

Let $\mathbf{b} = \mathbf{b}_1 + \mathbf{b}_2$, where $\mathbf{b}_1 \in \mathcal{R}(A)$ and $\mathbf{b}_2 \in \mathcal{R}(A)^\perp = \mathcal{N}(A^*)$, and denote $A^{(1,3)}$ as a matrix satisfying the Penrose conditions (1) and (3). Since

$$AA^{(1,3)}\mathbf{b}_1 = AA^{(1,3)}A\mathbf{y} = A\mathbf{y} = \mathbf{b}_1$$

and

$$AA^{(1,3)}\mathbf{b}_2 = (AA^{(1,3)})^*\mathbf{b}_2 = A^{(1,3)^*}A^*\mathbf{b}_2 = \mathbf{0},$$

we have

$$AA^{(1,3)}\mathbf{b} = AA^{(1,3)}A\mathbf{y} = A\mathbf{y} = \mathbf{b}_1.$$

From

$$\begin{aligned} \|AX\mathbf{b} - \mathbf{b}\|^2 &= \|(AX\mathbf{b} - \mathbf{b}_1) + (-\mathbf{b}_2)\|^2 \\ &= \|AX\mathbf{b} - \mathbf{b}_1\|^2 + \|\mathbf{b}_2\|^2, \end{aligned}$$

we can see that $\|AX\mathbf{b} - \mathbf{b}\|$ is minimized when and only when $AX\mathbf{b} = \mathbf{b}_1 = AA^{(1,3)}\mathbf{b}$ for all $\mathbf{b} \in \mathbb{C}^m$. It is clear that

$$AX = AA^{(1,3)},$$

which is equivalent to the Penrose conditions (1) and (3):

$$AXA = A, \quad (AX)^* = AX.$$

The next theorem is a formal statement of the above discussion.

Theorem 1.2.3 *For $A \in \mathbb{C}^{m \times n}$, $\mathbf{b} \in \mathbb{C}^m$, and $\mathbf{b} \notin \mathcal{R}(A)$, the vector $X\mathbf{b}$ is a least-squares solution of* (1.1.2) *if and only if X satisfies*

$$AXA = A \quad and \quad (AX)^* = AX.$$

Definition 1.2.3 A matrix X satisfying the Penrose conditions (1) and (3) is called the generalized inverse for solving the least-squares solution of the inconsistent system of linear equations (1.1.2) or the $\{1, 3\}$ inverse of A, and is denoted by $X = A^{(1,3)}$ or $X \in A\{1, 3\}$, where $A\{1, 3\}$ denotes the set of all the $\{1, 3\}$ inverses of A.

Since $A^{(1,3)}\mathbf{b}$ is a particular least-squares solution and $(I - A^{(1,3)}A)\mathbf{z} \in \mathcal{N}(A)$, for all $\mathbf{z} \in \mathbb{C}^n$, the general solution of the least-squares solution for (1.1.2) is

$$\mathbf{x} = A^{(1,3)}\mathbf{b} + (I - A^{(1,3)}A)\mathbf{z}, \quad \text{where } \mathbf{z} \in \mathbb{C}^n.$$

From Theorems 1.2.1, 1.2.2, and 1.2.3, one can see that each of the different types of X matrices discussed above can be characterized as a set of matrices satisfying some subset of the Penrose conditions (1) to (4). To simplify our nomenclature we make the following definition.

Definition 1.2.4 For any $A \in \mathbb{C}^{m \times n}$, a matrix $X \in \mathbb{C}^{n \times m}$ is called an $\{i, j, k\}$ inverse of A if X satisfies the ith, jth, and kth Penrose conditions, and is denoted by $X = A^{(i,j,k)}$ or $X \in A\{i, j, k\}$, where $A\{i, j, k\}$ denotes the set of all the $\{i, j, k\}$ inverses of A.

As shown above, the important types of the $\{i, j, k\}$ inverses, such as the $\{1, 3\}$ inverse, $\{1, 4\}$ inverse, and M-P inverse, are members of $A\{1\}$. They are all the equation solving generalized inverses of A. Therefore, we give the basic properties of the $\{1\}$ inverse.

Theorem 1.2.4 *For $A \in \mathbb{C}^{m \times n}$,*

(1) $(A^{(1)})^* \in A^*\{1\}$;
(2) $\lambda^\dagger A^{(1)} \in (\lambda A)\{1\}$, $\lambda \in \mathbb{C}$;
(3) $\text{rank}(A^{(1)}) \geq \text{rank}(A)$;
(4) *For nonsingular P and Q, $Q^{-1}A^{(1)}P^{-1} \in (PAQ)\{1\}$;*
(5) *If P is of full column rank, Q is of full row rank, then $Q^{(1)}A^{(1)}P^{(1)} \in (PAQ)\{1\}$;*
(6) *If A is of full column rank, then $A^{(1)}A = I_n$;*
   If A is of full row rank, then $AA^{(1)} = I_m$;
(7) *$AA^{(1)}$ and $A^{(1)}A$ are idempotent, and $\text{rank}(AA^{(1)}) = \text{rank}(A) = \text{rank}(A^{(1)}A)$;*
(8) *If A is nonsingular, then $A^{(1)} = A^{-1}$;*
(9) *If $A^* = A$, then there exists a matrix $X \in A\{1\}$ and $X^* = X$;*
(10) $\mathcal{R}(AA^{(1)}) = \mathcal{R}(A)$, $\mathcal{N}(A^{(1)}A) = \mathcal{N}(A)$, $\mathcal{R}((A^{(1)}A)^*) = \mathcal{R}(A^*)$.

The proof is left as an exercise.

1.2.4   The {1} Inverse and the Solution of the Matrix Equation $AXB = D$

From the above discussion, we have seen that all the important types of equation solving inverses are the {1} inverse. For the rest of this section, we focus our discussion on the {1} inverse as an equation solving inverse. In this subsection, we consider the problem of solving the matrix equation $AXB = D$.

Theorem 1.2.5 *Let $A \in \mathbb{C}^{m \times n}$, $B \in \mathbb{C}^{p \times q}$, and $D \in \mathbb{C}^{m \times q}$, then the matrix equation*

$$AXB = D \tag{1.2.1}$$

is consistent if and only if for some $A^{(1)}$ and $B^{(1)}$,

$$AA^{(1)}DB^{(1)}B = D. \tag{1.2.2}$$

In which case the general solution is

$$X = A^{(1)}DB^{(1)} + Y - A^{(1)}AYBB^{(1)} \tag{1.2.3}$$

for any $Y \in \mathbb{C}^{n \times p}$.

Proof If (1.2.2) holds, then $X = A^{(1)}DB^{(1)}$ is a solution of (1.2.1). Conversely, if X is any solution of (1.2.1), then

$$D = AXB = AA^{(1)}AXBB^{(1)}B = AA^{(1)}DB^{(1)}B.$$

Moreover, it follows from (1.2.2) and the definitions of $A^{(1)}$ and $B^{(1)}$ that every matrix X of the form (1.2.3) satisfies (1.2.1). On the other hand, let X be any solution of (1.2.1), then clearly

$$X = A^{(1)}DB^{(1)} + X - A^{(1)}AXBB^{(1)}$$

which is of the form (1.2.3). $\qquad\qquad\Box$

In the special case of ordinary system of linear equations, Theorem 1.2.5 gives:

Corollary 1.2.1 *Let $A \in \mathbb{C}^{m \times n}$, $\mathbf{b} \in \mathbb{C}^m$, then the system of linear equations $A\mathbf{x} = \mathbf{b}$ is consistent if and only if for some $A^{(1)}$,*

$$AA^{(1)}\mathbf{b} = \mathbf{b}.$$

In which case the general solution of $A\mathbf{x} = \mathbf{b}$ is

$$\mathbf{x} = A^{(1)}\mathbf{b} + (I - A^{(1)}A)\mathbf{y}, \ \forall \, \mathbf{y} \in \mathbb{C}^n. \tag{1.2.4}$$

1.2.5 The {1} Inverse and the Common Solution of $Ax = a$ and $Bx = b$

Let $A \in \mathbb{C}^{m \times n}$, $B \in \mathbb{C}^{p \times n}$, $\mathbf{a} \in \mathbb{C}^m$, and $\mathbf{b} \in \mathbb{C}^p$, we consider the two systems

$$A\mathbf{x} = \mathbf{a} \quad \text{and} \quad B\mathbf{x} = \mathbf{b} \tag{1.2.5}$$

of linear equations. The problem is to find an $\mathbf{x} \in \mathbb{C}^n$ satisfying the two systems in (1.2.5) simultaneously. It is clearly equivalent to solving the partitioned system

$$\begin{bmatrix} A \\ B \end{bmatrix} \mathbf{x} = \begin{bmatrix} \mathbf{a} \\ \mathbf{b} \end{bmatrix}. \tag{1.2.6}$$

First of all, we find a {1} inverse of the above partitioned matrix.

Theorem 1.2.6 *Let $A \in \mathbb{C}^{m \times n}$ and $B \in \mathbb{C}^{p \times n}$, then a {1} inverse for the row partitioned matrix*

$$M = \begin{bmatrix} A \\ B \end{bmatrix}$$

is given by

$$X_M = [Y \ Z] \tag{1.2.7}$$

where

$$Y = (I - (I - A^{(1)}A)(B(I - A^{(1)}A))^{(1)}B)A^{(1)}$$

and

$$Z = (I - A^{(1)}A)(B(I - A^{(1)}A))^{(1)}.$$

Let $C \in \mathbb{C}^{m \times r}$, then a {1} inverse for the column partitioned matrix

$$N = [A \ C]$$

is given by

$$X_N = \begin{bmatrix} A^{(1)}(I - C((I - AA^{(1)})C)^{(1)}(I - AA^{(1)})) \\ ((I - AA^{(1)})C)^{(1)}(I - AA^{(1)}) \end{bmatrix}. \tag{1.2.8}$$

Proof From (1.2.7), we have

$$MX_M M = \begin{bmatrix} A \\ B \end{bmatrix} [Y \ Z] \begin{bmatrix} A \\ B \end{bmatrix}$$

$$= \begin{bmatrix} A \\ B \end{bmatrix} (A^{(1)}A - (I - A^{(1)}A)(B(I - A^{(1)}A))^{(1)}BA^{(1)}A$$

$$+ (I - A^{(1)}A)(B(I - A^{(1)}A))^{(1)}B).$$

Since

$$AA^{(1)}A = A, \quad A(I - A^{(1)}A) = O,$$

and

$$
\begin{aligned}
&-B(I - A^{(1)}A)(B(I - A^{(1)}A))^{(1)}BA^{(1)}A \\
&+ B(I - A^{(1)}A)(B(I - A^{(1)}A))^{(1)}B \\
&= B(I - A^{(1)}A)(B(I - A^{(1)}A))^{(1)}B(I - A^{(1)}A) \\
&= B(I - A^{(1)}A),
\end{aligned}
$$

we have

$$MX_M M = \begin{bmatrix} A \\ BA^{(1)}A + B(I - A^{(1)}A) \end{bmatrix} = \begin{bmatrix} A \\ B \end{bmatrix} = M.$$

The proof of (1.2.8) is left as an exercise. □

The following theorem provides the common solution of the systems $Ax = \mathbf{a}$ and $Bx = \mathbf{b}$.

Theorem 1.2.7 *Let $A \in \mathbb{C}^{m \times n}$, $B \in \mathbb{C}^{p \times n}$, $\mathbf{a} \in \mathcal{R}(A)$ and $\mathbf{b} \in \mathcal{R}(B)$. Suppose that $\mathbf{x_a}$ and $\mathbf{x_b}$ are any two particular solutions for the two systems in (1.2.5) respectively. Denote $F = B(I - A^{(1)}A)$, then the following three statements are equivalent:*

$$\text{The two systems in (1.2.5) possess a common solution;} \qquad (1.2.9)$$

$$B\mathbf{x_a} - \mathbf{b} \in \mathcal{R}(F) = B\mathcal{N}(A); \qquad (1.2.10)$$

$$\mathbf{x_a} - \mathbf{x_b} \in \mathcal{N}(A) + \mathcal{N}(B). \qquad (1.2.11)$$

Furthermore, when a common solution exists, a particular common solution is given by

$$\mathbf{x_c} = (I - (I - A^{(1)}A)F^{(1)}B)\mathbf{x_a} + (I - A^{(1)}A)F^{(1)}\mathbf{b} \qquad (1.2.12)$$

and the set of all common solutions can be written as

$$\left\{ \mathbf{x_c} + (I - A^{(1)}A)(I - F^{(1)}F)\mathbf{h} : \mathbf{h} \in \mathbb{C}^n \right\}. \qquad (1.2.13)$$

Proof The chain of the implications to be proven is $(1.2.11) \Rightarrow (1.2.10) \Rightarrow (1.2.9) \Rightarrow (1.2.11)$.

Equation $(1.2.11) \Rightarrow (1.2.10)$: Suppose that $(1.2.11)$ holds, then

$$\mathbf{x_a} - \mathbf{x_b} = \mathbf{n_a} + \mathbf{n_b}, \quad \text{where} \quad \mathbf{n_a} \in \mathcal{N}(A) \text{ and } \mathbf{n_b} \in \mathcal{N}(B).$$

So
$$B\mathbf{x_a} - \mathbf{b} = B(\mathbf{x_a} - \mathbf{x_b}) = B\mathbf{n_a} \in \mathcal{R}(F),$$

which gives (1.2.10).

Equation (1.2.10) $\Rightarrow$ (1.2.9): If (1.2.10) holds, then the vector $\mathbf{x_c}$ of (1.2.12) is a common solution of (1.2.5). Since $A\mathbf{x_c} = A\mathbf{x_a} = \mathbf{a}$ and

$$B\mathbf{x_c} = B\mathbf{x_a} - FF^{(1)}B\mathbf{x_a} + FF^{(1)}\mathbf{b}, \qquad (1.2.14)$$

the statement (1.2.10) implies that

$$FF^{(1)}(B\mathbf{x_a} - \mathbf{b}) = B\mathbf{x_a} - \mathbf{b},$$

or

$$B\mathbf{x_a} - FF^{(1)}B\mathbf{x_a} = \mathbf{b} - FF^{(1)}\mathbf{b}.$$

Therefore (1.2.14) becomes $B\mathbf{x_c} = \mathbf{b}$. Thus $\mathbf{x_c}$ is a common solution for the two systems in (1.2.5).

Equation (1.2.9) $\Rightarrow$ (1.2.11): If there exists a common solution for the two systems in (1.2.5), then the two solution sets must intersect, that is,

$$\{\mathbf{x_a} + \mathcal{N}(A)\} \cap \{\mathbf{x_b} + \mathcal{N}(B)\} \neq \emptyset.$$

Thus there exist vectors $\mathbf{n_a} \in \mathcal{N}(A)$ and $\mathbf{n_b} \in \mathcal{N}(B)$ such that

$$\mathbf{x_a} + \mathbf{n_a} = \mathbf{x_b} + \mathbf{n_b},$$

and (1.2.11) follows.

To obtain the set of all solutions for (1.2.5), we rewrite the system, which is equivalent to (1.2.6), as

$$\{\mathbf{x_c} + \mathcal{N}(M)\} = \{\mathbf{x_c} + (I - X_M M)\mathbf{h} : \mathbf{h} \in \mathbb{C}^n\},$$

where

$$M = \begin{bmatrix} A \\ B \end{bmatrix} \quad \text{and} \quad X_M = \begin{bmatrix} A \\ B \end{bmatrix}^{(1)}$$

is given in (1.2.7). Now

$$\begin{aligned} I - X_M M &= (I - A^{(1)}A)(I - F^{(1)}B + F^{(1)}BA^{(1)}A) \\ &= (I - A^{(1)}A)(I - F^{(1)}F) \end{aligned}$$

which gives (1.2.13). $\qquad\qquad\qquad\qquad\qquad\qquad\qquad\qquad\qquad\qquad\qquad\qquad\Box$

1.2.6   The {1} Inverse and the Common Solution
of $AX = B$ and $XD = E$

We consider the common solution of the two matrix equations

$$AX = B \quad \text{and} \quad XD = E. \tag{1.2.15}$$

Theorem 1.2.8 *The matrix Eq. (1.2.15) have a common solution if and only if each equation has a solution and $AE = BD$. If there exists a common solution, then a particular common solution is*

$$X_c = A^{(1)}B + ED^{(1)} - A^{(1)}AED^{(1)}$$

and the general common solution is

$$X_c + (I - A^{(1)}A)Y(I - DD^{(1)}) \tag{1.2.16}$$

for any $A^{(1)} \in A\{1\}$, $D^{(1)} \in D\{1\}$, and Y of the same size as X.

Proof IF: If each equation has a solution and $AE = BD$, by Theorem 1.2.5, we have

$$AA^{(1)}B = B, \quad ED^{(1)}D = E,$$

and

$$\begin{aligned}
AX_c &= A(A^{(1)}B + ED^{(1)} - A^{(1)}AED^{(1)}) \\
&= B + AED^{(1)} - AED^{(1)} \\
&= B
\end{aligned}$$

and

$$\begin{aligned}
X_cD &= (A^{(1)}B + ED^{(1)} - A^{(1)}AED^{(1)})D \\
&= A^{(1)}BD + E - A^{(1)}AED^{(1)}D \\
&= A^{(1)}BD + E - A^{(1)}BDD^{(1)}D \\
&= E.
\end{aligned}$$

Therefore X_c is a common solution of (1.2.15).

Since

$$A(X_c + (I - A^{(1)}A)Y(I - DD^{(1)})) = B$$

and

$$(X_c + (I - A^{(1)}A)Y(I - DD^{(1)}))D = E,$$

the matrix (1.2.16) is a common solution of (1.2.15). Suppose X_c is a common solution of (1.2.15), it is easy to verify that

$$X = X_c + (I - A^{(1)}A)(X - X_c)(I - DD^{(1)})$$

is also a common solution. If we set $Y = X - X_c$, then X is of the form (1.2.16). Thus the general common solution is (1.2.16).

ONLY IF: If X is a common solution, then $AX = B$ and $XD = E$. Postmultiplying $AX = B$ with D and premultiplying $XD = E$ with A, we get $AE = BD$. $\quad\square$

Exercises 1.2

1. Prove Theorem 1.2.4.
2. If $X \in A\{1\}$, then the following three statements are equivalent:

    (1) $X \in A\{1, 2\}$;
    (2) $\text{rank}(A) = \text{rank}(X)$;
    (3) There exist $X_1 \in A\{1\}$ and $X_2 \in A\{1\}$ such that $X_1 A X_2 = X$.

3. If $\text{rank}(A^* V A) = \text{rank}(A)$, then

$$A(A^* V A)^{(1)}(A^* V A) = A \text{ and } (A^* V A)(A^* V A)^{(1)} A^* = A^*.$$

4. Prove that $A(A^* A)^{(1)} A^* = A A^\dagger$.
5. For every $A \in \mathbb{C}^{m \times n}$ and $B \in \mathbb{C}^{n \times p}$, there exist $G \in A\{1\}$ and $F \in B\{1\}$ such that

$$FG \in (AB)\{1\}.$$

6. For $A \in \mathbb{C}^{m \times n}$, prove that

$$A^{(1)} = A^\dagger + H(I - AA^\dagger) + (I - A^\dagger A)K$$

for arbitrary $H, K \in \mathbb{C}^{n \times m}$.

1.3 The Generalized Inverses With Prescribed Range and Null Space

Theorem 1.1.3 shows some properties of the range and null space of the Moore-Penrose inverse. For example, $\mathcal{R}(A^\dagger) = \mathcal{R}(A^*)$ and $\mathcal{N}(A^\dagger) = \mathcal{N}(A^*)$. In this section, we study the generalized inverses with prescribed range and null space.

1.3.1   Idempotent Matrices and Projectors

A projector is associated with subspaces, so we begin with projectors and related
idempotent matrices. We will present their properties and establish a one-to-one
correspondence between them.

Definition 1.3.1 For $E \in \mathbb{C}^{n \times n}$, if $E^2 = E$, then E is called an idempotent matrix.

Lemma 1.3.1 *Let $E \in \mathbb{C}^{n \times n}$ be idempotent, then*

(1) E^* *and* $I - E$ *are idempotent;*
(2) *The eigenvalues of E are 0 and 1, and the multiplicity of the eigenvalue 1 is*
     rank(E);
(3) rank$(E) = \text{tr}(E)$;
(4) $E(I - E) = (I - E)E = O$;
(5) $E\mathbf{x} = \mathbf{x}$ *if and only if* $\mathbf{x} \in \mathcal{R}(E)$;
(6) $E \in E\{1, 2\}$;
(7) $\mathcal{N}(E) = \mathcal{R}(I - E)$.

Proof Properties (1)–(6) are immediate consequences of Definition 1.3.1; (3) follows
from (2) and the fact that the trace of any square matrix is the sum of its eigenvalues
counting multiplicities; (7) is obtained by applying Corollary 1.2.1 to $E\mathbf{x} = \mathbf{0}$.   □

Lemma 1.3.2 *Let $E = FG$ be a full-rank factorization of E, then E is idempotent
if and only if $GF = I$.*

Proof If $GF = I$, then clearly

$$(FG)^2 = FGFG = FG.$$

On the other hand, since F is of full column rank and G is of full row rank, by
Theorem 1.2.4,

$$F^{(1)}F = GG^{(1)} = I.$$

If $FGFG = FG$, multiplying on the left by $F^{(1)}$ and on the right by $G^{(1)}$ gives
$GF = I$.   □

Now, we turn to projectors. Two subspaces L and M of $\mathbb{C}^n$ are called comple-
mentary if $\mathbb{C}^n = L \oplus M$. In this case, every $\mathbf{x} \in \mathbb{C}^n$ can be expressed uniquely as the
sum

$$\mathbf{x} = \mathbf{y} + \mathbf{z} \quad (\mathbf{y} \in L, \mathbf{z} \in M). \tag{1.3.1}$$

We shall then call $\mathbf{y}$ the projection of $\mathbf{x}$ on L along M.

Definition 1.3.2 Let $P_{L,M}$ denote the transformation that maps any $\mathbf{x} \in \mathbb{C}^n$ into its
projection on L along M. It is easy to verify that this transformation is linear. This
linear transformation can be represented by a matrix, which is uniquely determined
by the linear transformation and the standard basis of unit vectors. We denote $P_{L,M}$

as both the linear transformation and its matrix representation. It is easy to verify that this transformation is idempotent. The linear transformation $P_{L,M}$ is called the projector on L along M and $P_{L,M}\mathbf{x} = \mathbf{y}$.

The next theorem establishes a one-to-one correspondence between an idempotent matrix of order n and a projector $P_{L,M}$ where $L \oplus M = \mathbb{C}^n$. Moreover, for any two complementary subspaces L and M, a method for the construction of $P_{L,M}$ is given by (1.3.3).

Theorem 1.3.1 *For every idempotent matrix $E \in \mathbb{C}^{n \times n}$, $\mathcal{R}(E)$ and $\mathcal{N}(E)$ are complementary subspaces and*

$$E = P_{\mathcal{R}(E),\mathcal{N}(E)}. \tag{1.3.2}$$

Conversely, if L and M are complementary subspaces, then $P_{L,M}$ is the unique idempotent matrix such that

$$\mathcal{R}(P_{L,M}) = L, \quad and \quad \mathcal{N}(P_{L,M}) = M.$$

Proof Let E be idempotent and of order n, then it follows from (5) and (7) in Lemma 1.3.1 and

$$\mathbf{x} = E\mathbf{x} + (I - E)\mathbf{x}$$

that $\mathbb{C}^n$ is the sum of $\mathcal{R}(E)$ and $\mathcal{N}(E)$. Moreover, $\mathcal{R}(E) \cap \mathcal{N}(E) = \{\mathbf{0}\}$. Since $\mathbf{x} \in \mathcal{R}(E)$, $\mathbf{x} = E\mathbf{x}$ by (5) of Lemma 1.3.1, also $\mathbf{x} \in \mathcal{N}(E)$ implies $E\mathbf{x} = \mathbf{0}$, then $\mathbf{x} = \mathbf{0}$. Thus $\mathbb{C}^n = \mathcal{R}(E) \oplus \mathcal{N}(E)$. It follows from $\mathbf{x} = E\mathbf{x} + (I - E)\mathbf{x}$ that for every $\mathbf{x}$, $E\mathbf{x}$ is the projection of $\mathbf{x}$ on $\mathcal{R}(E)$ along $\mathcal{N}(E)$. This establishes (1.3.2).

On the other hand, if L and M are complementary subspaces, suppose that $\{\mathbf{x}_1, \mathbf{x}_2, \cdots, \mathbf{x}_l\}$ and $\{\mathbf{y}_1, \mathbf{y}_2, \cdots, \mathbf{y}_m\}$ are any bases for L and M, respectively. If there exists $P_{L,M}$ such that $\mathcal{R}(P_{L,M}) = L$ and $\mathcal{N}(P_{L,M}) = M$, then

$$P_{L,M}\mathbf{x}_i = \mathbf{x}_i, \ i = 1, 2, ..., l,$$
$$P_{L,M}\mathbf{y}_i = \mathbf{0}_i, \ i = 1, 2, ..., m.$$

Let $X = [\mathbf{x}_1 \ \mathbf{x}_2 \ \cdots \ \mathbf{x}_l]$ and $Y = [\mathbf{y}_1 \ \mathbf{y}_2 \ \cdots \ \mathbf{y}_m]$, then

$$P_{L,M}[X \ Y] = [X \ O].$$

Since $\{\mathbf{x}_1, \mathbf{x}_2, ..., \mathbf{x}_l, \mathbf{y}_1, \mathbf{y}_2, ..., \mathbf{y}_m\}$ is a basis for $\mathbb{C}^n$, the matrix $[X \ Y]$ is nonsingular. Thus

$$P_{L,M} = [X \ O][X \ Y]^{-1}. \tag{1.3.3}$$

Since $P_{L,M}[X \ Y] = [X \ O]$, we have

$$P_{L,M}^2 = P_{L,M}[X \ O][X \ Y]^{-1} = [X \ O][X \ Y]^{-1} = P_{L,M},$$

showing that $P_{L,M}$ given by (1.3.3) is idempotent.

The proof for the uniqueness of $P_{L,M}$ is as follows.

If there exists another idempotent matrix E such that $\mathcal{R}(E) = L$ and $\mathcal{N}(E) = M$, then

$$E\mathbf{x}_i = \mathbf{x}_i, \ i = 1, 2, ..., l,$$
$$E\mathbf{y}_i = \mathbf{0}, \ i = 1, 2, ..., m.$$

It is clear that $E = [X \ O][X \ Y]^{-1}$ and $E = P_{L,M}$.   □

Corollary 1.3.1 *Let* $\mathbb{C}^n = L \oplus M$, *then for every* $\mathbf{x} \in \mathbb{C}^n$, *the unique decomposition* (1.3.1) *is given by*

$$\mathbf{y} = P_{L,M}\mathbf{x} \quad and \quad \mathbf{z} = (I - P_{L,M})\mathbf{x}.$$

The above corollary shows a relation between the projector $P_{L,M}$ and the direct sum $\mathbb{C}^n = L \oplus M$. The following corollary shows a relation between the $\{1, 2\}$ inverse and projectors.

Corollary 1.3.2 *If A and X are $\{1, 2\}$ inverses of each other, then AX is the projector on $\mathcal{R}(A)$ along $\mathcal{N}(X)$ and XA is the projector on $\mathcal{R}(X)$ along $\mathcal{N}(A)$, i.e.,*

$$AX = P_{\mathcal{R}(A),\mathcal{N}(X)} \quad and \quad XA = P_{\mathcal{R}(X),\mathcal{N}(A)}. \tag{1.3.4}$$

Proof The equations in (1.3.4) can be obtained by Theorems 1.3.1 and 1.2.4.   □

Let L and M be complementary subspaces of $\mathbb{C}^n$ and consider the matrix $P_{L,M}^*$. By (1) of Lemma 1.3.1, it is idempotent and therefore a projector by Theorem 1.3.1. Since $\mathcal{N}(A^*) = \mathcal{R}(A)^\perp$,

$$\mathcal{R}(P_{L,M}^*) = \mathcal{N}(P_{L,M})^\perp = M^\perp$$

and

$$\mathcal{N}(P_{L,M}^*) = \mathcal{R}(P_{L,M})^\perp = L^\perp.$$

Thus, by Theorem 1.3.1,

$$P_{L,M}^* = P_{M^\perp,L^\perp},$$

from which the next corollary follows easily.

Corollary 1.3.3 *Let* $\mathbb{C}^n = L \oplus M$, *then* $M = L^\perp$ *if and only if* $P_{L,M}$ *is Hermitian.*

Proof If $P_{L,M}^* = P_{L,M}$, then

$$\mathcal{N}(P_{L,M}) = \mathcal{N}(P_{L,M}^*) \quad and \quad M = L^\perp.$$

Conversely, if $M = L^\perp$, then

$$P_{L,M}^* = P_{M^\perp,L^\perp} = P_{L,M}.$$

This completes the proof.   □

Just as there is a one-to-one correspondence between projectors and idempotent matrices, Corollary 1.3.3 shows that there is a one-to-one correspondence between orthogonal projectors and Hermitian idempotent matrices. The projector on L along $L^\perp$ is called the orthogonal projector on L and denoted by P_L.

Now we discuss the conditions under which the sum, difference, and product of two projectors are also projectors.

Theorem 1.3.2 *Let P_1 be the projector on R_1 along N_1, P_2 the projector on R_2 along N_2, then $P = P_1 + P_2$ is a projector if and only if*

$$P_1 P_2 = P_2 P_1 = O.$$

In this case, P is a projector on $R = R_1 \oplus R_2$ along $N = N_1 \cap N_2$.

Proof IF: Let $P_1^2 = P_1$ and $P_2^2 = P_2$. If $P^2 = P$, then

$$P_1 P_2 + P_2 P_1 = O.$$

Multiplying the above on the left by P_1 gives

$$P_1(P_1 P_2 + P_2 P_1) = P_1 P_2 + P_1 P_2 P_1 = O.$$

Multiplying on the right by P_1 gives

$$(P_1 P_2 + P_1 P_2 P_1) P_1 = 2 P_1 P_2 P_1 = O.$$

Hence $P_1 P_2 P_1 = O$. Substituting it into the previous equation, we have $P_1 P_2 = O$ and $P_2 P_1 = O$.

ONLY IF: If $P_1 P_2 = P_2 P_1 = O$, then

$$P^2 = (P_1 + P_2)^2 = P_1^2 + P_2^2 = P_1 + P_2 = P$$

and P is a projector by Theorem 1.3.1.

Now we prove $R = R_1 \oplus R_2$. Let $\mathbf{u} \in R_i$ ($i = 1, 2$), then

$$P_i \mathbf{u} = \mathbf{u}, \quad P\mathbf{u} = P P_i \mathbf{u} = P_i^2 \mathbf{u} = P_i \mathbf{u} = \mathbf{u}, \quad \mathbf{u} \in R.$$

Thus $R_i \subset R$. Let $\mathbf{u} \in R_1 \cap R_2$, then

$$P_1 \mathbf{u} = \mathbf{u}, \quad P_2 \mathbf{u} = \mathbf{u}, \quad \mathbf{u} = P_1 \mathbf{u} = P_2 \mathbf{u} = P_2 P_1 \mathbf{u} = \mathbf{0}.$$

Thus $R_1 \cap R_2 = \{\mathbf{0}\}$. Any vector $\mathbf{u} \in R$ can be expressed as the sum

$$\mathbf{u} = P\mathbf{u} = P_1 \mathbf{u} + P_2 \mathbf{u}, \quad P_1 \mathbf{u} \in R_1, \quad P_2 \mathbf{u} \in R_2,$$

so $R = R_1 \oplus R_2$.

Next we prove $N = N_1 \cap N_2$. Let $\mathbf{u} \in N$, then $P\mathbf{u} = \mathbf{0}$ and

$$\mathbf{0} = P_1 P\mathbf{u} = P_1^2 \mathbf{u} = P_1\mathbf{u}.$$

Thus $\mathbf{u} \in N_1$. The proof of $\mathbf{u} \in N_2$ is similar. Hence

$$N \subset N_1 \cap N_2.$$

Conversely, let $\mathbf{u} \in N_1 \cap N_2$, then

$$P_1\mathbf{u} = \mathbf{0}, \ \ P_2\mathbf{u} = \mathbf{0}, \ \ P\mathbf{u} = P_1\mathbf{u} + P_2\mathbf{u} = \mathbf{0}.$$

Thus $\mathbf{u} \in N$ and

$$N_1 \cap N_2 \subset N.$$

The proof is completed. $\qquad\qquad\qquad\qquad\qquad\qquad\qquad\qquad\qquad\qquad\quad\square$

Theorem 1.3.3 *Under the assumptions in Theorem 1.3.2, $P = P_1 - P_2$ is a projector if and only if*

$$P_1 P_2 = P_2 P_1 = P_2. \tag{1.3.5}$$

In this case, P is a projector on $R = R_1 \cap N_2$ along $N = N_1 \oplus R_2$.

Proof Noting that P is a projector if and only if $I - P$ is a complementary projector. Since $I - P = (I - P_1) + P_2$, by Theorem 1.3.2, $I - P$ is a projector if and only if $(I - P_1)P_2 = P_2(I - P_1) = O$. Thus (1.3.5) holds.

Next we prove that $N = N_1 \oplus R_2$ and $R = R_1 \cap N_2$. By Lemma 1.3.1 and Theorem 1.3.2,

$$\begin{aligned} N = \mathcal{N}(P) = \mathcal{R}(I - P) &= \mathcal{R}(I - P_1) \oplus \mathcal{R}(P_2) \\ &= \mathcal{N}(P_1) \oplus \mathcal{R}(P_2) = N_1 \oplus R_2 \end{aligned}$$

and

$$\begin{aligned} R = \mathcal{R}(P) = \mathcal{N}(I - P) &= \mathcal{N}(I - P_1) \cap \mathcal{N}(P_2) \\ &= \mathcal{R}(P_1) \cap \mathcal{N}(P_2) = R_1 \cap N_2. \end{aligned}$$

This completes the proof. $\qquad\qquad\qquad\qquad\qquad\qquad\qquad\qquad\qquad\qquad\quad\square$

Theorem 1.3.4 *Under the assumptions in Theorem 1.3.2, if*

$$P_1 P_2 = P_2 P_1, \tag{1.3.6}$$

then $P = P_1 P_2$ is a projector on $R = R_1 \cap R_2$ along $N = N_1 + N_2$.

Proof IF: If $P_1 P_2 = P_2 P_1$, then $P^2 = P$, so P is a projector. Now we prove $R = R_1 \cap R_2$. Let $\mathbf{u} \in R$, then

$$P_1 P_2 \mathbf{u} = P \mathbf{u} = \mathbf{u}.$$

Multiplying the above on the left by P_1 gives

$$P_1 \mathbf{u} = P_1^2 P_2 \mathbf{u} = P_1 P_2 \mathbf{u} = \mathbf{u},$$

thus $\mathbf{u} \in R_1$. The proof of $\mathbf{u} \in R_2$ is similar, thus $\mathbf{u} \in R_1 \cap R_2$. Conversely, let $\mathbf{u} \in R_1 \cap R_2$, then

$$P_1 \mathbf{u} = \mathbf{u} \quad \text{and} \quad P_2 \mathbf{u} = \mathbf{u},$$

thus

$$P \mathbf{u} = P_1 P_2 \mathbf{u} = P_1 \mathbf{u} = \mathbf{u}.$$

Therefore $\mathbf{u} \in R$. Thus $R = R_1 \cap R_2$.

Next we prove $N = N_1 + N_2$. If $\mathbf{u} \in N$, then

$$P_1 P_2 \mathbf{u} = P \mathbf{u} = \mathbf{0},$$

thus $P_2 \mathbf{u} \in N_1$. Moreover, $P_2 (I - P_2) \mathbf{u} = \mathbf{0}$, thus $(I - P_2) \mathbf{u} \in N_2$. Since

$$\mathbf{u} = P_2 \mathbf{u} + (I - P_2) \mathbf{u},$$

we have $N \subset N_1 + N_2$.

Conversely, if $\mathbf{u} \in N_1 + N_2$, then $\mathbf{u}$ can be expressed as the sum

$$\mathbf{u} = \mathbf{u}_1 + \mathbf{u}_2, \quad \mathbf{u}_1 \in N_1, \ \mathbf{u}_2 \in N_2.$$

Since

$$P \mathbf{u} = P_1 P_2 \mathbf{u} = P_1 P_2 \mathbf{u}_1 + P_1 P_2 \mathbf{u}_2 = P_1 P_2 \mathbf{u}_1 = P_2 P_1 \mathbf{u}_1 = \mathbf{0},$$

we have $\mathbf{u} \in N$. Thus $N = N_1 + N_2$. $\square$

1.3.2 Generalized Inverse $A_{T,S}^{(1,2)}$

Now we are ready for the generalized inverses with prescribed range and null space. Let $A \in \mathbb{C}^{m \times n}$ and $A^{(1)}$ be an element of $A\{1\}$. Suppose that $\mathcal{R}(A) = L$ and $\mathcal{N}(A) = M$, and $L \oplus S = \mathbb{C}^m$ and $T \oplus M = \mathbb{C}^n$, then $AA^{(1)}$ and $A^{(1)}A$ are idempotent and, by Theorems 1.3.1 and 1.2.4,

$$AA^{(1)} = P_{L,S} \quad \text{and} \quad A^{(1)}A = P_{T,M}.$$

Next, we introduce a generalized inverse X which is the unique matrix satisfying the following three equations

$$AX = P_{L,S}, \quad XA = P_{T,M}, \quad \text{and} \quad XAX = X.$$

First, we show the following lemma.

Lemma 1.3.3 *There exists at most one matrix X satisfying the three equations:*

$$AX = B, \quad XA = D, \quad \text{and} \quad XAX = X. \tag{1.3.7}$$

Proof The Eq. (1.3.7) may have no common solution. Now we suppose (1.3.7) have a common solution. Let both X_1 and X_2 satisfy (1.3.7) and $U = X_1 - X_2$. Then $AU = O, UA = O, UB = U$, and $DU = U$ by (1.3.7). Thus

$$U^*U = U^*D^*UB = U^*A^*X_i^*UAX_i = O, \quad i = 1, 2.$$

Therefore $U = O$, i.e., $X_1 = X_2$.   □

The following theorem gives an explicit expression for the $\{1, 2\}$ inverse with prescribed range and null space.

Theorem 1.3.5 *Let $A \in \mathbb{C}^{m \times n}$, $\mathcal{R}(A) = L$, $\mathcal{N}(A) = M$, $L \oplus S = \mathbb{C}^m$, and $T \oplus M = \mathbb{C}^n$, then*

(1) *X is a $\{1\}$ inverse of A such that $\mathcal{R}(XA) = T, \mathcal{N}(AX) = S$ if and only if*

$$AX = P_{L,S}, \quad XA = P_{T,M}; \tag{1.3.8}$$

(2) *The general solution of (1.3.8) is*

$$X = P_{T,M}A^{(1)}P_{L,S} + (I_n - A^{(1)}A)Y(I_m - AA^{(1)}),$$

where $A^{(1)}$ is a fixed (but arbitrary) element of $A\{1\}$ and Y is an arbitrary $n \times m$ matrix;

(3) *$A_{T,S}^{(1,2)} = P_{T,M}A^{(1)}P_{L,S}$ is the unique $\{1, 2\}$ inverse of A having range T and null space S.*

Proof (1) IF: By the assumptions $AX = P_{L,S}$ and $\mathcal{R}(A) = L$, and Exercise 1.3.2, we have

$$AXA = P_{L,S}A = A.$$

Thus $X \in A\{1\}$. Moreover,

$$\mathcal{N}(AX) = \mathcal{N}(P_{L,S}) = S \quad \text{and} \quad \mathcal{R}(XA) = \mathcal{R}(P_{T,M}) = T.$$

ONLY IF: By Lemma 1.3.1, AX and XA are idempotent. By Theorems 1.3.1 and 1.2.4,

$$AX = P_{\mathcal{R}(AX),\mathcal{N}(AX)} = P_{L,S} \quad \text{and} \quad XA = P_{\mathcal{R}(XA),\mathcal{N}(XA)} = P_{T,M}.$$

(2) Set $X_0 = P_{T,M} A^{(1)} P_{L,S}$. By $\mathcal{R}(P_{L,S}) = L = \mathcal{R}(A)$, there exists Y such that $P_{L,S} = AY$. By Exercise 1.3.2,

$$AX_0 = AP_{T,M} A^{(1)} P_{L,S} = AA^{(1)} P_{L,S} = AA^{(1)} AY = AY = P_{L,S}.$$

The proof of $X_0 A = P_{T,M}$ is similar. Thus X_0 is a common solution of (1.3.8). By using Theorem 1.2.8, the general solution of (1.3.8) is

$$X = P_{T,M} A^{(1)} P_{L,S} + (I_n - A^{(1)} A) Y (I_m - AA^{(1)}), \quad \forall\, Y \in \mathbb{C}^{n \times m}.$$

(3) Set $X = P_{T,M} A^{(1)} P_{L,S}$. It follows from (2) that X satisfies

$$AXA = P_{L,S} A = A.$$

Thus $X \in A\{1\}$ and $\text{rank}(X) \geq \text{rank}(A)$. Since

$$\text{rank}(X) = \text{rank}(P_{T,M} A^{(1)} P_{L,S}) \leq \text{rank}(P_{L,S}) \leq \text{rank}(A),$$

$\text{rank}(X) = \text{rank}(A)$. It is obvious that $\mathcal{R}(XA) \subset \mathcal{R}(X)$. By Theorem 1.2.4, $\text{rank}(XA) = \text{rank}(A)$. Thus $\mathcal{R}(XA) = \mathcal{R}(X)$ and there exists a Y such that $XAY = X$. Multiplying $XAY = X$ on the left by A gives

$$AX = AXAY = AY$$

and multiplying it on the left by X gives

$$XAX = XAY = X.$$

Therefore $X \in A\{2\}$.

Next, using (1.3.4), we have

$$\mathcal{R}(X) = \mathcal{R}(XA) = T \quad \text{and} \quad \mathcal{N}(X) = \mathcal{N}(AX) = S.$$

Thus $X = A_{T,S}^{(1,2)}$. Since X satisfies

$$AX = P_{L,S}, \quad XA = P_{T,M}, \quad \text{and} \quad XAX = X,$$

by Lemma 1.3.3, there exists at most one matrix X satisfying the above three equations. Therefore $A_{T,S}^{(1,2)}$ is the unique $\{1,2\}$ inverse of A having range T and null space S. $\qquad\square$

Finally, we establish a relation between $A_{T,S}^{(1,2)}$ and $A^{\dagger}$.

Theorem 1.3.6 *Let $A \in \mathbb{C}^{m \times n}$, $\mathcal{R}(A) = L$, and $\mathcal{N}(A) = M$, then*

$$A^\dagger = A^{(1,2)}_{\mathcal{R}(A^*), \mathcal{N}(A^*)}. \tag{1.3.9}$$

Proof Clearly $A^\dagger \in A\{1, 2\}$. By (2) and (4) of Theorem 1.1.3,

$$\mathcal{R}(A^\dagger) = \mathcal{R}(A^*) \quad \text{and} \quad \mathcal{N}(A^\dagger) = \mathcal{N}(A^*).$$

It then follows from the uniqueness of $A^{(1,2)}_{\mathcal{R}(A^*), \mathcal{N}(A^*)}$ that (1.3.9) holds. $\square$

This result means that the M-P inverse $A^\dagger$ is the $\{1, 2\}$ inverse of A having range $\mathcal{R}(A^*)$ and null space $\mathcal{N}(A^*)$.

1.3.3   Urquhart Formula

The formulas in Theorem 1.3.5 are not convenient for computational purposes. Using the results in [3], a useful formula for $A^{(1,2)}_{T,S}$ is given as follows.

Theorem 1.3.7 *Let $A \in \mathbb{C}^{m \times n}_r$, $U \in \mathbb{C}^{n \times p}$, $V \in \mathbb{C}^{q \times m}$, and*

$$X = U(VAU)^{(1)}V,$$

where $(VAU)^{(1)}$ is a fixed but arbitrary element of $(VAU)\{1\}$, then

(1) $X \in A\{1\}$ *if and only if* $\mathrm{rank}(VAU) = r$;
(2) $X \in A\{2\}$ *and* $\mathcal{R}(X) = \mathcal{R}(U)$ *if and only if* $\mathrm{rank}(VAU) = \mathrm{rank}(U)$;
(3) $X \in A\{2\}$ *and* $\mathcal{N}(X) = \mathcal{N}(V)$ *if and only if* $\mathrm{rank}(VAU) = \mathrm{rank}(V)$;
(4) $X = A^{(1,2)}_{\mathcal{R}(U), \mathcal{N}(V)}$ *if and only if* $\mathrm{rank}(VAU) = \mathrm{rank}(U) = \mathrm{rank}(V) = r$;
(5) $X = A^{(2)}_{\mathcal{R}(U), \mathcal{N}(V)}$ *if and only if* $\mathrm{rank}(VAU) = \mathrm{rank}(U) = \mathrm{rank}(V)$.

Proof (1) IF: Since

$$r = \mathrm{rank}(VAU) \leq \mathrm{rank}(AU) \leq \mathrm{rank}(A) = r,$$

we have
$$\mathrm{rank}(AU) = r = \mathrm{rank}(A),$$

thus $\mathcal{R}(AU) = \mathcal{R}(A)$ and there exists a matrix Y such that $A = AUY$. Moreover,

$$\mathrm{rank}(VAU) = \mathrm{rank}(AU) = r,$$

by Exercise 1.3.3,

$$AU(VAU)^{(1)}VAU = AU.$$

Thus

$$AXA = AU(VAU)^{(1)}VAUY = AUY = A,$$

i.e., $X \in A\{1\}$.

ONLY IF: Suppose that $X \in A\{1\}$, then

$$A = AXAXA = AU(VAU)^{(1)}VAU(VAU)^{(1)}VA,$$

thus $\text{rank}(VAU) \geq \text{rank}(A)$. It is clear that $\text{rank}(VAU) \leq \text{rank}(A)$. Therefore

$$\text{rank}(VAU) = \text{rank}(A) = r.$$

(2) IF: By Exercise 1.3.3,

$$XAU = U(VAU)^{(1)}VAU = U,$$

from which it follows that

$$XAX = XAU(VAU)^{(1)}V = U(VAU)^{(1)}V = X,$$

thus $X \in A\{2\}$. By $XAU = U$,

$$\text{rank}(U) \leq \text{rank}(X) \quad \text{and} \quad \mathcal{R}(U) \subset \mathcal{R}(X).$$

From $X = U(VAU)^{(1)}V$,

$$\text{rank}(X) \leq \text{rank}(U) \quad \text{and} \quad \mathcal{R}(X) \subset \mathcal{R}(U).$$

Thus

$$\text{rank}(X) = \text{rank}(U) \quad \text{and} \quad \mathcal{R}(U) = \mathcal{R}(X).$$

ONLY IF: Since $\mathcal{R}(U) = \mathcal{R}(X)$ and $\text{rank}(X) = \text{rank}(U)$. By $X \in A\{2\}$,

$$X = XAX = U(VAU)^{(1)}VAU(VAU)^{(1)}V.$$

Therefore

$$\text{rank}(U) = \text{rank}(X) \leq \text{rank}(VAU) \leq \text{rank}(U).$$

Thus

$$\text{rank}(VAU) = \text{rank}(U).$$

(3) IF: By Exercise 1.3.3,

$$VAX = VAU(VAU)^{(1)}V = V,$$

from which it follows that

$$XAX = U(VAU)^{(1)}VAX = U(VAU)^{(1)}V = X,$$

thus $X \in A\{2\}$. By $V = VAX$,

$$\text{rank}(V) \leq \text{rank}(X) \quad \text{and} \quad \mathcal{N}(X) \subset \mathcal{N}(V).$$

Since $X = U(VAU)^{(1)}V$,

$$\text{rank}(X) \leq \text{rank}(V) \quad \text{and} \quad \mathcal{N}(V) \subset \mathcal{N}(X).$$

Thus

$$\text{rank}(X) = \text{rank}(V) \quad \text{and} \quad \mathcal{N}(V) = \mathcal{N}(X).$$

ONLY IF: Since $\mathcal{N}(X) = \mathcal{N}(V)$ and the number of the columns of both X and V is m, we have $\text{rank}(V) = \text{rank}(X)$. By $X \in A\{2\}$,

$$X = XAX = U(VAU)^{(1)}VAU(VAU)^{(1)}V,$$

thus

$$\text{rank}(V) = \text{rank}(X) \leq \text{rank}(VAU) \leq \text{rank}(V).$$

Therefore

$$\text{rank}(VAU) = \text{rank}(V).$$

(4) Follows from (1), (2) and (3).
(5) Follows from (2) and (3).   □

By Theorem 1.3.7, we can derive the formula of Zlobec [4].

$$A^\dagger = A^*YA^*, \tag{1.3.10}$$

where $Y \in (A^*AA^*)\{1\}$. Indeed, since

$$\text{rank}(A^*AA^*) = \text{rank}(AA^*) = \text{rank}(A) = \text{rank}(A^*) = r,$$

by (4) in Theorems 1.3.6 and 1.3.7,

$$A^*(A^*AA^*)^{(1)}A^* = A^{(1,2)}_{\mathcal{R}(A^*),\mathcal{N}(A^*)} = A^\dagger.$$

Using Theorem 1.3.7, we can construct not only $A^{(1,2)}_{\mathcal{R}(U),\mathcal{N}(V)}$ but also the $\{2\}$ inverse of A with prescribed range $\mathcal{R}(U)$ and null space $\mathcal{N}(V)$. We discuss this kind of generalized inverses in the following subsection.

1.3.4 Generalized Inverse $A_{T,S}^{(2)}$

In this subsection, we present a necessary and sufficient condition for the existence of $A_{T,S}^{(2)}$ and then a condition for $A_{T,S}^{(2)}$ to be $A_{T,S}^{(1,2)}$.

Theorem 1.3.8 *Let $A \in \mathbb{C}_r^{m \times n}$, T be a subspace of $\mathbb{C}^n$ of dimension $t \le r$, and S a subspace of $\mathbb{C}^m$ of dimension $m - t$, then A has a $\{2\}$ inverse X such that $\mathcal{R}(X) = T$ and $\mathcal{N}(X) = S$ if and only if*

$$AT \oplus S = \mathbb{C}^m, \tag{1.3.11}$$

in which case X is unique and denoted by $A_{T,S}^{(2)}$.

Proof IF: Let the columns of $U \in \mathbb{C}_t^{n \times t}$ form a basis for T and the columns of $V^* \in \mathbb{C}_t^{m \times t}$ form a basis for $S^\perp$, that is, $\mathcal{R}(U) = T$ and $\mathcal{N}(V) = S$, then the columns of AU span AT. It follows from (1.3.11) that $\dim(AT) = t$, so

$$\text{rank}(AU) = \dim(\mathcal{R}(AU)) = \dim(AT) = t. \tag{1.3.12}$$

Another consequence of (1.3.11) is

$$AT \cap S = \{0\}. \tag{1.3.13}$$

Moreover, the $t \times t$ matrix VAU is nonsingular. If $VAU\mathbf{y} = \mathbf{0}$, then $AU\mathbf{y} \in \mathcal{N}(V) = S$ and $AU\mathbf{y} \in \mathcal{R}(AU) = AT$, thus $AU\mathbf{y} = \mathbf{0}$ by (1.3.13). It follows from (1.3.12) that AU is of full column rank, thus $\mathbf{y} = \mathbf{0}$. Therefore

$$\text{rank}(VAU) = \text{rank}(U) = \text{rank}(V) = t \ (\le r). \tag{1.3.14}$$

By (5) of Theorem 1.3.7,

$$X = U(VAU)^{-1}V \tag{1.3.15}$$

is a $\{2\}$ inverse of A having range $\mathcal{R}(X) = \mathcal{R}(U) = T$ and null space $\mathcal{N}(X) = \mathcal{N}(V) = S$.

ONLY IF: Since $X \in A\{2\}$, i.e., $A \in X\{1\}$, AX is idempotent. By Theorem 1.3.1,

$$\mathcal{R}(AX) \oplus \mathcal{N}(AX) = \mathbb{C}^m.$$

Moreover, by (10) of Theorem 1.2.4,

$$\mathcal{R}(AX) = A\mathcal{R}(X) = AT \quad \text{and} \quad \mathcal{N}(AX) = \mathcal{N}(X) = S.$$

Thus $AT \oplus S = \mathbb{C}^m$ holds.

UNIQUENESS: Let X_1 and X_2 be $\{2\}$ inverses of A having range T and null space S, then $A \in X_1\{1\}$, $A \in X_2\{1\}$,

$$X_1 A = P_{\mathcal{R}(X_1 A), \mathcal{N}(X_1 A)} = P_{\mathcal{R}(X_1), \mathcal{N}(X_1 A)} = P_{T, \mathcal{N}(X_1 A)},$$

and

$$AX_2 = P_{\mathcal{R}(AX_2), \mathcal{N}(AX_2)} = P_{\mathcal{R}(AX_2), \mathcal{N}(X_2)} = P_{\mathcal{R}(AX_2), S}.$$

Since $\mathcal{R}(X_2) = T$ and $\mathcal{N}(X_1) = S$, by Exercise 1.3.2,

$$X_2 = P_{T, \mathcal{N}(X_1 A)} X_2 = X_1 A X_2 = X_1 P_{\mathcal{R}(AX_2), S} = X_1,$$

which completes the proof.   □

The following corollary shows that if $t = r$, then $A_{T,S}^{(2)} = A_{T,S}^{(1,2)}$ in Theorem 1.3.8.

Corollary 1.3.4 *Let $A \in \mathbb{C}_r^{m \times n}$, T be a subspace of $\mathbb{C}^n$ of dimension r, and S a subspace of $\mathbb{C}^m$ of dimension $m - r$, then the following three statements are equivalent:*

(1) $AT \oplus S = \mathbb{C}^m$;
(2) $\mathcal{R}(A) \oplus S = \mathbb{C}^m$ *and* $\mathcal{N}(A) \oplus T = \mathbb{C}^n$;
(3) *There exists an $X \in A\{1, 2\}$ such that $\mathcal{R}(X) = T$ and $\mathcal{N}(X) = S$.*

Proof (1)$\Rightarrow$(3): If (1) holds, by (1.3.14),

$$\mathrm{rank}(VAU) = \mathrm{rank}(U) = \mathrm{rank}(V) = r.$$

It follows from (4) of Theorem 1.3.7 that $X = U(VAU)^{-1}V$ is a $\{1, 2\}$ inverse of A having range $\mathcal{R}(X) = T$ and null space $\mathcal{N}(X) = S$.

(3)$\Rightarrow$(1): (1) is obtained by applying Theorem 1.3.8.

(1)$\Rightarrow$(2): If (1) holds, then $\mathrm{rank}(VAU) = r$. By Theorem 1.3.7, $X \in A\{1\}$. It follows from Theorem 1.2.4 that

$$\mathcal{R}(AX) = \mathcal{R}(A)$$

and

$$AT = A\mathcal{R}(X) = \mathcal{R}(AX) = \mathcal{R}(A).$$

Therefore $\mathcal{R}(A) \oplus S = \mathbb{C}^m$ by (1).

On the other hand, since (1)$\Leftrightarrow$(3), $X \in A\{1\}$ and $X \in A\{2\}$, i.e., $A \in X\{1\}$. It follows from Theorem 1.2.4 that $\mathcal{R}(XA) = \mathcal{R}(X) = T$ and $\mathcal{N}(XA) = \mathcal{N}(A)$. Thus

$$XA = P_{\mathcal{R}(XA), \mathcal{N}(XA)} = P_{T, \mathcal{N}(A)}$$

and $T \oplus \mathcal{N}(A) = \mathbb{C}^n$ by Theorem 1.3.1.

$(2) \Rightarrow (1)$: Since $\mathbb{C}^n = \mathcal{N}(A) \oplus T$, it is easy to verify that $\mathcal{R}(A) = AT$. In fact, it is clear that $\mathcal{R}(A) \supset AT$. Conversely, if $\mathbf{x}$ is any vector in $\mathcal{R}(A)$, then $\mathbf{x} = A\mathbf{y}$, $\mathbf{y} \in \mathbb{C}^n$. Setting $\mathbf{y} = \mathbf{y}_1 + \mathbf{y}_2$, where $\mathbf{y}_1 \in \mathcal{N}(A)$ and $\mathbf{y}_2 \in T$, we have

$$\mathbf{x} = A\mathbf{y} = A\mathbf{y}_1 + A\mathbf{y}_2 = A\mathbf{y}_2 \in AT.$$

Thus $\mathcal{R}(A) = AT$. Therefore $AT \oplus S = \mathbb{C}^m$ holds by (2). $\square$

Exercises 1.3

1. Let L and M be complementary subspaces of $\mathbb{C}^n$, and $P_{L,M}$ denote the transformation that carries any $\mathbf{x} \in \mathbb{C}^n$ into its projection on L along M. Prove that this transformation is linear.
2. Let L and M be complementary subspaces of $\mathbb{C}^n$, prove that

    (1) $P_{L,M} A = A \Leftrightarrow \mathcal{R}(A) \subset L$;
    (2) $A P_{L,M} = A \Leftrightarrow \mathcal{N}(A) \supset M$.

3. Prove that

    (1) $AB(AB)^{(1)}A = A \Leftrightarrow \operatorname{rank}(AB) = \operatorname{rank}(A)$;
    (2) $B(AB)^{(1)}AB = B \Leftrightarrow \operatorname{rank}(AB) = \operatorname{rank}(B)$.

4. Prove that $I - P_{L,M} = P_{M,L}$.
5. Prove that (1.3.11) is equivalent to $A^* S^\perp \oplus T^\perp = \mathbb{C}^n$.
6. Let L be a subspace of $\mathbb{C}^n$ and the columns of F form a basis for L. Show that

    $$P_L = F F^\dagger = F(F^*F)^{-1}F^*.$$

7. Prove that $A_{T,S}^{(2)} = (P_{S^\perp} A P_T)^\dagger$.
8. Prove that $(A_{T,S}^{(2)})^* = (A^*)_{S^\perp, T^\perp}^{(2)}$.
9. Prove that

    (1) $A(A_{T,S}^{(2)}) = P_{AT,S}$;
    (2) $(A_{T,S}^{(2)})A = P_{T,(A^*S^\perp)^\perp}$.

1.4 Weighted Moore-Penrose Inverse

In Sects. 1.1 and 1.2, the relations between the generalized inverses $A^{(1,4)}$, $A^{(1,3)}$ and $A^\dagger$, and the minimum-norm solution, least-squares solution and minimum-norm least-squares solution are discussed. There, the vector 2-norm $\|\mathbf{x}\|_2$ was used. Let $\mathbf{x}, \mathbf{y} \in \mathbb{C}^n$, then the inner product in $\mathbb{C}^n$ is defined by

$$(\mathbf{x}, \mathbf{y}) = \mathbf{y}^* \mathbf{x}.$$

Moreover ,

$$\|\mathbf{x}\|_2 = (\mathbf{x}, \mathbf{x})^{1/2} = (\mathbf{x}^*\mathbf{x})^{1/2}.$$

A matrix norm can be induced from a vector norm. For example, the 2-norm of a matrix A can be defined by

$$\|A\|_2 = \max_{\|\mathbf{x}\|_2=1} \|A\mathbf{x}\|_2.$$

In the vector 2-norm defined above, all the components of the vector are equally important. In practice, we may want to give different weights to the components of the residual of the linear system $A\mathbf{x} = \mathbf{b}$. A generalization of the standard least-squares is the minimization of the weighted norm

$$\|A\mathbf{x} - \mathbf{b}\|_M^2 \equiv (A\mathbf{x} - \mathbf{b})^* M (A\mathbf{x} - \mathbf{b}),$$

where M is a given Hermitian positive definite matrix. Now we discuss the weighted norm and some related generalized inverses.

1.4.1 Weighted Norm and Weighted Conjugate Transpose Matrix

Let M and N be Hermitian positive definite matrices of orders m and n respectively. The weighted inner products in $\mathbb{C}^m$ and $\mathbb{C}^n$ are defined by

$$(\mathbf{x}, \mathbf{y})_M = \mathbf{y}^* M \mathbf{x}, \quad \mathbf{x}, \mathbf{y} \in \mathbb{C}^m$$

and

$$(\mathbf{x}, \mathbf{y})_N = \mathbf{y}^* N \mathbf{x}, \quad \mathbf{x}, \mathbf{y} \in \mathbb{C}^n$$

respectively. Correspondingly, the definitions of weighted vector norms are

$$\|\mathbf{x}\|_M = (\mathbf{x}, \mathbf{x})_M^{1/2} = (\mathbf{x}^* M \mathbf{x})^{1/2} = \|M^{1/2}\mathbf{x}\|_2, \quad \mathbf{x} \in \mathbb{C}^m$$

and

$$\|\mathbf{x}\|_N = (\mathbf{x}, \mathbf{x})_N^{1/2} = (\mathbf{x}^* N \mathbf{x})^{1/2} = \|N^{1/2}\mathbf{x}\|_2, \quad \mathbf{x} \in \mathbb{C}^n$$

respectively. Let $\mathbf{x}, \mathbf{y} \in \mathbb{C}^m$ and $(\mathbf{x}, \mathbf{y})_M = 0$, then $\mathbf{x}$ and $\mathbf{y}$ are called M-orthogonal, i.e., $M^{1/2}\mathbf{x}$ and $M^{1/2}\mathbf{y}$ are orthogonal. It is easy to verify that

$$\|\mathbf{x} + \mathbf{y}\|_M^2 = \|\mathbf{x}\|_M^2 + \|\mathbf{y}\|_M^2, \quad \mathbf{x}, \mathbf{y} \in \mathbb{C}^m, \tag{1.4.1}$$

which is called the weighted Pythagorean theorem.

The definitions of weighted matrix norm are:

$$\|A\|_{MN} = \max_{\|\mathbf{x}\|_N=1} \|A\mathbf{x}\|_M, \quad A \in \mathbb{C}^{m\times n}, \ \mathbf{x} \in \mathbb{C}^n$$

and

$$\|B\|_{NM} = \max_{\|\mathbf{x}\|_M=1} \|B\mathbf{x}\|_N, \quad B \in \mathbb{C}^{n\times m}, \ \mathbf{x} \in \mathbb{C}^m$$

respectively. Such a norm is sometimes called an operator norm subordinate to vector norms. It is easy to verify that

$$\|A\|_{MN} = \|M^{1/2}AN^{-1/2}\|_2$$

and

$$\|B\|_{NM} = \|N^{1/2}BM^{-1/2}\|_2.$$

The next lemma shows the consistent property of the weighted matrix norm induced from a weighted vector norm.

Lemma 1.4.1 *Let* $A \in \mathbb{C}^{m\times n}$, $B \in \mathbb{C}^{n\times m}$, $\mathbf{x} \in \mathbb{C}^n$, *and* $\mathbf{y} \in \mathbb{C}^m$, *then*

$$\|A\mathbf{x}\|_M \le \|A\|_{MN}\|\mathbf{x}\|_N; \tag{1.4.2}$$

$$\|B\mathbf{y}\|_N \le \|B\|_{NM}\|\mathbf{y}\|_M; \tag{1.4.3}$$

$$\|AB\|_{MM} \le \|A\|_{MN}\|B\|_{NM}.$$

Proof We give only the proof of last inequality and leave the rest for Exercise 1.4.2. From the relation between the weighted matrix norm and 2-norm, we get

$$\begin{aligned}
\|AB\|_{MM} &= \|M^{1/2}AN^{-1/2}N^{1/2}BM^{-1/2}\|_2 \\
&\le \|M^{1/2}AN^{-1/2}\|_2 \|N^{1/2}BM^{-1/2}\|_2 \\
&= \|A\|_{MN}\|B\|_{NM}.
\end{aligned}$$

Let $A \in \mathbb{C}^{n\times n}$, $\mathbf{x}, \mathbf{y} \in \mathbb{C}^n$, and A^* be the conjugate transpose matrix of A, then

$$(A\mathbf{x}, \mathbf{y}) = \mathbf{y}^*A\mathbf{x} = (A^*\mathbf{y})^*\mathbf{x} = (\mathbf{x}, A^*\mathbf{y}).$$

This shows a relation between the inner product and the conjugate transpose matrix. The weighted conjugate transpose matrix of A is the generalization of the conjugate transpose matrix of A.

Definition 1.4.1 Let $A \in \mathbb{C}^{m\times n}$, and M and N be Hermitian positive definite matrices of orders m and n respectively. The matrix $X \in \mathbb{C}^{n\times m}$ satisfying

$$(A\mathbf{x}, \mathbf{y})_M = (\mathbf{x}, X\mathbf{y})_N, \quad \text{for all } \mathbf{x} \in \mathbb{C}^n, \mathbf{y} \in \mathbb{C}^m$$

is called the weighted conjugate transpose matrix and denoted by $X = A^\#$.

From the above definition,

$$A^\# = N^{-1}A^*M, \quad A \in \mathbb{C}^{m \times n}$$
$$B^\# = M^{-1}B^*N, \quad B \in \mathbb{C}^{n \times m}. \tag{1.4.4}$$

In the special case when $A \in \mathbb{C}^{n \times n}$, $\mathbf{x}, \mathbf{y} \in \mathbb{C}^n$, and N is a Hermitian positive definite matrix of order n, then

$$(A\mathbf{x}, \mathbf{y})_N = (\mathbf{x}, A^\#\mathbf{y})_N,$$

where

$$A^\# = N^{-1}A^*N. \tag{1.4.5}$$

If $(A\mathbf{x}, \mathbf{y})_N = (\mathbf{x}, A\mathbf{y})_N$, then $A^\# = A$ and A is called the weighted self-conjugate matrix. The following lemmas list some useful properties of the weighted conjugate transpose matrix and the weighted matrix norm.

Lemma 1.4.2

$$(A + B)^\# = A^\# + B^\#, \quad A, B \in \mathbb{C}^{m \times n}; \tag{1.4.6}$$
$$(AB)^\# = B^\# A^\#, \quad A \in \mathbb{C}^{m \times n}, B \in \mathbb{C}^{n \times m}; \tag{1.4.7}$$
$$(A^\#)^\# = A, \quad A \in \mathbb{C}^{m \times n}; \tag{1.4.8}$$
$$(A^\#)^{-1} = (A^{-1})^\#, \quad A \in \mathbb{C}_n^{n \times n}. \tag{1.4.9}$$

Proof These conclusions can be easily verified by using Definition 1.4.1, Eqs. (1.4.4), and (1.4.5), and are left as exercises.   □

Lemma 1.4.3 Let $A \in \mathbb{C}^{m \times n}$, then

$$\|A\|_{MN} = \|A^\#\|_{NM}; \tag{1.4.10}$$
$$\|A\|_{MN}^2 = \|AA^\#\|_{MM} = \|A^\# A\|_{NN}. \tag{1.4.11}$$

Proof

$$\|A^\#\|_{NM} = \|N^{1/2}A^\# M^{-1/2}\|_2 = \|N^{-1/2}A^* M^{1/2}\|_2$$
$$= \|M^{1/2}AN^{-1/2}\|_2 = \|A\|_{MN},$$
$$\|AA^\#\|_{MM} = \|M^{1/2}AA^\# M^{-1/2}\|_2$$
$$= \|(M^{1/2}AN^{-1/2})(M^{1/2}AN^{-1/2})^*\|_2$$
$$= \|M^{1/2}AN^{-1/2}\|_2^2 = \|A\|_{MN}^2.$$

The proof of $\|A^\# A\|_{NN} = \|A\|_{MN}^2$ is similar.   □

The relations between the weighted generalized inverses and the solutions of linear equations are discussed in the following subsections.

1.4.2 The $\{1, 4N\}$ Inverse and the Minimum-Norm (N) Solution of a Consistent System of Linear Equations

Now that we have introduced the weighted norm in the previous subsection, we consider a generalized inverse in terms of finding the minimal weighted norm solution for a consistent system of linear equations. The following theorem gives a sufficient and necessary condition for such solution.

Theorem 1.4.1 Let $A \in \mathbb{C}^{m \times n}$, $\mathbf{b} \in \mathcal{R}(A)$, and N be a Hermitian positive definite matrix of order n, then $\mathbf{x} = X\mathbf{b}$ is the minimum-norm (N) solution of the consistent system of linear equations (1.1.1) if and only if X satisfies

$$
\begin{aligned}
&(1) \ AXA = A, \\
&(4N) \ (NXA)^* = NXA.
\end{aligned}
\tag{1.4.12}
$$

Proof By (1.2.4), the general solution of (1.1.1) is

$$\mathbf{x} = X\mathbf{b} + (I - XA)\mathbf{y}, \quad \forall \, \mathbf{y} \in \mathbb{C}^n,$$

where $X \in A\{1\}$. If $X\mathbf{b}$ is the minimum-norm (N) solution, then

$$\|X\mathbf{b}\|_N \le \|X\mathbf{b} + (I - XA)\mathbf{y}\|_N, \quad \forall \, \mathbf{b} \in \mathcal{R}(A), \ \mathbf{y} \in \mathbb{C}^n.$$

By Exercise 1.4.5,

$$
\begin{aligned}
&\|XA\mathbf{u}\|_N \le \|XA\mathbf{u} + (I - XA)\mathbf{y}\|_N, \ \forall \, \mathbf{u}, \mathbf{y} \in \mathbb{C}^n \\
\Leftrightarrow \ &(XA\mathbf{u}, \ (I - XA)\mathbf{y})_N = 0, && \forall \, \mathbf{u}, \mathbf{y} \in \mathbb{C}^n \\
\Leftrightarrow \ &(\mathbf{u}, \ (XA)^{\#}(I - XA)\mathbf{y})_N = 0, && \forall \, \mathbf{u}, \mathbf{y} \in \mathbb{C}^n \\
\Leftrightarrow \ &(XA)^{\#}(I - XA) = O \\
\Leftrightarrow \ &(XA)^{\#} = XA \\
\Leftrightarrow \ &(NXA)^* = NXA,
\end{aligned}
$$

which completes the proof. □

A matrix X satisfying (1.4.12) is called the generalized inverse for the minimum-norm (N) solution of the consistent system of linear equations (1.1.1), and is denoted by $X = A^{(1,4N)}$ or $X \in A\{1, 4N\}$, where $A\{1, 4N\}$ denotes the set of all the $\{1, 4N\}$ inverses of A.

1.4.3   The {1, 3M} Inverse and the Least-Squares (M) Solution of An Inconsistent System of Linear Equations

Previously, we introduced weights to the solution norm and considered its associated generalized inverse. We can also introduce weights to the residual norm, that is, we consider the problem of finding the least-squares solution in a weighted residual norm and its associated generalized inverse.

Theorem 1.4.2 *Let $A \in \mathbb{C}^{m \times n}$, $\mathbf{b} \in \mathbb{C}^m$, and M be a Hermitian positive definite matrix of order m, then $\mathbf{x} = X\mathbf{b}$ is the least-squares (M) solution of the inconsistent system of linear equations (1.1.2) if and only if X satisfies*

$$\begin{aligned}(1)\ & AXA = A, \\ (3M)\ & (MAX)^* = MAX.\end{aligned} \qquad (1.4.13)$$

Proof IF:

$$\begin{aligned}\|AX\mathbf{b} - \mathbf{b}\|_M &\le \|A\mathbf{x} - \mathbf{b}\|_M, \quad \forall\, \mathbf{x} \in \mathbb{C}^n,\ \mathbf{b} \in \mathbb{C}^m \\ &= \|AX\mathbf{b} - \mathbf{b} + A\mathbf{x} - AX\mathbf{b}\|_M, \quad \forall\, \mathbf{x} \in \mathbb{C}^n,\ \mathbf{b} \in \mathbb{C}^m \\ &= \|AX\mathbf{b} - \mathbf{b} + A\mathbf{w}\|_M, \quad \forall\, \mathbf{b} \in \mathbb{C}^m,\ \mathbf{w} = \mathbf{x} - X\mathbf{b} \in \mathbb{C}^n.\end{aligned}$$

By Exercise 1.4.5, the above inequality holds if and only if

$$\begin{aligned}&(A\mathbf{w},\ (AX - I)\mathbf{b})_M = 0, \quad \forall\, \mathbf{b} \in \mathbb{C}^m,\ \mathbf{w} \in \mathbb{C}^n \\ \Leftrightarrow\ & A^{\#}(AX - I) = O \\ \Leftrightarrow\ & A^{\#}AX = A^{\#}.\end{aligned} \qquad (1.4.14)$$

From (1.4.14), we have

$$(AX)^{\#} = X^{\#}A^{\#} = X^{\#}A^{\#}AX = (AX)^{\#}AX.$$

Thus $(AX)^{\#} = AX$, which is equivalent to $(MAX)^* = MAX$, and

$$A = (A^{\#}AX)^{\#} = X^{\#}A^{\#}A = (AX)^{\#}A = AXA.$$

Therefore (1.4.13) holds.
ONLY IF: If (1.4.13) holds, then

$$A^{\#}AX = A^{\#}(AX)^{\#} = (AXA)^{\#} = A^{\#},$$

i.e., (1.4.14) holds, equivalently, $X\mathbf{b}$ is the least-squares (M) solution of (1.1.2). $\square$

A matrix X satisfying (1.4.13) is called the generalized inverse for the least-squares (M) solution of the inconsistent system of linear equations (1.1.2), and is denoted by $X = A^{(1,3M)}$ or $X \in A\{1, 3M\}$, where $A\{1, 3M\}$ denotes the set of all the $\{1, 3M\}$ inverses of A.

1.4.4 Weighted Moore-Penrose Inverse and The Minimum-Norm (N) and Least-Squares (M) Solution of An Inconsistent System of Linear Equations

Finally, we introduce weights to both the solution norm and the residual norm and its associated generalized inverse, that is, the weighted Moore-Penrose inverse for expressing the minimal weighted norm and weighted least-squares solution.

Theorem 1.4.3 *Let $A \in \mathbb{C}^{m \times n}$ and $\mathbf{b} \in \mathbb{C}^m$, M and N be Hermitian positive definite matrices of orders m and n respectively, then $\mathbf{x} = X\mathbf{b}$ is the minimum-norm (N) and least-squares (M) solution if and only if X satisfies*

$$\begin{align}
&(1)\ AXA = A, \\
&(2)\ XAX = X, \\
&(3M)\ (MAX)^* = MAX, \\
&(4N)\ (NXA)^* = NXA.
\end{align} \tag{1.4.15}$$

Proof ONLY IF: Since $X\mathbf{b}$ is a least-squares (M) solution, (1) and (3M) of (1.4.15) hold by Theorem 1.4.2. It is clear that the general solution of the least-squares (M) solution of (1.1.2) is

$$X\mathbf{b} + (I - XA)\mathbf{z}, \quad X \in A\{1, 3M\}, \ \forall \mathbf{z} \in \mathbb{C}^n.$$

From the meaning of the minimum-norm (N) and least-squares (M) solution,

$$\begin{align}
&\|X\mathbf{b}\|_N \le \|X\mathbf{b} + (I - XA)\mathbf{z}\|_N, \quad \forall \mathbf{b} \in \mathbb{C}^m, \ \mathbf{z} \in \mathbb{C}^n \\
&\Leftrightarrow (X\mathbf{b}, \ (I - XA)\mathbf{z})_N = 0, \quad \forall \mathbf{b} \in \mathbb{C}^m, \ \mathbf{z} \in \mathbb{C}^n \\
&\Leftrightarrow X^\#(I - XA) = O \\
&\Leftrightarrow X^\# = X^\# XA.
\end{align} \tag{1.4.16}$$

By (1.4.16),

$$(XA)^\# = A^\# X^\# = A^\# X^\# XA = (XA)^\# XA.$$

Thus $(XA)^\# = XA$, which is equivalent to $(NXA)^* = NXA$, and

$$X = (X^\# XA)^\# = (XA)^\# X = XAX.$$

Therefore (2) and (4N) of (1.4.15) hold.

IF: If (1.4.15) holds, (1) and ($3M$) show that $X\mathbf{b}$ is a least-squares (M) solution of (1.1.2) by Theorem 1.4.2. Equations (2) and ($4N$) in (1.4.15) imply

$$X^{\#}XA = X^{\#}(XA)^{\#} = (XAX)^{\#} = X^{\#}.$$

Thus (1.4.16) holds, i.e., $X\mathbf{b}$ is the minimum-norm (N) and least-squares (M) solution of (1.1.2). $\qquad\square$

A matrix X satisfying (1.4.15) is called the generalized inverse for the minimum-norm (N) and least-squares (M) solution or the weighted Moore-Penrose inverse, and is denoted by $X = A_{MN}^{\dagger}$. It is readily found that X is unique.

The weighted least-squares and weighted minimum-norm problem can be reduced to the standard least-squares and minimum-norm problem by some simple transformations [5].

Let $A \in \mathbb{C}^{m \times n}$, $\mathbf{b} \in \mathbb{C}^m$, $\mathbf{x} \in \mathbb{C}^n$, and M and N be Hermitian positive definite matrices of orders m and n respectively. Set

$$\widetilde{A} = M^{1/2}AN^{-1/2}, \quad \widetilde{\mathbf{x}} = N^{1/2}\mathbf{x}, \quad \text{and} \quad \widetilde{\mathbf{b}} = M^{1/2}\mathbf{b},$$

then it is easy to verify that

$$\|A\mathbf{x} - \mathbf{b}\|_M = \|M^{1/2}A\mathbf{x} - M^{1/2}\mathbf{b}\|_2 = \|\widetilde{A}\widetilde{\mathbf{x}} - \widetilde{\mathbf{b}}\|_2$$

and

$$\|\mathbf{x}\|_N = \|\widetilde{\mathbf{x}}\|_2.$$

Thus we reduce the weighted least-squares problem to the standard least-squares problem:

$$\min_{\mathbf{x}} \|A\mathbf{x} - \mathbf{b}\|_M = \min_{\widetilde{\mathbf{x}}} \|\widetilde{A}\widetilde{\mathbf{x}} - \widetilde{\mathbf{b}}\|_2.$$

Moreover, there exists the least-squares solution generalized inverse $\widetilde{X}$ satisfying

$$\widetilde{A}\widetilde{X}\widetilde{A} = \widetilde{A}, \quad (\widetilde{A}\widetilde{X})^* = \widetilde{A}\widetilde{X}$$

such that $\widetilde{X}\mathbf{b}$ is the least-squares solution of $\widetilde{A}\widetilde{\mathbf{x}} = \widetilde{\mathbf{b}}$. Let

$$X = N^{-1/2}\widetilde{X}M^{1/2} \quad \text{or} \quad \widetilde{X} = N^{1/2}XM^{-1/2},$$

then

$$N^{1/2}\mathbf{x} = \widetilde{\mathbf{x}} = \widetilde{X}\widetilde{\mathbf{b}} = N^{1/2}XM^{-1/2}M^{1/2}\mathbf{b} = N^{1/2}X\mathbf{b}.$$

Thus

$$\tilde{\mathbf{x}} = \tilde{X}\tilde{\mathbf{b}} \Leftrightarrow \mathbf{x} = X\mathbf{b}$$
$$\tilde{A}\tilde{X}\tilde{A} = \tilde{A} \Leftrightarrow AXA = A,$$
$$(\tilde{A}\tilde{X})^* = \tilde{A}\tilde{X} \Leftrightarrow (MAX)^* = MAX.$$

It shows that there exists the least-squares solution generalized inverse X satisfying

$$AXA = A \quad \text{and} \quad (MAX)^* = MAX$$

such that $X\mathbf{b}$ is a least-squares (M) solution of $A\mathbf{x} = \mathbf{b}$.

This result is the same as Theorem 1.4.2. Theorems 1.4.1 and 1.4.3 can be obtained by the same method. It is omitted here.

The weighted Moore-Penrose inverse $A_{MN}^{\dagger}$ is a generalization of the Moore-Penrose inverse $A^{\dagger}$. Specifically, when $M = I_m$ and $N = I_n$, $A_{I_m,I_n}^{\dagger} = A^{\dagger}$.

Some properties of $A_{MN}^{\dagger}$ are given as follows.

Theorem 1.4.4 *Let $A \in \mathbb{C}^{m \times n}$. If M and N are Hermitian positive definite matrices of orders m and n respectively, then*

(1) $(A_{MN}^{\dagger})_{NM}^{\dagger} = A;$

(2) $(A_{MN}^{\dagger})^* = (A^*)_{N^{-1},M^{-1}}^{\dagger};$

(3) $A_{MN}^{\dagger} = (A^*MA)_{IN}^{\dagger}A^*M = N^{-1}A^*(AN^{-1}A^*)_{MI}^{\dagger};$

(4) $\mathcal{R}(A_{MN}^{\dagger}) = N^{-1}\mathcal{R}(A^*) = \mathcal{R}(A^{\#}),$
$\quad \mathcal{N}(A_{MN}^{\dagger}) = M^{-1}\mathcal{N}(A^*) = \mathcal{N}(A^{\#});$

(5) $\mathcal{R}(AA_{MN}^{\dagger}) = \mathcal{R}(A),$
$\quad \mathcal{N}(AA_{MN}^{\dagger}) = M^{-1}\mathcal{N}(A^*) = \mathcal{N}(A^{\#}),$
$\quad \mathcal{R}(A_{MN}^{\dagger}A) = N^{-1}\mathcal{R}(A^*) = \mathcal{R}(A^{\#}),$
$\quad \mathcal{N}(A_{MN}^{\dagger}A) = \mathcal{N}(A);$

(6) *If $A = FG$ is a full-rank factorization of A, then*

$$A_{MN}^{\dagger} = N^{-1}G^*(F^*MAN^{-1}G^*)^{-1}F^*M;$$

(7) $A_{MN}^{\dagger} = A_{N^{-1}\mathcal{R}(A^*),M^{-1}\mathcal{N}(A^*)}^{(1,2)} = A_{\mathcal{R}(A^{\#}),\mathcal{N}(A^{\#})}^{(1,2)};$

(8) $A_{MN}^{\dagger} = N^{-1/2}(M^{1/2}AN^{-1/2})^{\dagger}M^{1/2};$

(9) $A_{MN}^{\dagger} = A^{\#}YA^{\#}$, *where $Y \in (A^{\#}AA^{\#})\{1\}$.*

The proof is left to the reader as an exercise.

Exercises 1.4

1. Prove (1.4.1).
2. Prove (1.4.2) and (1.4.3).
3. Prove (1.4.6)–(1.4.9).

4. Prove Theorem 1.4.4.
5. Let N be a Hermitian positive definite matrix of order n, L a subspace of $\mathbb{C}^n$, and
   $\mathcal{R}(B) = L$, show that

$$\|\mathbf{x}\|_N \le \|\mathbf{x} + \mathbf{y}\|_N, \ \forall\, \mathbf{y} \in L$$
$$\Leftrightarrow \mathbf{x}^* N \mathbf{y} = 0, \qquad\qquad \forall\, \mathbf{y} \in L$$
$$\Leftrightarrow \mathbf{x}^* N B = \mathbf{0}^T.$$

1.5   Bott-Duffin Inverse and Its Generalization

To conclude this chapter on the equation solving generalized inverses, we introduce
another type of generalized inverse called Bott-Duffin inverse in terms of solving
constrained linear systems.

1.5.1   Bott-Duffin Inverse and the Solution of Constrained Linear Equations

Let $A \in \mathbb{C}^{n \times n}$, $\mathbf{b} \in \mathbb{C}^n$ and a subspace $L \subset \mathbb{C}^n$, the constrained linear equations

$$A\mathbf{x} + \mathbf{y} = \mathbf{b}, \quad \mathbf{x} \in L, \ \mathbf{y} \in L^{\perp} \tag{1.5.1}$$

arise in electrical network theory. It is readily found that the consistency of (1.5.1)
is equivalent to the consistency of the following linear equations

$$(AP_L + P_{L^{\perp}})\mathbf{z} = \mathbf{b}. \tag{1.5.2}$$

Also, the pair $(\mathbf{x}, \mathbf{y})$ is a solution of (1.5.1) if and only if

$$\mathbf{x} = P_L \mathbf{z}, \quad \mathbf{y} = P_{L^{\perp}} \mathbf{z} = \mathbf{b} - A P_L \mathbf{z}, \tag{1.5.3}$$

where $\mathbf{z}$ is a solution of (1.5.2), and P_L and $P_{L^{\perp}}$ are projectors on L and $L^{\perp}$ respectively. If the matrix $AP_L + P_{L^{\perp}}$ is nonsingular, then (1.5.2) is consistent for all $\mathbf{b} \in \mathbb{C}^n$
and the solution

$$\mathbf{x} = P_L(AP_L + P_{L^{\perp}})^{-1}\mathbf{b}, \quad \mathbf{y} = \mathbf{b} - A\mathbf{x}$$

is unique. This leads to another equation solving generalized inverses.

Definition 1.5.1 Let $A \in \mathbb{C}^{n \times n}$ and L be a subspace of $\mathbb{C}^n$. If $AP_L + P_{L^{\perp}}$ is nonsingular, then the Bott-Duffin inverse of A with respect to L, denoted by $A_{(L)}^{(-1)}$, is
defined by

$$A_{(L)}^{(-1)} = P_L(AP_L + P_{L^{\perp}})^{-1}. \tag{1.5.4}$$

The basic properties of $A_{(L)}^{(-1)}$ are given in the following theorem.

Theorem 1.5.1 *Suppose that* $AP_L + P_{L^\perp}$ *is nonsingular, then*

(1) *The constrained linear equations (1.5.1) has a unique solution*

$$\mathbf{x} = A_{(L)}^{(-1)}\mathbf{b}, \quad \mathbf{y} = (I - AA_{(L)}^{(-1)})\mathbf{b} \tag{1.5.5}$$

for any $\mathbf{b} \in \mathbb{C}^n$;

(2)

$$P_L = A_{(L)}^{(-1)}AP_L = P_L AA_{(L)}^{(-1)}, \tag{1.5.6}$$

$$A_{(L)}^{(-1)} = P_L A_{(L)}^{(-1)} = A_{(L)}^{(-1)}P_L; \tag{1.5.7}$$

(3)

$$\mathcal{R}(A_{(L)}^{(-1)}) = L, \quad \mathcal{N}(A_{(L)}^{(-1)}) = L^\perp; \tag{1.5.8}$$

(4)

$$A_{(L)}^{(-1)} = (AP_L)_{L,L^\perp}^{(1,2)} = (P_L A)_{L,L^\perp}^{(1,2)} = (P_L A P_L)_{L,L^\perp}^{(1,2)}; \tag{1.5.9}$$

(5)

$$(A_{(L)}^{(-1)})_{(L)}^{(-1)} = P_L A P_L.$$

Proof (1) This follows from the equivalence of (1.5.4) and (1.5.1)–(1.5.3).

(2) Premultiplying (1.5.4) with P_L and using $P_L^2 = P_L$, we have

$$P_L A_{(L)}^{(-1)} = A_{(L)}^{(-1)}. \tag{1.5.10}$$

From (1.5.4),

$$A_{(L)}^{(-1)}(AP_L + P_{L^\perp}) = P_L. \tag{1.5.11}$$

Postmultiplying (1.5.11) with P_L and using $P_{L^\perp}P_L = O$, we get

$$A_{(L)}^{(-1)}AP_L = P_L. \tag{1.5.12}$$

By (1.5.11) and (1.5.12), we have $A_{(L)}^{(-1)}P_{L^\perp} = O$, thus

$$A_{(L)}^{(-1)}P_L = A_{(L)}^{(-1)}(I - P_{L^\perp}) = A_{(L)}^{(-1)} \tag{1.5.13}$$

and

$$P_L A A^{(-1)}_{(L)} = P_L A P_L (A P_L + P_{L^\perp})^{-1}$$
$$= P_L (A P_L + P_{L^\perp})(A P_L + P_{L^\perp})^{-1}$$
$$= P_L. \tag{1.5.14}$$

It follows from (1.5.12), (1.5.14), (1.5.10) and (1.5.13) that (1.5.6) and (1.5.7) hold.
(3) From (1.5.6) and (1.5.7),

$$\dim(L) \le \operatorname{rank}(A^{(-1)}_{(L)}) \le \dim(L),$$

$$\mathcal{R}(A^{(-1)}_{(L)}) \subset \mathcal{R}(P_L) = L,$$

and

$$\mathcal{N}(A^{(-1)}_{(L)}) \supset \mathcal{N}(P_L) = L^\perp.$$

Therefore

$$\operatorname{rank}(A^{(-1)}_{(L)}) = \dim(L) \tag{1.5.15}$$

and

$$\mathcal{R}(A^{(-1)}_{(L)}) = L, \quad \mathcal{N}(A^{(-1)}_{(L)}) = L^\perp. \tag{1.5.16}$$

(4) Now, $A^{(-1)}_{(L)}$ is a $\{1, 2\}$ inverse of $A P_L$. By (1.5.6) and (1.5.7),

$$A^{(-1)}_{(L)} = A^{(-1)}_{(L)} A P_L A^{(-1)}_{(L)}. \tag{1.5.17}$$

Premultiplying the first equality in (1.5.6) with $A P_L$ gives

$$A P_L = A P_L A^{(-1)}_{(L)} A P_L. \tag{1.5.18}$$

From (1.5.16)–(1.5.18), we have

$$A^{(-1)}_{(L)} = (A P_L)^{(1,2)}_{L, L^\perp}.$$

The proof of

$$A^{(-1)}_{(L)} = (P_L A)^{(1,2)}_{L, L^\perp} = (P_L A P_L)^{(1,2)}_{L, L^\perp}$$

is similar.
(5) Firstly, we show that $(A^{(-1)}_{(L)})^{(-1)}_{(L)}$ is defined, i.e., $A^{(-1)}_{(L)} P_L + P_{L^\perp}$ is nonsingular.
From (1.5.7),

$$A^{(-1)}_{(L)} P_L + P_{L^\perp} = A^{(-1)}_{(L)} + P_{L^\perp}.$$

If $(A^{(-1)}_{(L)} + P_{L^\perp})\mathbf{x} = \mathbf{0}$, then

$$A^{(-1)}_{(L)}\mathbf{x} = -P_{L^\perp}\mathbf{x} \in L \cap L^\perp = \{\mathbf{0}\},$$

thus

$$A_{(L)}^{(-1)}\mathbf{x} = P_{L^\perp}\mathbf{x} = \mathbf{0} \quad \text{and} \quad \mathbf{x} \in L \cap L^\perp = \{\mathbf{0}\}.$$

Therefore $\mathbf{x} = \mathbf{0}$, which implies that $A_{(L)}^{(-1)} + P_{L^\perp}$ is nonsingular. Moreover, $A_{(L)}^{(-1)}P_L + P_{L^\perp}$ is also nonsingular.

Secondly, by (1.5.9), $P_L A P_L$ and $A_{(L)}^{(-1)}$ are $\{1, 2\}$ inverses of each other, and

$$\text{rank}(P_L A P_L) = \text{rank}(A_{(L)}^{(-1)}) = \dim(L)$$

by (1.5.15), which, together with

$$\mathcal{R}(P_L A P_L) \subset \mathcal{R}(P_L) = L \quad \text{and} \quad \mathcal{N}(P_L A P_L) \supset \mathcal{N}(P_L) = L^\perp,$$

shows that

$$\mathcal{R}(P_L A P_L) = L \quad \text{and} \quad \mathcal{N}(P_L A P_L) = L^\perp. \tag{1.5.19}$$

It then follows from (1.5.7) and (1.5.9) that

$$P_L A P_L = (A_{(L)}^{(-1)})_{L,L^\perp}^{(1,2)} = (A_{(L)}^{(-1)} P_L)_{L,L^\perp}^{(1,2)} = (A_{(L)}^{(-1)})_{(L)}^{(-1)}.$$

This completes the proof. $\qquad\qquad\qquad\qquad\qquad\qquad\qquad\qquad\qquad\square$

1.5.2 The Necessary and Sufficient Conditions for the Existence of the Bott-Duffin Inverse

In this subsection, we present some conditions equivalent to the nonsingularity of $A P_L + P_{L^\perp}$. Based on the equivalent conditions, we give more properties of the Bott-Duffin inverse.

Theorem 1.5.2 Let $A \in \mathbb{C}^{n \times n}$ and a subspace $L \subset \mathbb{C}^n$, then the following statements are equivalent:

(a) $A P_L + P_{L^\perp}$ is nonsingular;
(b) $AL \oplus L^\perp = \mathbb{C}^n$;
(c) $P_L AL \oplus L^\perp = \mathbb{C}^n$;
(d) $\text{rank}(P_L A P_L) = \text{rank}(P_L) = \dim(L)$;
(e) $\text{rank}(P_L A^* P_L) = \text{rank}(P_L) = \dim(L)$;
(f) $P_L A^* L \oplus L^\perp = \mathbb{C}^n$;
(g) $A^* L \oplus L^\perp = \mathbb{C}^n$;
(h) $L \oplus (A^* L)^\perp = \mathbb{C}^n$;
(i) $A^* P_L + P_{L^\perp}$ is nonsingular;

(j) $P_L A + P_{L^\perp}$ is nonsingular;

(k) $P_L A P_L + P_{L^\perp}$ is nonsingular.

Thus, each of the above conditions is necessary and sufficient for the existence of $A_{(L)}^{(-1)}$, the Bott-Duffin inverse of A with respect to L.

Proof (a) $\Leftrightarrow$ (b): If $AP_L + P_{L^\perp}$ is nonsingular, then $A_{(L)}^{(-1)}$ exists. By (1.5.7) and (1.5.6),

$$A_{(L)}^{(-1)} A A_{(L)}^{(-1)} = A_{(L)}^{(-1)} A P_L A_{(L)}^{(-1)} = P_L A_{(L)}^{(-1)} = A_{(L)}^{(-1)},$$

thus $A \in A_{(L)}^{(-1)}\{1\}$ and $A A_{(L)}^{(-1)} = P_{\mathcal{R}(A A_{(L)}^{(-1)}),\mathcal{N}(A A_{(L)}^{(-1)})}$. Since

$$\mathcal{R}(A A_{(L)}^{(-1)}) = A\mathcal{R}(A_{(L)}^{(-1)}) = AL$$

and

$$\mathcal{N}(A A_{(L)}^{(-1)}) = \mathcal{N}(A_{(L)}^{(-1)}) = L^\perp,$$

$AL \oplus L^\perp = \mathbb{C}^n$ holds.

Conversely, if $AL \oplus L^\perp = \mathbb{C}^n$ but $AP_L + P_{L^\perp}$ is singular, then $(AP_L + P_{L^\perp})$ $\mathbf{x} = \mathbf{0}$ for some nonzero $\mathbf{x} \in \mathbb{C}^n$, i.e., $AP_L\mathbf{x} + P_{L^\perp}\mathbf{x} = \mathbf{0}$. Since $AP_L\mathbf{x} \in AL$, $P_{L^\perp}\mathbf{x} \in L^\perp$, and $AL \oplus L^\perp = \mathbb{C}^n$, $AP_L\mathbf{x} = \mathbf{0}$ and $P_{L^\perp}\mathbf{x} = \mathbf{0}$, thus $\mathbf{x} \in L$ and $A\mathbf{x} = \mathbf{0}$. Hence $\dim(AL) \le \dim(L) - 1$. The contradiction establishes (a).

(a) $\Leftrightarrow$ (c): If $AP_L + P_{L^\perp}$ is nonsingular, then $A_{(L)}^{(-1)}$ exists. By (1.5.9), $A_{(L)}^{(-1)} = (P_L A)_{L,L^\perp}^{(1,2)}$, thus

$$P_L A A_{(L)}^{(-1)} = P_{\mathcal{R}(P_L A A_{(L)}^{(-1)}),\mathcal{N}(P_L A A_{(L)}^{(-1)})}$$
$$= P_{P_L A \mathcal{R}(A_{(L)}^{(-1)}),\mathcal{N}(A_{(L)}^{(-1)})}$$
$$= P_{P_L AL, L^\perp}$$

and $P_L AL \oplus L^\perp = \mathbb{C}^n$.

Conversely, the proof of (c) $\Rightarrow$ (a) is similar to that of (b) $\Rightarrow$ (a).

(c) $\Leftrightarrow$ (d): For any $\mathbf{x} \in P_L AL$, $\mathbf{x} = P_L A\mathbf{w}$, where $\mathbf{w} \in L$. Thus

$$\mathbf{x} = P_L A P_L \mathbf{z} \in \mathcal{R}(P_L A P_L).$$

Conversely, for all $\mathbf{x} \in \mathcal{R}(P_L A P_L)$, $\mathbf{x} = P_L A P_L\mathbf{z}$. Thus

$$\mathbf{x} = P_L A\mathbf{w}, \ \mathbf{w} \in L, \ \text{and} \ \mathbf{x} \in P_L AL.$$

Therefore $\mathcal{R}(P_L A P_L) = P_L AL$. Thus

$$P_L AL \oplus L^\perp = \mathbb{C}^n$$
$$\Leftrightarrow \text{rank}(P_L A P_L) = \dim(P_L AL) = \dim(L) = \text{rank}(P_L).$$

(d) $\Leftrightarrow$ (e): $\text{rank}(P_L A P_L) = \text{rank}((P_L A P_L)^*) = \text{rank}(P_L A^* P_L)$.
(e) $\Leftrightarrow$ (f): The proof is similar to (c) $\Leftrightarrow$ (d).
(f) $\Leftrightarrow$ (g): The proof is similar to (b) $\Leftrightarrow$ (c).
(g) $\Leftrightarrow$ (h): By orthogonal complement.
(g) $\Leftrightarrow$ (i): The proof is similar to (a) $\Leftrightarrow$ (b).
(i) $\Leftrightarrow$ (j): By conjugate transposition.
(c) $\Leftrightarrow$ (k): The proof is similar to (a) $\Leftrightarrow$ (b). $\square$

We show more properties of $A_{(L)}^{(-1)}$.

Theorem 1.5.3 *If $A_{(L)}^{(-1)}$ exists, then*

(1)

$$
\begin{aligned}
A_{(L)}^{(-1)} &= P_L(AP_L + P_{L^\perp})^{-1} \\
&= (P_L A + P_{L^\perp})^{-1} P_L && (1.5.20) \\
&= (P_L A P_L)^\dagger && (1.5.21) \\
&= P_L(P_L A P_L + P_{L^\perp})^{-1} \\
&= (P_L A P_L + P_{L^\perp})^{-1} P_L && (1.5.22) \\
&= (P_L A P_L + P_{L^\perp})^{-1} - P_{L^\perp} && (1.5.23) \\
&= A_{L,L^\perp}^{(2)}; && (1.5.24)
\end{aligned}
$$

(2) *Let U be a matrix whose columns form a basis for L, then $A_{(L)}^{(-1)}$ exists if and only if $U^* A U$ is nonsingular, in which case*

$$A_{(L)}^{(-1)} = U(U^* A U)^{-1} U^*; \qquad (1.5.25)$$

(3)

$$
\begin{aligned}
(A_{(L)}^{(-1)})^* &= (A^*)_{(L)}^{(-1)}, && (1.5.26) \\
A A_{(L)}^{(-1)} &= P_{AL, L^\perp}, && (1.5.27) \\
A_{(L)}^{(-1)} A &= P_{L,(A^*L)^\perp}. && (1.5.28)
\end{aligned}
$$

Proof (1) From (1.5.4), $A_{(L)}^{(-1)} = P_L(AP_L + P_{L^\perp})^{-1}$. Since

$$(P_L A + P_{L^\perp})P_L = P_L(AP_L + P_{L^\perp})$$

and $AP_L + P_{L^\perp}$ is nonsingular by (j) of Theorem 1.5.2,

$$A_{(L)}^{(-1)} = P_L(AP_L + P_{L^\perp})^{-1} = (P_L A + P_{L^\perp})^{-1} P_L,$$

i.e., (1.5.20) holds. From (1.5.9), (1.5.19) and (1.5.12),

$$A_{(L)}^{(-1)} = (P_L A P_L)_{L,L^\perp}^{(1,2)} = (P_L A P_L)^\dagger,$$

thus (1.5.21) holds.

Since

$$A_{(L)}^{(-1)} = (P_L(P_L A P_L)P_L)^\dagger = (P_L A P_L)_{(L)}^{(-1)} = P_L(P_L A P_L + P_{L^\perp})^{-1}$$

and

$$(P_L A P_L + P_{L^\perp})P_L = P_L(P_L A P_L + P_{L^\perp}),$$

Eq. (1.5.22) holds.

Since

$$\begin{aligned}
&(P_L A P_L + P_{L^\perp})((P_L A P_L)^\dagger + P_{L^\perp})\\
&= (P_L A P_L)(P_L A P_L)^\dagger + P_{L^\perp}\\
&= P_{\mathcal{R}(P_L A P_L)} + P_{L^\perp}\\
&= P_L + P_{L^\perp}\\
&= I,
\end{aligned}$$

we have

$$(P_L A P_L + P_{L^\perp})^{-1} = (P_L A P_L)^\dagger + P_{L^\perp}$$

and

$$A_{(L)}^{(-1)} = (P_L A P_L)^\dagger = (P_L A P_L + P_{L^\perp})^{-1} - P_{L^\perp}.$$

So (1.5.23) holds.

By (1.5.7) and (1.5.6),

$$A_{(L)}^{(-1)} A A_{(L)}^{(-1)} = A_{(L)}^{(-1)} A P_L A_{(L)}^{(-1)} = P_L A_{(L)}^{(-1)} = A_{(L)}^{(-1)}, \quad A_{(L)}^{(-1)} \in A\{2\}.$$

From (1.5.16): $\mathcal{R}(A_{(L)}^{(-1)}) = L$ and $\mathcal{N}(A_{(L)}^{(-1)}) = L^\perp$, $A_{(L)}^{(-1)} = A_{L,L^\perp}^{(2)}$, that is, (1.5.24) holds.

(2) If $A_{(L)}^{(-1)}$ exists, from (b) of Theorem 1.5.2, $AL \oplus L^\perp = \mathbb{C}^n$. Let U be a matrix whose columns form a basis for L, then $L = \mathcal{R}(U)$ and $L^\perp = \mathcal{N}(U^*)$. Now we prove that $U^* A U$ is nonsingular. Suppose $U^* A U \mathbf{x} = \mathbf{0}$, then

$$A U \mathbf{x} \in \mathcal{N}(U^*) = L^\perp$$

and

$$A U \mathbf{x} \in \mathcal{R}(AU) = A\mathcal{R}(U) = AL,$$

thus

$$A U \mathbf{x} \in AL \cap L^\perp = \{\mathbf{0}\}.$$

It is clear that the columns of AU span AL and AU is of full column rank, thus $\mathbf{x} = \mathbf{0}$. Therefore U^*AU is nonsingular. In this case,

$$\text{rank}(U^*AU) = \text{rank}(U) = \text{rank}(U^*).$$

By (5) of Theorem 1.3.7 and (1.5.24),

$$U(U^*AU)^{-1}U^* = A^{(2)}_{\mathcal{R}(U),\mathcal{N}(U^*)} = A^{(2)}_{L,L^\perp} = A^{(-1)}_{(L)}.$$

(3) From (1.5.24),

$$(A^{(-1)}_{(L)})^* = U(U^*A^*U)^{-1}U^* = (A^*)^{(-1)}_{(L)},$$

thus (1.5.26) holds. Since

$$\mathcal{R}(AA^{(-1)}_{(L)}) = A\mathcal{R}(A^{(-1)}_{(L)}) = AL \quad \text{and} \quad \mathcal{N}(AA^{(-1)}_{(L)}) = \mathcal{N}(A^{(-1)}_{(L)}) = L^\perp,$$

we get $AA^{(-1)}_{(L)} = P_{AL,L^\perp}$, which is (1.5.27). Since

$$\mathcal{R}(A^{(-1)}_{(L)}A) = \mathcal{R}(A^{(-1)}_{(L)}) = L$$

and

$$\mathcal{N}(A^{(-1)}_{(L)}A)^\perp = \mathcal{R}(A^*(A^{(-1)}_{(L)})^*) = A^*\mathcal{R}((A^{(-1)}_{(L)})^*) = A^*\mathcal{R}((A^*)^{(-1)}_{(L)}) = A^*L,$$

we have

$$\mathcal{N}(A^{(-1)}_{(L)}A) = (A^*L)^\perp.$$

Therefore $A^{(-1)}_{(L)}A = P_{L,(A^*L)^\perp}$, i.e., (1.5.28) holds. □

1.5.3 Generalized Bott-Duffin Inverse and Its Properties

Definition 1.5.1 of the Bott-Duffin inverse requires the nonsingularity of $AP_L + P_{L^\perp}$. In this subsection, we generalize the Bott-Duffin inverse by extending it to the case when $AP_L + P_{L^\perp}$ is singular. We first give some slightly extended concepts related to Hermitian positive definiteness and Hermitian nonnegative definiteness [6].

Definition 1.5.2 Let $A^* = A$. If A satisfies the condition

$$\mathbf{x}^*A\mathbf{x} > 0, \quad \text{for all } \mathbf{x} \in L \text{ and } \mathbf{x} \neq \mathbf{0},$$

then A is called an L-positive definite (L-p.d.) matrix.

Definition 1.5.3 Let $A^* = A$. If A satisfies the two conditions:
(1) $\mathbf{x}^* A \mathbf{x} \geq 0$ for all $\mathbf{x} \in L$ and
(2) $\mathbf{x}^* A \mathbf{x} = 0$, where $\mathbf{x} \in L$, implies $A \mathbf{x} = \mathbf{0}$,
then A is called an L-positive semidefinite (L-p.s.d.) matrix.

Now we introduce the generalized Bott-Duffin inverse.

Definition 1.5.4 Let $A \in \mathbb{C}^{n \times n}$ and a subspace $L \subset \mathbb{C}^n$, then

$$A_{(L)}^{(\dagger)} = P_L (A P_L + P_{L^\perp})^\dagger$$

is called the generalized Bott-Duffin inverse of A with respect to L.

When $A P_L + P_{L^\perp}$ is nonsingular, $A_{(L)}^{(\dagger)}$ exists and equals the Bott-Duffin inverse. Naturally, before studying the generalized Bott-Duffin inverse, we investigate $A P_L + P_{L^\perp}$. Some expressions of the range and null space of $A P_L + P_{L^\perp}$ are given in the following lemmas. We first consider the case of a general A.

Lemma 1.5.1 *For any $A \in \mathbb{C}^{n \times n}$ and a subspace $L \subset \mathbb{C}^n$, we have*
(1)

$$\mathcal{N}(P_L A + P_{L^\perp}) = \mathcal{N}(P_L A P_L + P_{L^\perp}) = (A^* L)^\perp \cap L = \mathcal{N}(P_L A P_L) \cap L,$$

(2)

$$\mathcal{R}(A P_L + P_{L^\perp}) = \mathcal{R}(P_L A P_L + P_{L^\perp}) = AL + L^\perp = P_L AL \oplus L^\perp.$$

Proof (1) Firstly,

$$\mathbf{x} \in \mathcal{N}(P_L A P_L + P_{L^\perp}) \Leftrightarrow P_L A P_L \mathbf{x} = -P_{L^\perp} \mathbf{x} = \mathbf{0}$$
$$\Leftrightarrow \mathbf{x} \in \mathcal{N}(P_L A P_L) \cap L.$$

Secondly,

$$\mathbf{x} \in \mathcal{N}(P_L A + P_{L^\perp}) \Leftrightarrow P_L A \mathbf{x} = -P_{L^\perp} \mathbf{x} = \mathbf{0}$$
$$\Leftrightarrow P_L A \mathbf{x} = \mathbf{0} \text{ and } \mathbf{x} \in L$$
$$\Leftrightarrow P_L \mathbf{x} = \mathbf{x}, \ \mathbf{x} \in L \text{ and } P_L A P_L \mathbf{x} = P_L A \mathbf{x} = \mathbf{0}$$
$$\Leftrightarrow \mathbf{x} \in \mathcal{N}(P_L A P_L) \cap L$$
$$\Leftrightarrow \mathbf{x} \in \mathcal{N}(A P_L P_L + P_{L^\perp}).$$

Thirdly, since
$$\mathcal{N}(P_L A)^\perp = \mathcal{R}(A^* P_L) = A^* \mathcal{R}(P_L) = A^* L,$$

$\mathcal{N}(P_L A) = (A^* L)^\perp$. Thus

$$\mathbf{x} \in \mathcal{N}(P_L A + P_{L^\perp}) \Leftrightarrow \mathbf{x} \in \mathcal{N}(P_L A) \text{ and } \mathbf{x} \in L$$
$$\Leftrightarrow \mathbf{x} \in (A^* L)^\perp \cap L.$$

(2) Firstly, by (1) above,

$$(\mathcal{R}(A P_L + P_{L^\perp}))^\perp = \mathcal{N}(P_L A^* + P_{L^\perp})$$
$$= \mathcal{N}(P_L A^* P_L + P_{L^\perp})$$
$$= (\mathcal{R}(P_L A P_L + P_{L^\perp}))^\perp.$$

Thus

$$\mathcal{R}(A P_L + P_{L^\perp}) = \mathcal{R}(P_L A P_L + P_{L^\perp}).$$

Secondly,

$$\mathbf{x} \in \mathcal{R}(A P_L + P_{L^\perp}) \quad \Leftrightarrow \quad \mathbf{x} = (A P_L + P_{L^\perp})\mathbf{z} = A P_L \mathbf{z} + P_{L^\perp}\mathbf{z} \in AL + L^\perp.$$

Thirdly,

$$\mathbf{x} \in \mathcal{R}(P_L A P_L + P_{L^\perp}) \quad \Leftrightarrow \quad \mathbf{x} = P_L A P_L \mathbf{z} + P_{L^\perp}\mathbf{z} \in P_L AL \oplus L^\perp.$$

The proof is completed. $\qquad\qquad\qquad\qquad\qquad\qquad\qquad\qquad\qquad\qquad$ □

Then we consider the case of an L-positive semidefinite A.

Lemma 1.5.2 *Let A be L-p.s.d. (including p.d., p.s.d., and L-p.d.), then we have*
(1)

$$\mathcal{N}(A P_L + P_{L^\perp}) = \mathcal{N}(P_L A + P_{L^\perp}) = \mathcal{N}(P_L A P_L + P_{L^\perp})$$
$$= \mathcal{N}(A) \cap L = \mathcal{N}(A P_L) \cap L,$$

(2)

$$\mathcal{R}(A P_L + P_{L^\perp}) = \mathcal{R}(P_L A + P_{L^\perp}) = \mathcal{R}(P_L A P_L + P_{L^\perp})$$
$$= \mathcal{R}(A) + L^\perp = P_L \mathcal{R}(A) \oplus L^\perp.$$

Proof (1) Firstly,

$$\mathbf{x} \in \mathcal{N}(A P_L + P_{L^\perp}) \quad \Leftrightarrow \quad A P_L \mathbf{x} = -P_{L^\perp}\mathbf{x}.$$

Multiplying the above equation on the left with P_L, we get

$$P_L A P_L \mathbf{x} = P_L(-P_{L^\perp}\mathbf{x}) = \mathbf{0}.$$

Thus

$$\mathbf{x}^* P_L A P_L \mathbf{x} = 0,$$

equivalently,

$$(P_L \mathbf{x})^* A (P_L \mathbf{x}) = 0.$$

Since A is L-p.s.d, we have

$$A P_L \mathbf{x} = \mathbf{0}$$

and

$$P_{L^\perp} \mathbf{x} = -A P_L \mathbf{x} = \mathbf{0},$$

thus

$$\mathbf{x} \in \mathcal{N}(A P_L + P_{L^\perp}) \Leftrightarrow A P_L \mathbf{x} = -P_{L^\perp} \mathbf{x} = \mathbf{0}$$
$$\Leftrightarrow \mathbf{x} \in \mathcal{N}(A P_L) \cap L.$$

Secondly,

$$\mathbf{x} \in \mathcal{N}(A P_L) \cap L \Leftrightarrow \mathbf{x} \in \mathcal{N}(A P_L) \text{ and } P_L \mathbf{x} = \mathbf{x}, \ \mathbf{x} \in L$$
$$\Leftrightarrow \mathbf{0} = A P_L \mathbf{x} = A \mathbf{x}, \ \mathbf{x} \in L$$
$$\Leftrightarrow \mathbf{x} \in \mathcal{N}(A) \cap L.$$

Thirdly,

$$\mathbf{x} \in \mathcal{N}(A) \cap L \Leftrightarrow \mathbf{x} \in \mathcal{N}(P_L A) \text{ and } \mathbf{x} \in L$$
$$\Leftrightarrow P_L A \mathbf{x} = \mathbf{0} \text{ and } P_{L^\perp} \mathbf{x} = \mathbf{0}$$
$$\Leftrightarrow \mathbf{x} \in \mathcal{N}(P_L A + P_{L^\perp}).$$

Lastly, it is clear that

$$\mathcal{N}(P_L A + P_{L^\perp}) = \mathcal{N}(P_L A P_L + P_{L^\perp})$$

by (1) of Lemma 1.5.1.

(2) Let A be L-p.s.d. and

$$\mathcal{N}(A P_L + P_{L^\perp})^\perp = \mathcal{N}(P_L A + P_{L^\perp})^\perp = \mathcal{N}(P_L A P_L + P_{L^\perp})^\perp,$$

then we have

$$\mathcal{R}(P_L A + P_{L^\perp}) = \mathcal{R}(A P_L + P_{L^\perp}) = \mathcal{R}(P_L A P_L + P_{L^\perp}).$$

On the other hand,

$$\mathbf{x} \in \mathcal{R}(P_L A + P_{L^\perp}) \quad \Leftrightarrow \quad \mathbf{x} = P_L A \mathbf{z} + P_{L^\perp} \mathbf{z} \in P_L \mathcal{R}(A) \oplus L^\perp$$

and

$$\mathbf{x} \in \mathcal{R}(A P_L + P_{L^\perp}) \quad \Leftrightarrow \quad \mathbf{x} = A P_L \mathbf{z} + P_{L^\perp} \mathbf{z} \in \mathcal{R}(A) + L^\perp,$$

which completes the proof. $\qquad\qquad\qquad\qquad\qquad\qquad\qquad\qquad\qquad\qquad$ □

Lemma 1.5.3 *Let A be L-p.s.d. and $S = \mathcal{R}(P_L A)$, then we have the following properties:*

(1)

$$(A P_L + P_{L^\perp})(A P_L + P_{L^\perp})^\dagger = (A P_L + P_{L^\perp})^\dagger (A P_L + P_{L^\perp})$$
$$= P_S + P_{L^\perp};$$
$$(P_L A + P_{L^\perp})(P_L A + P_{L^\perp})^\dagger = (P_L A + P_{L^\perp})^\dagger (P_L A + P_{L^\perp})$$
$$= P_S + P_{L^\perp}.$$

(2)

$$P_S = P_L A (P_L A + P_{L^\perp})^\dagger = (A P_L + P_{L^\perp})^\dagger A P_L;$$
$$P_{L^\perp} = P_{L^\perp} (P_L A + P_{L^\perp})^\dagger = (A P_L + P_{L^\perp})^\dagger P_{L^\perp}.$$

(3)

$$P_L (A P_L + P_{L^\perp})^\dagger P_{L^\perp} = O;$$
$$P_{L^\perp} (A P_L + P_{L^\perp})^\dagger A P_L = O.$$

Proof (1) By Lemma 1.5.2 and Theorem 1.3.2,

$$(A P_L + P_{L^\perp})(A P_L + P_{L^\perp})^\dagger = P_{\mathcal{R}(A P_L + P_{L^\perp})} = P_{\mathcal{R}(P_L A + P_{L^\perp})}$$
$$= P_{S \oplus L^\perp} = P_S + P_{L^\perp},$$

$$(A P_L + P_{L^\perp})^\dagger (A P_L + P_{L^\perp}) = P_{\mathcal{R}((A P_L + P_{L^\perp})^*)} = P_{\mathcal{R}(P_L A + P_{L^\perp})}$$
$$= P_{S \oplus L^\perp} = P_S + P_{L^\perp},$$

$$(P_L A + P_{L^\perp})(P_L A + P_{L^\perp})^\dagger = P_{\mathcal{R}(P_L A + P_{L^\perp})} = P_S + P_{L^\perp},$$

and

$$(P_L A + P_{L^\perp})^\dagger (P_L A + P_{L^\perp}) = P_{\mathcal{R}((P_L A + P_{L^\perp})^*)} = P_{\mathcal{R}(A P_L + P_{L^\perp})}$$
$$= P_S + P_{L^\perp}.$$

(2) Since
$$P_S P_L = P_S = P_L P_S \quad \text{and} \quad P_S P_{L^\perp} = O = P_{L^\perp} P_S,$$

multiplying the first equation in (1) on the right with P_L and $P_{L^\perp}$ gives

$$(AP_L + P_{L\perp})^\dagger AP_L = P_S \quad \text{and} \quad (AP_L + P_{L\perp})^\dagger P_{L\perp} = P_{L\perp}$$

respectively. Multiplying the second equation in (1) on the left with P_L and $P_{L\perp}$ gives

$$P_L A(P_L A + P_{L\perp})^\dagger = P_S \quad \text{and} \quad P_{L\perp}(P_L A + P_{L\perp})^\dagger = P_{L\perp}$$

respectively.

(3) Multiplying the first and second equations in (2) on the left with $P_{L\perp}$ and P_L respectively gives

$$P_{L\perp}(AP_L + P_{L\perp})^\dagger AP_L = P_{L\perp} P_S = O$$

and

$$P_L(AP_L + P_{L\perp})^\dagger P_{L\perp} = P_L P_{L\perp} = O.$$

This completes the proof.   □

The properties of $A_{(L)}^{(\dagger)}$ are given as follows:

Theorem 1.5.4 *Let A be L-p.s.d. and $S = \mathcal{R}(P_L A)$ and $T = \mathcal{R}(AP_L)$, then $A_{(L)}^{(\dagger)}$ has the following properties:*

(1)

$$A_{(L)}^{(\dagger)} = P_L A_{(L)}^{(\dagger)} = A_{(L)}^{(\dagger)} P_L = P_L A_{(L)}^{(\dagger)} P_L; \tag{1.5.29}$$

$$A_{(L)}^{(\dagger)} AP_L = P_L AA_{(L)}^{(\dagger)} = P_S; \tag{1.5.30}$$

$$\mathcal{R}(A_{(L)}^{(\dagger)}) = S \quad \text{and} \quad \mathcal{N}(A_{(L)}^{(\dagger)}) = S^\perp; \tag{1.5.31}$$

$$AP_S = AP_L \quad \text{and} \quad P_S A = P_L A; \tag{1.5.32}$$

$$P_L(A - AA_{(L)}^{(\dagger)} A) = (A - AA_{(L)}^{(\dagger)} A)P_L = O. \tag{1.5.33}$$

(2)

$$A_{(L)}^{(\dagger)} = A_{S,S^\perp}^{(2)} = (AP_L)_{S,S^\perp}^{(1,2)} = (P_L A)_{S,S^\perp}^{(1,2)}$$
$$= (P_L AP_L)_{S,S^\perp}^{(1,2)} = (P_L AP_L)^\dagger; \tag{1.5.34}$$

$$AA_{(L)}^{(\dagger)} = P_{T,S^\perp} \quad \text{and} \quad A_{(L)}^{(\dagger)} A = P_{S,T^\perp}. \tag{1.5.35}$$

(3)

$$A_{(L)}^{(\dagger)} = P_L(AP_L + P_{L\perp})^\dagger = (P_L A + P_{L\perp})^\dagger P_L \tag{1.5.36}$$
$$= P_L(P_L AP_L + P_{L\perp})^\dagger$$
$$= (P_L AP_L + P_{L\perp})^\dagger P_L \tag{1.5.37}$$
$$= (P_L AP_L + P_{L\perp})^\dagger - P_{L\perp}. \tag{1.5.38}$$

(4)

$$A_{(L)}^{(\dagger)} = A_{(S)}^{(-1)}$$
$$= P_S(AP_S + P_{S^\perp})^{-1}$$
$$= (P_S A + P_{S^\perp})^{-1} P_S. \tag{1.5.39}$$

Proof (1) From Definition 1.5.4, $A_{(L)}^{(\dagger)} = P_L(AP_L + P_{L^\perp})^\dagger$. The premultiplication of P_L gives $P_L A_{(L)}^{(\dagger)} = A_{(L)}^{(\dagger)}$. Also, by the first equation in (3) of Lemma 1.5.3, the postmultiplication of P_L gives

$$A_{(L)}^{(\dagger)} P_L = P_L(AP_L + P_{L^\perp})^\dagger P_L + P_L(AP_L + P_{L^\perp})^\dagger P_{L^\perp}$$
$$= P_L(AP_L + P_{L^\perp})^\dagger (P_L + P_{L^\perp})$$
$$= A_{(L)}^{(\dagger)}.$$

Consequently,

$$P_L A_{(L)}^{(\dagger)} P_L = P_L A_{(L)}^{(\dagger)} = A_{(L)}^{(\dagger)}.$$

Thus (1.5.29) holds.

From the first equation in (1) of Lemma 1.5.3, we have

$$(AP_L + P_{L^\perp})(AP_L + P_{L^\perp})^\dagger = P_S + P_{L^\perp}.$$

Since $P_L P_{L^\perp} = O$ and $P_L P_S = P_S$, premultiplying the above equation with P_L gives

$$P_L AP_L(AP_L + P_{L^\perp})^\dagger = P_S.$$

Thus $P_L A A_{(L)}^{(\dagger)} = P_S$. Again, by the first equation in (1) of Lemma 1.5.3, multiplying

$$(AP_L + P_{L^\perp})^\dagger (AP_L + P_{L^\perp}) = P_S + P_{L^\perp}$$

on both the left and right by P_L gives $A_{(L)}^{(\dagger)} AP_L = P_S$. Thus (1.5.30) holds.

By (2) of Lemma 1.5.2,

$$\mathcal{R}(A_{(L)}^{(\dagger)}) = P_L \mathcal{R}(AP_L + P_{L^\perp})$$
$$= P_L(\mathcal{R}(A) + L^\perp)$$
$$= P_L \mathcal{R}(A)$$
$$= \mathcal{R}(P_L A)$$
$$= S.$$

From (1.5.30), we have $\mathcal{N}(A_{(L)}^{(\dagger)}) \subset S^\perp$. Since

$$\dim(\mathcal{N}(A_{(L)}^{(\dagger)})) = n - \dim(\mathcal{R}(A_{(L)}^{(\dagger)})) = n - \dim(S) = \dim(S^{\perp}),$$

$\mathcal{N}(A_{(L)}^{(\dagger)}) = S^{\perp}$. Thus (1.5.31) holds.

Since $P_L A (P_L A)^{\dagger} P_L A = P_L A$,

$$(P_L A)^* ((P_L A)(P_L A)^{\dagger})^* = (P_L A)^*;$$
$$\Rightarrow A P_L (P_L A)(P_L A)^{\dagger} = A P_L;$$
$$\Rightarrow A P_L A (P_L A)^{\dagger} = A P_L;$$
$$\Rightarrow A P_{\mathcal{R}(P_L A)} = A P_L.$$

Thus $A P_S = A P_L$.

The proof of $P_L A = P_S A$ is similar to $A P_L (A P_L)^{\dagger} A P_L = A P_L$. Thus (1.5.32) holds.

It follows from (1.5.29) and (1.5.30) that

$$A P_L = A P_S = A A_{(L)}^{(\dagger)} A P_L \quad \text{and} \quad P_L A = P_S A = P_L A A_{(L)}^{(\dagger)} A,$$

therefore, we have

$$(A - A A_{(L)}^{(\dagger)} A) P_L = O \quad \text{and} \quad P_L (A - A A_{(L)}^{(\dagger)} A) = O.$$

Thus (1.5.33) holds.

(2) From (1.5.29), (1.5.30), and (1.5.31), we have

$$A_{(L)}^{(\dagger)} A A_{(L)}^{(\dagger)} = A_{(L)}^{(\dagger)} A P_L A_{(L)}^{(\dagger)} = P_S A_{(L)}^{(\dagger)} = A_{(L)}^{(\dagger)},$$

which implies that $A_{(L)}^{(\dagger)} \in A\{2\}$ and $A_{(L)}^{(\dagger)} = A_{S,S^{\perp}}^{(2)}$.

From (1.5.30) and (1.5.32),

$$A P_L A_{(L)}^{(\dagger)} A P_L = A P_L P_S = A P_S = A P_L,$$

therefore $A_{(L)}^{(\dagger)} \in A P_L\{1\}$.

From (1.5.29) and and $A_{(L)}^{(\dagger)} \in A\{2\}$,

$$A_{(L)}^{(\dagger)} A P_L A_{(L)}^{(\dagger)} = A_{(L)}^{(\dagger)} A A_{(L)}^{(\dagger)} = A_{(L)}^{(\dagger)},$$

therefore $A_{(L)}^{(\dagger)} \in (A P_L)\{2\}$. Thus $A_{(L)}^{(\dagger)} = (A P_L)_{S,S^{\perp}}^{(1,2)}$.

Similarly, we can prove $A_{(L)}^{(\dagger)} = (P_L A)_{S,S^{\perp}}^{(1,2)}$ and $A_{(L)}^{(\dagger)} = (P_L A P_L)_{S,S^{\perp}}^{(1,2)}$.

Finally, since A is L-p.s.d.,

$$\mathbf{x} \in \mathcal{N}(P_L A P_L) \Leftrightarrow P_L A P_L \mathbf{x} = \mathbf{0}$$
$$\Leftrightarrow \mathbf{x}^* P_L A P_L \mathbf{x} = \mathbf{0}$$
$$\Leftrightarrow (P_L \mathbf{x})^* A (P_L \mathbf{x}) = \mathbf{0}$$
$$\Leftrightarrow A P_L \mathbf{x} = \mathbf{0}$$
$$\Leftrightarrow \mathbf{x} \in \mathcal{N}(A P_L) = S^\perp,$$

implying that $\mathcal{N}(P_L A P_L) = \mathcal{N}(A P_L) = S^\perp$. Moreover, $\mathcal{R}(P_L A P_L) = \mathcal{N}(P_L A P_L)^\perp = S$. It then follows that

$$A_{(L)}^{(\dagger)} = (P_L A P_L)_{S,S^\perp}^{(1,2)}$$
$$= (P_L A P_L)_{\mathcal{R}(P_L A P_L),\mathcal{N}(P_L A P_L)}^{(1,2)}$$
$$= (P_L A P_L)^\dagger.$$

Thus (1.5.34) holds.

Noting that

$$\mathcal{R}(A A_{(L)}^{(\dagger)}) = A\mathcal{R}(A_{(L)}^{(\dagger)}) = AS$$
$$= A\mathcal{R}(P_L A) = \mathcal{R}(A P_L P_L A)$$
$$= \mathcal{R}(A P_L (A P_L)^*) = \mathcal{R}(A P_L)$$
$$= T$$

and

$$\mathcal{N}(A A_{(L)}^{(\dagger)}) = \mathcal{N}(A_{(L)}^{(\dagger)}) = S^\perp,$$

we have $A A_{(L)}^{(\dagger)} = P_{T,S^\perp}$.

The proof of $A_{(L)}^{(\dagger)} A = P_{S,T^\perp}$ is similar. Thus (1.5.35) holds.

(3) From (1.5.34), we have $(A_{(L)}^{(\dagger)})^* = (P_L A P_L)^\dagger = A_{(L)}^{(\dagger)}$ and

$$A_{(L)}^{(\dagger)} = (P_L A P_L)^\dagger = (P_L (P_L A P_L) P_L)^\dagger$$
$$= (P_L A P_L)_{(L)}^{(\dagger)} = P_L (P_L A P_L + P_{L^\perp})^\dagger$$
$$= (P_L A P_L + P_{L^\perp})^\dagger P_L,$$

thus (1.5.37) holds. Since

$$A_{(L)}^{(\dagger)} = (P_L A P_L)^\dagger = (P_L A P_L P_L)^\dagger$$
$$= (A P_L)_{(L)}^{(\dagger)} = P_L (A P_L + P_{L^\perp})^\dagger$$
$$= (P_L A + P_{L^\perp})^\dagger P_L,$$

(1.5.36) holds.

Note that

$$(P_L A P_L + P_{L^\perp})^\dagger = (P_L A P_L)^\dagger + P_{L^\perp} = A^{(\dagger)}_{(L)} + P_{L^\perp}.$$

Thus (1.5.38) holds.

(4) Since

$$AS = A\mathcal{R}(A^{(\dagger)}_{(L)}) = \mathcal{R}(AA^{(\dagger)}_{(L)}) \quad \text{and} \quad S^\perp = \mathcal{N}(A^{(\dagger)}_{(L)}) = \mathcal{N}(AA^{(\dagger)}_{(L)}),$$

we have

$$AS \oplus S^\perp = \mathcal{R}(AA^{(\dagger)}_{(L)}) \oplus \mathcal{N}(AA^{(\dagger)}_{(L)}) = \mathbb{C}^n.$$

By Theorem 1.5.2, $AP_S + P_{S^\perp}$ is nonsingular, then $A^{(-1)}_{(S)}$ exists and

$$A^{(-1)}_{(S)} = A^{(2)}_{S,S^\perp} = (P_S A P_S)^\dagger = (P_L A P_L)^\dagger = A^{(\dagger)}_{(L)}$$

and

$$A^{(-1)}_{(S)} = P_S (AP_S + P_{S^\perp})^{-1} = (P_S A + P_{S^\perp})^{-1} P_S.$$

Thus (1.5.39) holds.   □

1.5.4 The Generalized Bott-Duffin Inverse and the Solution of Linear Equations

The solution of the constrained linear equations (1.5.1) can be expressed by the Bott-Duffin inverse $A^{(-1)}_{(L)}$. The relations between the general solution of the linear system

$$\begin{cases} Ax + B^*y = b, \\ Bx \quad\quad\; = d \end{cases} \tag{1.5.40}$$

and the generalized Bott-Duffin inverse are discussed in this subsection.

The consistency of (1.5.40) is given in following theorem.

Theorem 1.5.5 Let $A \in \mathbb{C}^{n \times n}$, $B \in \mathbb{C}^{m \times n}$, $L = \mathcal{N}(B)$, $\mathbf{b} \in \mathbb{C}^n$, and $\mathbf{d} \in \mathbb{C}^m$, then (1.5.40) and the system

$$(AP_L + P_{L^\perp})\mathbf{u} = \mathbf{b} - AB^{(1)}\mathbf{d}, \quad \mathbf{d} \in \mathcal{R}(B) \tag{1.5.41}$$

have the same consistency. A necessary and sufficient condition for the consistency of (1.5.40) is

$$\mathbf{d} \in \mathcal{R}(B) \quad \text{and} \quad \mathbf{b} - AB^{(1)}\mathbf{d} \in AL + L^\perp. \tag{1.5.42}$$

If the condition (1.5.42) *is satisfied, then the vector pair* $(\mathbf{x}, \mathbf{y})$ *is a solution of* (1.5.40) *if and only if* $\mathbf{x}$ *and* $\mathbf{y}$ *can be expressed as*

$$\mathbf{x} = B^{(1)}\mathbf{d} + P_L \mathbf{u} \tag{1.5.43}$$

$$\mathbf{y} = (B^{(1)})^* P_{L^\perp}\mathbf{u} + P_{\mathcal{N}(B^*)}\mathbf{v}, \quad \textit{for an arbitrary } \mathbf{v}, \tag{1.5.44}$$

where $\mathbf{u}$ *is a solution of* (1.5.41).

Proof Let $(\mathbf{x}, \mathbf{y})$ be a solution pair of (1.5.40), then $\mathbf{d} \in \mathcal{R}(B)$, $BB^{(1)}\mathbf{d} = \mathbf{d}$, and $\mathbf{x}$ and $\mathbf{y}$ satisfy

$$A(\mathbf{x} - B^{(1)}\mathbf{d}) + B^*\mathbf{y} = \mathbf{b} - AB^{(1)}\mathbf{d}, \tag{1.5.45}$$

$$B(\mathbf{x} - B^{(1)}\mathbf{d}) = \mathbf{0}. \tag{1.5.46}$$

Set $\mathbf{u} = (\mathbf{x} - B^{(1)}\mathbf{d}) + B^*\mathbf{y}$. Now, from (1.5.46), $\mathbf{x} - B^{(1)}\mathbf{d} \in \mathcal{N}(B) = L$, and $B^*\mathbf{y} \in \mathcal{R}(B^*) = \mathcal{N}(B)^\perp = L^\perp$, we have

$$\mathbf{x} - B^{(1)}\mathbf{d} = P_L\mathbf{u}, \quad \text{i.e., } \mathbf{x} = B^{(1)}\mathbf{d} + P_L\mathbf{u},$$

which is (1.5.43). Since $B^*\mathbf{y} = P_{L^\perp}\mathbf{u}$, we have

$$\mathbf{y} = (B^{(1)})^* P_{L^\perp}\mathbf{u} + P_{\mathcal{N}(B^*)}\mathbf{v},$$

for an arbitrary $\mathbf{v}$, which is (1.5.44). From (1.5.45), we obtain

$$AP_L\mathbf{u} + P_{L^\perp}\mathbf{u} = \mathbf{b} - AB^{(1)}\mathbf{d}.$$

This shows that $\mathbf{u}$ is a solution of (1.5.41) and by Lemma 1.5.1

$$\mathbf{b} - AB^{(1)}\mathbf{d} \in \mathcal{R}(AP_L + P_{L^\perp}) = AL + L^\perp,$$

which is (1.5.42).

On the other hand, let $(\mathbf{x}_0, \mathbf{y}_0)$ be a solution pair of (1.5.40), then

$$\begin{aligned} \mathbf{x}_0 &= B^{(1)}\mathbf{d} + (I - BB^{(1)})\mathbf{x}_0 \\ &= B^{(1)}\mathbf{d} + P_L(I - BB^{(1)})\mathbf{x}_0 \\ &= B^{(1)}\mathbf{d} + P_L((I - BB^{(1)})\mathbf{x}_0 + B^*\mathbf{y}_0) \\ &= B^{(1)}\mathbf{d} + P_L\mathbf{u}, \end{aligned}$$

which is (1.5.43), and

$$B^* \mathbf{y}_0 = \mathbf{b} - A\mathbf{x}_0$$
$$= \mathbf{b} - A(B^{(1)}\mathbf{d} + P_L\mathbf{u})$$
$$= \mathbf{b} - AB^{(1)}\mathbf{d} - AP_L\mathbf{u}$$
$$= \mathbf{b} - AB^{(1)}\mathbf{d} - (\mathbf{b} - AB^{(1)}\mathbf{d} - P_{L^\perp}\mathbf{u})$$
$$= P_{L^\perp}\mathbf{u}$$

and

$$\mathbf{y}_0 = B^{(1)*} P_{L^\perp}\mathbf{u} + (I - B^{(1)*} B^*)\mathbf{y}_0$$
$$= B^{(1)*} P_{L^\perp}\mathbf{u} + P_{\mathcal{N}(B^*)}\mathbf{y}_0,$$

which is (1.5.44).

Conversely, if the condition (1.5.42) is satisfied, then the system (1.5.41) is consistent. Let $\mathbf{u}$ be a solution of (1.5.41) and $\mathbf{x}$ and $\mathbf{y}$ be expressed as in (1.5.43) and (1.5.44), then

$$B\mathbf{x} = B(B^{(1)}\mathbf{d} + P_L\mathbf{u}) = BB^{(1)}\mathbf{d} = \mathbf{d}.$$

Using (1.5.41) and

$$B^*(B^{(1)})^* = P_{\mathcal{R}(B^*(B^{(1)})^*), \mathcal{N}(B^*(B^{(1)})^*)} = P_{\mathcal{R}(B^*), M} = P_{L^\perp, M},$$

where $M = \mathcal{N}(B^*(B^{(1)})^*)$, we have

$$A\mathbf{x} + B^*\mathbf{y} = AB^{(1)}\mathbf{d} + AP_L\mathbf{u} + B^*(B^{(1)})^* P_{L^\perp}\mathbf{u} + B^* P_{\mathcal{N}(B^*)}\mathbf{v}$$
$$= AB^{(1)}\mathbf{d} + AP_L\mathbf{u} + P_{L^\perp}\mathbf{u}$$
$$= AB^{(1)}\mathbf{d} + \mathbf{b} - AB^{(1)}\mathbf{d}$$
$$= \mathbf{b}.$$

This shows that $(\mathbf{x}, \mathbf{y})$ is a solution pair of (1.5.40). $\qquad\square$

The general solution of (1.5.40) is given as follows.

Theorem 1.5.6 *Let* $L = \mathcal{N}(B)$. *If the condition* (1.5.42) *is satisfied, then the general solution of* (1.5.40) *is given by*

$$\mathbf{x} = A_{(L)}^{(\dagger)}\mathbf{b} + (I - A_{(L)}^{(\dagger)}A)B^{(1)}\mathbf{d} + P_L P_{\mathcal{N}(AP_L + P_{L^\perp})}\mathbf{z}, \tag{1.5.47}$$
$$\mathbf{y} = (B^{(1)})^*(I - AA_{(L)}^{(\dagger)})\mathbf{b} + (B^{(1)})^*(AA_{(L)}^{(\dagger)}A - A)B^{(1)}\mathbf{d}$$
$$\quad - (B^{(1)})^* AP_L P_{\mathcal{N}(AP_L + P_{L^\perp})}\mathbf{z} + P_{\mathcal{N}(B^*)}\mathbf{v}, \tag{1.5.48}$$

for arbitrary $\mathbf{z} \in \mathbb{C}^n$ *and* $\mathbf{v} \in \mathbb{C}^m$.

Proof From (1.2.4), the general solution of (1.5.41) is

$$\mathbf{u} = (AP_L + P_{L^\perp})^\dagger (\mathbf{b} - AB^{(1)}\mathbf{d}) + P_{\mathcal{N}(AP_L + P_{L^\perp})}\mathbf{z}, \tag{1.5.49}$$

for an arbitrary $\mathbf{z}$. Substituting $\mathbf{u}$ in (1.5.43) with the above equation, we obtain (1.5.47).

By (1.5.42), we have

$$(AP_L + P_{L^\perp})(AP_L + P_{L^\perp})^\dagger(\mathbf{b} - AB^{(1)}\mathbf{d}) = \mathbf{b} - AB^{(1)}\mathbf{d},$$

equivalently,

$$P_{L^\perp}(AP_L + P_{L^\perp})^\dagger(\mathbf{b} - AB^{(1)}\mathbf{d}) = (I - AA_{(L)}^{(\dagger)})(\mathbf{b} - AB^{(1)}\mathbf{d}). \tag{1.5.50}$$

From $(AP_L + P_{L^\perp})P_{\mathcal{N}(AP_L+P_{L^\perp})} = O$, we have

$$P_{L^\perp}P_{\mathcal{N}(AP_L+P_{L^\perp})} = -AP_L P_{\mathcal{N}(AP_L+P_{L^\perp})}. \tag{1.5.51}$$

Substituting $\mathbf{u}$ in (1.5.44) with (1.5.49) and using (1.5.50) and (1.5.51), we get

$$\begin{aligned}
\mathbf{y} &= (B^{(1)})^* P_{L^\perp}((AP_L + P_{L^\perp})^\dagger(\mathbf{b} - AB^{(1)}\mathbf{d}) + P_{\mathcal{N}(AP_L+P_{L^\perp})}\mathbf{z}) \\
&\quad + P_{\mathcal{N}(B^*)}\mathbf{v} \\
&= (B^{(1)})^*(I - AA_{(L)}^{(\dagger)})(\mathbf{b} - AB^{(1)}\mathbf{d}) \\
&\quad - (B^{(1)})^* AP_L P_{\mathcal{N}(AP_L+P_{L^\perp})}\mathbf{z} + P_{\mathcal{N}(B^*)}\mathbf{v} \\
&= (B^{(1)})^*(I - AA_{(L)}^{(\dagger)})\mathbf{b} + (B^{(1)})^*(AA_{(L)}^{(\dagger)}A - A)B^{(1)}\mathbf{d} \\
&\quad - (B^{(1)})^* AP_L P_{\mathcal{N}(AP_L+P_{L^\perp})}\mathbf{z} + P_{\mathcal{N}(B^*)}\mathbf{v}
\end{aligned}$$

for arbitrary $\mathbf{z} \in \mathbb{C}^n$ and $\mathbf{v} \in \mathbb{C}^m$. Thus (1.5.48) holds. $\square$

Corollary 1.5.1 *If A is L-p.s.d. in Theorem 1.5.6, then the general solution of (1.5.40) is simplified to*

$$\begin{aligned}
\mathbf{x} &= A_{(L)}^{(\dagger)}\mathbf{b} + (I - A_{(L)}^{(\dagger)}A)B^{(1)}\mathbf{d} + P_{\mathcal{N}(A)\cap L}\mathbf{z}, \\
\mathbf{y} &= (B^{(1)})^*(I - AA_{(L)}^{(\dagger)})\mathbf{b} + (B^{(1)})^*(AA_{(L)}^{(\dagger)}A - A)B^{(1)}\mathbf{d} + P_{\mathcal{N}(B^*)}\mathbf{v},
\end{aligned}$$

for arbitrary $\mathbf{z} \in \mathbb{C}^n$ and $\mathbf{v} \in \mathbb{C}^m$.

Proof If A is L-p.s.d., then

$$\mathcal{N}(AP_L + P_{L^\perp}) = \mathcal{N}(A) \cap L \subset L,$$

implying that

$$P_L P_{\mathcal{N}(AP_L+P_{L^\perp})} = P_{\mathcal{N}(AP_L+P_{L^\perp})} = P_{\mathcal{N}(A)\cap L}. \tag{1.5.52}$$

Since

$$\mathcal{R}(AB^{(1)}) \subset \mathcal{R}(A) \subset \mathcal{R}(A) + L^{\perp} = (\mathcal{N}(A) \cap L)^{\perp},$$

we have

$$
\begin{aligned}
(B^{(1)})^* A P_L P_{\mathcal{N}(AP_L + P_{L^{\perp}})} \mathbf{z} &= (B^{(1)})^* A P_{\mathcal{N}(A) \cap L} \mathbf{z} \\
&= (P_{\mathcal{N}(A) \cap L} A B^{(1)})^* \mathbf{z} \\
&= \mathbf{0}.
\end{aligned}
\tag{1.5.53}
$$

The proof is completed by substituting (1.5.52) and (1.5.53) into (1.5.47) and (1.5.48) respectively. $\square$

Exercises 1.5

1. Prove that
   (1) $P_{L^{\perp}} A_{(L)}^{(-1)} = O$;
   (2) $A_{(L)}^{(-1)} P_{L^{\perp}} = O$.
2. Let A be L-p.s.d. and $S = \mathcal{R}(P_L A)$, then
   (1) $P_S P_L = P_S = P_L P_S$;
   (2) $P_S P_{L^{\perp}} = O = P_{L^{\perp}} P_S$.

Remarks

This chapter surveys basic concepts and important results on the solution of linear equations and various generalized inverses. Sections 1.1, 1.2, 1.3 and 1.4 are based on [5, 7, 8], Sect. 1.5 is based on [6, 9].

As for the linear least squares problem, there are several excellent books and papers [10–13].

The ray uniqueness of the Moore-Penrose inverse is discussed in [14]. Some results on the weighted projector and weighted generalized inverse matrices can be found in [15].

The matrix Moore-Penrose inverse can be generalized to tensors [16, 17] and projectors [18].

Other types of generalized inverses are studied, for example, outer generalized inverse [19–22], MK-weighted generalized inverse [23], signed generalized inverse [24–26], scaled projections and generalized inverses [27, 28], core inverses [29, 30], and related randomized generalized SVD [31].

The analysis of a recursive least squares signal processing algorithm is given in [32]. Applications of the {2} inverse in statistics can be found in [33].

References

1. G.W. Stewart, *Introduction to Matrix Computation* (Academic Press, New York, 1973)
2. C.C. MacDuffe, *The Theory of Matrices* (Chelsea, New York, 1956)
3. N.S. Urquhart, Computation of generalized inverse matrtices which satisfy specified conditions. SIAM Rev. **10**, 216–218 (1968)
4. S. Zlobec, An explicit form of the Moore-Penrose inverse of an arbitrary complex matrix. SIAM Rev. **12**, 132–134 (1970)
5. A. Ben-Israel, T.N.E. Greville, *Generalized Inverses: Theory and Applications*, 2nd edn. (Springer Verlag, New York, 2003)
6. Y. Chen, Generalized Bott-Duffin inverse and its applications. Appl. Math. J. Chinese Univ. **4**, 247–257 (1989). in Chinese
7. S.L. Campbell, C.D. Meyer Jr., *Generalized Inverses of Linear Transformations* (Pitman, London, 1979)
8. X. He, W. Sun. *Introduction to Generalized Inverses of Matrices*. (Jiangsu Science and Technology Press, 1990). in Chinese
9. Y. Chen, The generalized Bott-Duffin inverse and its applications. Linear Algebra Appl. **134**, 71–91 (1990)
10. C.L. Lawson, R.J. Hanson, *Solving Least Squares Problems* (Prentice-Hall Inc, Englewood Cliffs, N.J., 1974)
11. Å. Björck, *Numerical Methods for Least Squares Problems* (SIAM, Philadelphia, 1996)
12. M. Wei, *Supremum and Stability of Weighted Pseudoinverses and Weighted Least Squares Problems Analysis and Computations* (Nova Science Publisher Inc, Huntington, NY, 2001)
13. L. Eldén, A weighted pseudoinverse, generalized singular values and constrained least squares problems. BIT **22**, 487–502 (1982)
14. C. Bu, W. Gu, J. Zhou, Y. Wei, On matrices whose Moore-Penrose inverses are ray unique. Linear Multilinear Algebra **64**(6), 1236–1243 (2016)
15. Y. Chen, On the weighted projector and weighted generalized inverse matrices. Acta Math. Appl. Sinica **6**, 282–291 (1983). in Chinese
16. L. Sun, B. Zheng, C. Bu, Y. Wei, Moore-Penrose inverse of tensors via Einstein product. Linear Multilinear Algebra **64**(4), 686–698 (2016)
17. J. Ji, Y. Wei, Weighted Moore-Penrose inverses and fundamental theorem of even-order tensors with Einstein product. Front. Math. China **12**(6), 1319–1337 (2017)
18. C. Deng, Y. Wei, Further results on the Moore-Penrose invertibility of projectors and its applications. Linear Multilinear Algebra **60**(1), 109–129 (2012)
19. D.S. Djordjević, P.S. Stanimirović, Y. Wei, The representation and approximations of outer generalized inverses. Acta Math. Hungar. **104**, 1–26 (2004)
20. Y. Wei, H. Wu, On the perturbation and subproper splittings for the generalized inverse $A_{T,S}^{(2)}$ of rectangular matrix A. J. Comput. Appl. Math. **137**, 317–329 (2001)
21. Y. Wei, H. Wu, $(T\text{-}S)$ splitting methods for computing the generalized inverse $A_{T,S}^{(2)}$ of rectangular systems. Int. J. Comput. Math. **77**, 401–424 (2001)
22. Y. Wei, N. Zhang, Condition number related with generalized inverse $A_{T,S}^{(2)}$ and constrained linear systems. J. Comput. Appl. Math. **157**, 57–72 (2003)
23. M. Wei, B. Zhang, Structures and uniqueness conditions of MK-weighted pseudoinverses. BIT **34**, 437–450 (1994)
24. J. Shao, H. Shan, The solution of a problem on matrices having signed generalized inverses. Linear Algebra Appl. **345**, 43–70 (2002)
25. J. Zhou, C. Bu, Y. Wei, Group inverse for block matrices and some related sign analysis. Linear Multilinear Algebra **60**, 669–681 (2012)
26. J. Zhou, C. Bu, Y. Wei, Some block matrices with signed Drazin inverses. Linear Algebra Appl. **437**, 1779–1792 (2012)
27. G.W. Stewart, On scaled projections and pseudoinverses. Linear Algebra Appl. **112**, 189–193 (1989)

28. M. Wei, Upper bound and stability of scaled pseudoinverses. Numer. Math. **72**(2), 285–293 (1995)
29. O.M. Baksalary, G. Trenkler, Core inverse of matrices. Linear Multilinear Algebra **58**(5–6), 681–697 (2010)
30. D. Rakić, N. Dinčić, D. Djordjević, Group, Moore-Penrose, core and dual core inverse in rings with involution. Linear Algebra Appl. **463**, 115–133 (2014)
31. Y. Wei, P. Xie, L. Zhang, Tikhonov regularization and randomized GSVD. SIAM J. Matrix Anal. Appl. **37**(2), 649–675 (2016)
32. F.T. Luk, S. Qiao, Analysis of a recursive least squares signal processing algorithm. SIAM J. Sci. Stat. Comput. **10**, 407–418 (1989)
33. F. Hsuan, P. Langenberg, A. Getson, The {2}-inverse with applications in statistics. Linear Algebra Appl. **70**, 241–248 (1985)

Chapter 2
Drazin Inverse

In Chap. 1, we discussed the Moore-Penrose inverse and the $\{i, j, k\}$ inverses which possess some "inverse-like" properties. The $\{i, j, k\}$ inverses provide some types of solution, or the least-square solution, for a system of linear equations just as the regular inverse provides a unique solution for a nonsingular system of linear equations. Hence the $\{i, j, k\}$ inverses are called equation solving inverses. However, there are some properties of the regular inverse matrix that the $\{i, j, k\}$ inverses do not possess. For example, if $A, B \in \mathbb{C}^{n \times n}$, then there is no class $\mathbb{C}\{i, j, k\}$ of $\{i, j, k\}$ inverses of A and B such that $A^-, B^- \in \mathbb{C}\{i, j, k\}$ implies any of the following properties:

(1) $AA^- = A^-A$;
(2) $(A^-)^p = (A^p)^-$ for positive integer p;
(3) $\lambda \in \lambda(A) \Leftrightarrow \lambda^\dagger \in \lambda(A^-)$, where $\lambda(A)$ denotes the set of the eigenvalues of an $n \times n$ matrix A;
(4) $A^{p+1}A^- = A^p$ for positive integer p;
(5) $P^{-1}AP = B \Rightarrow P^{-1}A^-P = B^-$.

The Drazin inverse and its special case of the group inverse introduced in Sects. 2.1 and 2.2 possess all of the above properties. Moreover, in some cases, the Drazin inverse and group inverse provide not only solutions of linear equations, but also solutions of linear differential equations and linear difference equations. Hence, they resemble the regular inverse more closely than the $\{i, j, k\}$ inverses. The weighted Drazin inverse presented in Sect. 2.3 is a generalization of the Drazin inverse for rectangular matrices, with some interesting applications.

© Springer Nature Singapore Pte Ltd. and Science Press 2018
G. Wang et al., *Generalized Inverses: Theory and Computations*,
Developments in Mathematics 53, https://doi.org/10.1007/978-981-13-0146-9_2

2.1   Drazin Inverse

The Drazin inverse is associated with matrix index, which is defined only for square matrices. In this section, we first introduce the index of a square matrix before defining the Drazin inverse. Then we present a matrix decomposition related to the Drazin inverse.

2.1.1   Matrix Index and Its Basic Properties

The index of a square matrix is defined as follows.

Definition 2.1.1 Let $A \in \mathbb{C}^{n \times n}$. If

$$\operatorname{rank}(A^{k+1}) = \operatorname{rank}(A^k), \tag{2.1.1}$$

then the smallest positive integer k for which (2.1.1) holds is called the index of A and is denoted by

$$\operatorname{Ind}(A) = k.$$

If A is nonsingular, then $\operatorname{Ind}(A) = 0$; if A is singular, then $\operatorname{Ind}(A) \geq 1$.

In the following, we only consider the singular case.

The basic properties of the index of a square matrix are summarized in the following theorem.

Theorem 2.1.1 *Let* $A \in \mathbb{C}^{n \times n}$.

(1) *If* $\operatorname{Ind}(A) = k$, *then*

$$\operatorname{rank}(A^l) = \operatorname{rank}(A^k), \quad l \geq k; \tag{2.1.2}$$
$$\mathcal{R}(A^l) = \mathcal{R}(A^k), \quad l \geq k; \tag{2.1.3}$$
$$\mathcal{N}(A^l) = \mathcal{N}(A^k), \quad l \geq k. \tag{2.1.4}$$

(2) $\operatorname{Ind}(A) = k$ *if and only if* k *is the smallest positive integer such that*

$$A^k = A^{k+1}X, \tag{2.1.5}$$

for some matrix X.

(3) $\operatorname{Ind}(A) = k$ *if and only if* $\mathcal{R}(A^k)$ *and* $\mathcal{N}(A^k)$ *are complementary subspaces, that is,*

$$\mathcal{R}(A^k) \oplus \mathcal{N}(A^k) = \mathbb{C}^n. \tag{2.1.6}$$

Proof (1) It follows from $\text{rank}(A^{k+1}) = \text{rank}(A^k)$ and Theorem 1.1.3 that

$$\mathcal{R}(A^{k+1}) = \mathcal{R}(A^k), \quad \mathcal{N}(A^{k+1}) = \mathcal{N}(A^k).$$

Therefore $A^k = A^{k+1}X$ holds for some matrix X, then multiplying on the left with A^{l-k} gives

$$A^l = A^{l+1}X, \quad l \geq k.$$

Hence $\text{rank}(A^l) = \text{rank}(A^{l+1})$. By Theorem 1.1.3, we have

$$\mathcal{R}(A^{l+1}) = \mathcal{R}(A^l) \quad \text{and} \quad \mathcal{N}(A^{l+1}) = \mathcal{N}(A^l), \quad l \geq k.$$

Thus (2.1.2)–(2.1.4) hold.

(2) Since $\text{rank}(A^k) = \text{rank}(A^{k+1})$ and $A^k = A^{k+1}X$ are equivalent, (2.1.5) holds.

(3) Suppose that $\text{rank}(A^k) > \text{rank}(A^{k+1})$, equivalently, there exists some $\mathbf{x} \in \mathbb{C}^n$ such that

$$A^{k+1}\mathbf{x} = \mathbf{0} \quad \text{and} \quad A^k\mathbf{x} \neq \mathbf{0}.$$

Let $\mathbf{y} = A^k\mathbf{x} \in \mathcal{R}(A^k)$, then $A^k\mathbf{y} = A^{2k}\mathbf{x} = \mathbf{0}$. Thus

$$\mathbf{y} \in \mathcal{N}(A^k) \quad \text{and} \quad \mathbf{0} \neq \mathbf{y} = A^k\mathbf{x} \in \mathcal{R}(A^k) \cap \mathcal{N}(A^k),$$

which completes the proof. $\qquad\square$

2.1.2 Drazin Inverse and Its Properties

In this section, we first define the Drazin inverse, then show its existence and uniqueness and study its basic properties.

Definition 2.1.2 Let $A \in \mathbb{C}^{n \times n}$ and $\text{Ind}(A) = k$, then the matrix $X \in \mathbb{C}^{n \times n}$ satisfying

$$\begin{aligned} (1^k) \quad & A^k X A = A^k, & (2.1.7) \\ (2) \quad & XAX = X, & (2.1.8) \\ (5) \quad & AX = XA & (2.1.9) \end{aligned}$$

is called the Drazin inverse of A and is denoted by $X = A_d$ or $X = A^{(1^k,2,5)}$.

It is easy to verify that (2.1.7)–(2.1.9) are equivalent to

$$\begin{aligned} A^{k+1}X &= A^k, \\ AX^2 &= X, \\ AX &= XA, \end{aligned}$$

and by Definition 2.1.1, we have

$$A^{l+1} X = A^l, \quad l \geq k.$$

If A is nonsingular, then $A_d = A^{-1}$.

The existence and uniqueness of the Drazin inverse are given in the following theorem.

Theorem 2.1.2 *Let $A \in \mathbb{C}^{n \times n}$ and* $\mathrm{Ind}(A) = k$, *then the Drazin inverse of A is existent and unique.*

Proof EXISTENCE: Since $\mathrm{Ind}(A) = k$, by (2.1.6), $\mathcal{R}(A^k) \oplus \mathcal{N}(A^k) = \mathbb{C}^n$. Let

$$P = [\mathbf{v}_1, \mathbf{v}_2, \ldots, \mathbf{v}_r, \mathbf{v}_{r+1}, \mathbf{v}_{r+2}, \ldots, \mathbf{v}_n],$$

where $\mathbf{v}_1, \mathbf{v}_2, \ldots, \mathbf{v}_r$ and $\mathbf{v}_{r+1}, \mathbf{v}_{r+2}, \ldots, \mathbf{v}_n$ form the bases for $\mathcal{R}(A^k)$ and $\mathcal{N}(A^k)$ respectively. Set

$$P = [P_1 \ P_2], \quad P_1 = [\mathbf{v}_1, \mathbf{v}_2, \ldots, \mathbf{v}_r], \quad P_2 = [\mathbf{v}_{r+1}, \mathbf{v}_{r+2}, \ldots, \mathbf{v}_n].$$

Since $\mathcal{R}(A^k)$ and $\mathcal{N}(A^k)$ are invariant subspaces for A, there exist $C \in \mathbb{C}^{r \times r}$ and $N \in \mathbb{C}^{(n-r) \times (n-r)}$ such that

$$A P_1 = P_1 C \quad \text{and} \quad A P_2 = P_2 N.$$

Thus A has the decomposition:

$$A = P \begin{bmatrix} C & O \\ O & N \end{bmatrix} P^{-1}. \tag{2.1.10}$$

Since $A^k \mathcal{N}(A^k) = O$, we have $O = A^k P_2 = P_2 N^k$. Thus $N^k = O$. Moreover

$$A^k = P \begin{bmatrix} C^k & O \\ O & O \end{bmatrix} P^{-1}$$

and

$$r = \mathrm{rank}(A^k) = \mathrm{rank}(C^k) \leq r.$$

Thus $\mathrm{rank}(C) = r$, that is, C is a nonsingular matrix of order r. Using (2.1.10), we set

$$X = P \begin{bmatrix} C^{-1} & O \\ O & O \end{bmatrix} P^{-1}. \tag{2.1.11}$$

It is easy to verify that X satisfies (2.1.7)–(2.1.9). Thus $X = A_d$.

UNIQUENESS: Suppose both X and Y are the Drazin inverses of A. Set

$$AX = XA = E \quad \text{and} \quad AY = YA = F.$$

It is clear that

$$E^2 = E \quad \text{and} \quad F^2 = F.$$

Thus

$$E = AX = A^k X^k = A^k Y A X^k = A Y A^k X^k = FAX = FE$$
$$F = YA = Y^k A^k = Y^k A^k XA = YAE = FE.$$

Therefore $E = F$ and

$$X = AX^2 = EX = FX = YAX = YE = YF = Y^2 A = AY^2 = Y,$$

meaning that the Drazin inverse is unique. $\square$

If A is nonsingular, then it is easy to show that A^{-1} can be expressed as a polynomial of A. Indeed, if $A \in \mathbb{C}_n^{n \times n}$, the characteristic polynomial of A is

$$f(\lambda) = \det(\lambda I - A)$$
$$= \lambda^n + a_1 \lambda^{n-1} + \cdots + a_{n-1} \lambda + a_n,$$

where I is the identity matrix of order n. By Cayley-Hamilton Theorem, $f(A) = O$ implies

$$A^{-1} = -(A^{n-1} + a_1 A^{n-2} + \cdots + a_{n-1} I)/a_n.$$

Hence A^{-1} is expressed as a polynomial of A. This property does not carry over to the $\{i, j, k\}$ inverses. However, the Drazin inverse of A is always expressible as a polynomial of A.

Theorem 2.1.3 *Let $A \in \mathbb{C}^{n \times n}$ and $\mathrm{Ind}(A) = k$. There exists a polynomial $q(x)$ such that*

$$A_d = A^l (q(A))^{l+1}, \quad l \geq k. \tag{2.1.12}$$

Proof By (2.1.10),

$$A = P \begin{bmatrix} C & O \\ O & N \end{bmatrix} P^{-1},$$

where P and C are nonsingular and N is nilpotent of index k.

Since C is nonsingular, there exists a polynomial $q(x)$ such that $C^{-1} = q(C)$. Thus

$$
\begin{aligned}
A^l(q(A))^{l+1} &= P \begin{bmatrix} C^l & O \\ O & O \end{bmatrix} P^{-1} P \begin{bmatrix} q(C) & O \\ O & q(N) \end{bmatrix}^{l+1} P^{-1} \\
&= P \begin{bmatrix} C^l(q(C))^{l+1} & O \\ O & O \end{bmatrix} P^{-1} \\
&= P \begin{bmatrix} C^{-1} & O \\ O & O \end{bmatrix} P^{-1} \\
&= A_d,
\end{aligned}
$$

which completes the proof.   □

The basic properties of the Drazin inverse are summarized in the following theorem.

Theorem 2.1.4 *Let $A \in \mathbb{C}^{n \times n}$ and $\mathrm{Ind}(A) = k$, then*

$$
\begin{aligned}
\mathcal{R}(A_d) &= \mathcal{R}(A^l), & l \geq k, & \qquad (2.1.13) \\
\mathcal{N}(A_d) &= \mathcal{N}(A^l), & l \geq k, & \qquad (2.1.14) \\
AA_d = A_d A &= P_{\mathcal{R}(A_d),\mathcal{N}(A_d)} = P_{\mathcal{R}(A^l),\mathcal{N}(A^l)}, & l \geq k, & \qquad (2.1.15) \\
I - AA_d = I - A_d A &= P_{\mathcal{N}(A^l),\mathcal{R}(A^l)}, & l \geq k. & \qquad (2.1.16)
\end{aligned}
$$

Proof (1) Let $A_d = X$. From the definition of the Drazin inverse,

$$
A^k = A^{k+1} X = X A^{k+1}.
$$

Multiplication on the right by A^{l-k} gives

$$
A^l = X A^{l+1}, \quad l \geq k.
$$

Thus $\mathcal{R}(A^l) \subset \mathcal{R}(X)$. From (2.1.12), $\mathcal{R}(X) \subset \mathcal{R}(A^l)$. Therefore, we have

$$
\mathcal{R}(X) = \mathcal{R}(A^l), \quad l \geq k.
$$

(2) From (2.1.13), $\mathrm{rank}(X) = \mathrm{rank}(A^l)$. By (2.1.12), $\mathcal{N}(A^l) \subset \mathcal{N}(X)$. Thus

$$
\mathcal{N}(X) = \mathcal{N}(A^l), \quad l \geq k.
$$

(3) Let $X = A_d \in A\{1^k, 2, 5\}$, then $XAX = X$, $X \in A\{1\}$ and $AX = XA$. Thus

$$
\mathcal{R}(AX) = \mathcal{R}(XA) = R(X) \quad \text{and} \quad \mathcal{N}(AX) = \mathcal{N}(X).
$$

Since $AX = XA$ is idempotent, it follows from Theorem 1.3.1 that

$$AX = XA = P_{\mathcal{R}(AX),\mathcal{N}(AX)} = P_{\mathcal{R}(X),\mathcal{N}(X)} = P_{\mathcal{R}(A^l),\mathcal{N}(A^l)}, \quad l \geq k.$$

(4) From (2.1.6), $\mathcal{R}(A^l) \oplus \mathcal{N}(A^l) = \mathbb{C}^n$. By Exercise 1.3.4,

$$P_{\mathcal{R}(A^l),\mathcal{N}(A^l)} + P_{\mathcal{N}(A^l),\mathcal{R}(A^l)} = I.$$

Thus

$$I - AA_d = I - A_dA = P_{\mathcal{N}(A^l),\mathcal{R}(A^l)}.$$

This completes the proof. □

From the above theorem, we have

$$A_d = A^{(2)}_{\mathcal{R}(A^l),\mathcal{N}(A^l)},$$

which shows that the Drazin inverse A_d is the $\{2\}$ inverse of A with range $\mathcal{R}(A^l)$ and null space $\mathcal{N}(A^l)$.

Recall the Zlobec formula (1.3.10):

$$A^\dagger = A^*(A^*AA^*)^{(1)}A^*$$

for $A^\dagger$, where the Moore-Penrose inverse $A^\dagger$ is expressed by any $\{1\}$ inverse of A^*AA^*. Similarly, it is possible to express the Drazin inverse in terms of any $\{1\}$ inverse of A^{2l+1}.

Theorem 2.1.5 *Let $A \in \mathbb{C}^{n \times n}$ and $\mathrm{Ind}(A) = k$, then for any $\{1\}$ inverse of A^{2l+1}, for each integer $l \geq k$,*

$$A_d = A^l(A^{2l+1})^{(1)}A^l, \tag{2.1.17}$$

in particular,

$$A_d = A^l(A^{2l+1})^\dagger A^l.$$

Proof Let

$$A = P \begin{bmatrix} C & O \\ O & N \end{bmatrix} P^{-1},$$

where P and C are nonsingular, N is nilpotent of index k, then

$$A^{2l+1} = P \begin{bmatrix} C^{2l+1} & O \\ O & O \end{bmatrix} P^{-1}.$$

If X is a $\{1\}$ inverse of A^{2l+1}, then it is easy to see that

$$X = P \begin{bmatrix} C^{-2l-1} & X_1 \\ X_2 & X_3 \end{bmatrix} P^{-1},$$

where X_1, X_2 and X_3 are arbitrary. It can be verified that

$$A_d = A^l X A^l$$

by multiplying the block matrices.   □

In Theorem 1.1.5, the Moore-Penrose inverse is expressed by applying the full rank factorization of a matrix. The Drazin inverse can also be obtained by applying the full rank factorization.

Theorem 2.1.6 *Let $A \in \mathbb{C}^{n \times n}$. We perform a sequence of full rank factorizations:*

$$A = B_1 C_1, \quad C_1 B_1 = B_2 C_2, \quad C_2 B_2 = B_3 C_3, \ldots$$

so that $B_i C_i$ are full rank factorizations of $C_{i-1} B_{i-1}$, for $i = 2, 3, \ldots$. Eventually, there will be a pair of factors, B_k and C_k, such that either $(C_k B_k)^{-1}$ exists or $C_k B_k = O$. If k is the smallest integer for which this occurs, then

$$\mathrm{Ind}(A) = \begin{cases} k & \text{when } (C_k B_k)^{-1} \text{ exists,} \\ k+1 & \text{when } C_k B_k = O. \end{cases}$$

When $C_k B_k$ is nonsingular,

$$\mathrm{rank}(A^k) = \text{number of columns of } B_k = \text{number of rows of } C_k$$

and

$$\mathcal{R}(A^k) = \mathcal{R}(B_1 B_2 \cdots B_k), \quad \mathcal{N}(A^k) = \mathcal{N}(C_k C_{k-1} \cdots C_1).$$

Moreover

$$A_d = \begin{cases} B_1 \cdots B_k (C_k B_k)^{-(k+1)} C_k \cdots C_1 & \text{when } (C_k B_k)^{-1} \text{ exists,} \\ O & \text{when } C_k B_k = O. \end{cases} \tag{2.1.18}$$

Proof If $C_i B_i$ is $p \times p$ and has rank $q < p$, then $C_{i+1} B_{i+1}$ will be $q \times q$. That is, the size of $C_{i+1} B_{i+1}$ must be strictly smaller than that of $C_i B_i$ when $C_i B_i$ is singular. It follows that there eventually must be a pair of factors B_k and C_k, such that $C_k B_k$ is either nonsingular or zero matrix. Let k be the smallest integer when this occurs and write

$$
\begin{aligned}
A^k &= (B_1 C_1)^k = B_1 (C_1 B_1)^{k-1} C_1 \\
&= B_1 (B_2 C_2)^{k-1} C_1 = B_1 B_2 (C_2 B_2)^{k-2} C_2 C_1 \\
&= \cdots \\
&= B_1 B_2 \cdots B_{k-1} (B_k C_k) C_{k-1} C_{k-2} \cdots C_1, \\
A^{k+1} &= B_1 B_2 \cdots B_k (C_k B_k) C_k C_{k-1} \cdots C_1.
\end{aligned}
\tag{2.1.19}
$$

Assume that $C_k B_k$ is nonsingular. If $B_k \in \mathbb{C}_r^{p \times r}$ and $C_k \in \mathbb{C}_r^{r \times p}$, then $\operatorname{rank}(B_k C_k) = r$. Since $C_k B_k$ is $r \times r$ and nonsingular, it follows that

$$
\operatorname{rank}(C_k B_k) = r = \operatorname{rank}(B_k C_k).
$$

Noting that B_i and C_i are full column rank and full row rank respectively, for $i = 1, 2, 3, \ldots$. It follows from (2.1.19) that

$$
\operatorname{rank}(A^{k+1}) = \operatorname{rank}(C_k B_k) = \operatorname{rank}(B_k C_k) = \operatorname{rank}(A^k).
$$

Since k is the smallest integer for this to hold, it must be the case that $\operatorname{Ind}(A) = k$.

We can clearly see that $\operatorname{rank}(A^k)$ equals the number of columns of B_k, which, in turn, equals the number of rows of C_k.

By using the fact that B_i and C_i are full rank factors, for $i = 1, 2, 3, \ldots$, it is not difficult to show that

$$
\mathcal{R}(A^k) = \mathcal{R}(B_1 B_2 \cdots B_k), \quad \mathcal{N}(A^k) = \mathcal{N}(C_k C_{k-1} \cdots C_1).
$$

If $C_k B_k = O$, then $A^{k+1} = O$, $\operatorname{rank}(A^{k+2}) = \operatorname{rank}(A^{k+1}) = 0$. Thus $\operatorname{Ind}(A) = k + 1$.

To prove the formula (2.1.18), one simply verifies the three conditions (2.1.7)–(2.1.9) of Definition 2.1.2. $\qquad \square$

2.1.3 Core-Nilpotent Decomposition

If $\operatorname{Ind}(A) = k \neq 1$, then A_d is not always a $\{1\}$ inverse of $A \in \mathbb{C}^{n \times n}$. Although $AA_d A \neq A$, the product $AA_d A = A^2 A_d$ still plays an important role in the theory of the Drazin inverse.

Definition 2.1.3 Let $A \in \mathbb{C}^{n \times n}$, then the product

$$
C_A = AA_d A = A^2 A_d = A_d A^2
$$

is called the core part of A.

Theorem 2.1.7 Let $A \in \mathbb{C}^{n \times n}$, $\operatorname{Ind}(A) = k$, and $N_A = A - C_A$, then N_A satisfies

$$
N_A^k = O \quad \text{and} \quad \operatorname{Ind}(N_A) = k.
$$

Proof The theorem is trivial when $\text{Ind}(A) = 0$. Thus, assuming $\text{Ind}(A) \geq 1$, we have

$$
\begin{aligned}
N_A^k &= (A - AA_dA)^k = A^k(I - AA_d)^k \\
&= A^k(I - AA_d) = A^k - A^k \\
&= O.
\end{aligned}
$$

Since $N_A^l = A^l - A^{l+1}A_d \neq O$ for $l < k$, we have $\text{Ind}(N_A) = k$. $\square$

Definition 2.1.4 Let $A \in \mathbb{C}^{n \times n}$, then the matrix $N_A = A - C_A = (I - AA_d)A$ is called the nilpotent part of A and the decomposition

$$
A = C_A + N_A
$$

is called the core-nilpotent decomposition of A.

In terms of the decomposition (2.1.10) of A, we have the following results.

Theorem 2.1.8 *Let $A \in \mathbb{C}^{n \times n}$ be written as (2.1.10):*

$$
A = P \begin{bmatrix} C & O \\ O & N \end{bmatrix} P^{-1},
$$

where P and C are nonsingular and N is nilpotent of index $k = \text{Ind}(A)$, then

$$
C_A = P \begin{bmatrix} C & O \\ O & O \end{bmatrix} P^{-1}
$$

and

$$
N_A = P \begin{bmatrix} O & O \\ O & N \end{bmatrix} P^{-1}.
$$

Proof The above two equations can be easily verified by using (2.1.11):

$$
A_d = P \begin{bmatrix} C^{-1} & O \\ O & O \end{bmatrix} P^{-1}
$$

and the definitions of C_A and N_A. $\square$

The next theorem summarizes some of the basic relationships between A, C_A, N_A and A_d.

Theorem 2.1.9 *Let $A \in \mathbb{C}^{n \times n}$, the following statements are true.*

(1) $\text{Ind}(A_d) = \text{Ind}(C_A) = \begin{cases} 1 \ \textit{if} \ \ \text{Ind}(A) \geq 1, \\ 0 \ \textit{if} \ \ \text{Ind}(A) = 0; \end{cases}$

(2) $N_A C_A = C_A N_A = O$;

(3) $N_A A_d = A_d N_A = O$;
(4) $C_A A A_d = A A_d C_A = C_A$;
(5) $(A_d)_d = C_A$;
(6) $A = C_A \Leftrightarrow \mathrm{Ind}(A) \leq 1$;
(7) $((A_d)_d)_d = A_d$;
(8) $A_d = (C_A)_d$;
(9) $(A_d)^* = (A^*)_d$.

The proof is left to the reader as an exercise.

Exercise 2.1

1. Let $T \in \mathbb{C}^{n \times n}$, $R \in \mathbb{C}_n^{r \times n}$, and $S \in \mathbb{C}_n^{n \times s}$. Prove

$$\mathrm{rank}(RTS) = \mathrm{rank}(T).$$

2. Prove Theorem 2.1.9.
3. Show that

$$(A^l)_d = (A_d)^l, \quad l = 1, 2, \ldots.$$

4. Prove that

$$\mathrm{rank}(A) = \mathrm{rank}(A_d) + \mathrm{rank}(N_A).$$

5. If $A, B \in \mathbb{C}^{n \times n}$ and $AB = BA$, then

$$(AB)_d = B_d A_d.$$

6. If $A, B \in \mathbb{C}^{n \times n}$, then

$$(AB)_d = A((BA)^2)_d B$$

even if $AB \neq BA$.
7. Let $A^{k+1}U = A^k$ and $VA^{l+1} = A^l$, prove that
   (1) $\mathrm{Ind}(A) \leq \min\{k, l\}$;
   (2) $A_d = A^k U^{k+1} = V^{l+1} A^l = VA^k U^k = V^l A^l U$;
   (3) $AA_d = A^{k+m} U^{k+m} = V^{l+n} A^{l+n}$ for all integers $m \geq 0$ and $n \geq 0$;
   (4) $VA^{k+1} = A^k$, if $k \leq l$,
       $A^{l+1}U = A^l$, if $k \geq l$.

2.2 Group Inverse

Let $A \in \mathbb{C}^{n \times n}$. If $\mathrm{Ind}(A) = 1$, then this special case of the Drazin inverse is known as the group inverse. Notice that in this case, the condition (1^k) becomes $AA_d A = A$.

2.2.1 Definition and Properties of the Group Inverse

We start with the definition of the group inverse.

Definition 2.2.1 Let $A \in \mathbb{C}^{n \times n}$. If $X \in \mathbb{C}^{n \times n}$ satisfies

(1) $AXA = A$,
(2) $XAX = X$,
(5) $AX = XA$,

then X is called the group inverse of A and denoted by $X = A_g$ or $X = A^{(1,2,5)}$.

From Sect. 2.1.2, for every $A \in \mathbb{C}^{n \times n}$, A_d always uniquely exists. However, the group inverse may not exist.

Theorem 2.2.1 Let $A \in \mathbb{C}^{n \times n}$ be singular, then A has a group inverse if and only if $\mathrm{Ind}(A) = 1$. When the group inverse exists, it is unique.

Proof Let $X = A^{(1,2,5)}$, then $X \in A\{1, 2\}$. From Sect. 1.3,

$$AX = P_{\mathcal{R}(A), \mathcal{N}(X)}, \quad \text{and} \quad XA = P_{\mathcal{R}(X), \mathcal{N}(A)}.$$

Since $AX = XA$, we have

$$\mathcal{R}(X) = \mathcal{R}(A), \quad \mathcal{N}(X) = \mathcal{N}(A),$$

and

$$X = A^{(1,2)}_{\mathcal{R}(A), \mathcal{N}(A)}. \tag{2.2.1}$$

By Theorem 1.3.5, there is at most one such inverse, and such an inverse exists if and only if $\mathcal{R}(A)$ and $\mathcal{N}(A)$ are complementary subspaces. By (2.1.6), $\mathcal{R}(A)$ and $\mathcal{N}(A)$ are complementary subspaces if and only if $\mathrm{Ind}(A) = 1$.   □

Another necessary and sufficient condition for the existence of the group inverse is given in the next theorem.

Theorem 2.2.2 Let $A \in \mathbb{C}^{n \times n}$, then A has a group inverse if and only if there exist nonsingular matrices P and C, such that

$$A = P \begin{bmatrix} C & O \\ O & O \end{bmatrix} P^{-1}.$$

Proof From Theorem 2.2.1, A has a group inverse A_g if and only if $\mathrm{Ind}(A) = 1$. By Theorem 2.1.9, $\mathrm{Ind}(A) = 1$ if and only if

$$A = C_A = P \begin{bmatrix} C & O \\ O & O \end{bmatrix} P^{-1}.$$

The proof is completed by verifying that

$$A_g = P \begin{bmatrix} C^{-1} & O \\ O & O \end{bmatrix} P^{-1}$$

is the group inverse of A. $\square$

From Theorem 2.1.3, the group inverse of A can be expressed as a polynomial of A.

Corollary 2.2.1 *Let $A \in \mathbb{C}^{n \times n}$ and $\mathrm{Ind}(A) = 1$, then there exists a polynomial $q(x)$ such that*

$$A_g = A(q(x))^2.$$

From Theorem 2.1.4, the basic properties of the group inverse are as follows.

Corollary 2.2.2 *Let $A \in \mathbb{C}^{n \times n}$ and $\mathrm{Ind}(A) = 1$, then*

$$\mathcal{R}(A_g) = \mathcal{R}(A),$$
$$\mathcal{N}(A_g) = \mathcal{N}(A),$$
$$AA_g = A_g A = P_{\mathcal{R}(A_g),\mathcal{N}(A_g)} = P_{\mathcal{R}(A),\mathcal{N}(A)},$$
$$I - AA_g = I - A_g A = P_{\mathcal{N}(A),\mathcal{R}(A)}.$$

From Theorem 2.1.5, the group inverse A_g can be expressed by any $\{1\}$ inverse of A^3.

Corollary 2.2.3 *Let $A \in \mathbb{C}^{n \times n}$ and $\mathrm{Ind}(A) = 1$, then*

$$A_g = A(A^3)^{(1)}A.$$

In particular,

$$A_g = A(A^3)^\dagger A.$$

From Theorem 2.1.6, the group inverse A_g can be expressed by applying the full rank factorization of A.

Corollary 2.2.4 *Let $A \in \mathbb{C}^{n \times n}$ and $\mathrm{Ind}(A) = 1$. If $A = BC$ is a full rank factorization, then*

$$A_g = B(CB)^{-2}C.$$

There are cases when the group inverse coincides with the Moore-Penrose inverse.

Definition 2.2.2 Suppose $A \in \mathbb{C}^{n \times n}$ and $\mathrm{Ind}(A) = r$. If $A^\dagger A = AA^\dagger$, then A is called an EP_r, or simply EP, matrix.

Theorem 2.2.3 *Let $A \in \mathbb{C}^{n \times n}$ and $\mathrm{Ind}(A) = r$, then $A_g = A^\dagger$ if and only if A is an EP matrix.*

Proof From (2.2.1) and (1.3.9), we have

$$A_g = A^{(1,2)}_{\mathcal{R}(A),\mathcal{N}(A)}, \quad \text{and} \quad A^\dagger = A^{(1,2)}_{\mathcal{R}(A^*),\mathcal{N}(A^*)}.$$

Thus

$$A_g = A^\dagger \Leftrightarrow \mathcal{R}(A) = \mathcal{R}(A^*)$$
$$\Leftrightarrow P_{\mathcal{R}(A)} = P_{\mathcal{R}(A^*)}$$
$$\Leftrightarrow A^\dagger A = AA^\dagger,$$

completing the proof.   □

2.2.2 Spectral Properties of the Drazin and Group Inverses

As we know, if A is nonsingular, then A does not have 0 as an eigenvalue and

$$A\mathbf{x} = \lambda\mathbf{x} \Leftrightarrow A^{-1}\mathbf{x} = \lambda^{-1}\mathbf{x},$$

that is, $\mathbf{x}$ is an eigenvector of A associated with the eigenvalue λ if and only if $\mathbf{x}$ is an eigenvector of A^{-1} associated with the eigenvalue λ^{-1}. Can we generalize this spectral property of the regular inverse to the generalized inverses?

We first introduce a generalization of eigenvectors.

Definition 2.2.3 The principal vector of grade p associated with an eigenvalue λ is a vector $\mathbf{x}$ such that

$$(A - \lambda I)^p \mathbf{x} = \mathbf{0} \quad \text{and} \quad (A - \lambda I)^{p-1}\mathbf{x} \neq \mathbf{0},$$

where p is some positive integer.

Evidently, principal vectors are a generalization of eigenvectors. In fact an eigenvector is a principal vector of grade 1. The term "principal vector of grade p associated with the eigenvalue λ" is abbreviated to "λ-vector of A of grade p".

It is not difficult to show that if A is nonsingular, a vector $\mathbf{x}$ is a λ^{-1}-vector of A^{-1} of grade p if and only if it is a λ-vector of A of grade p, i.e.,

$$\begin{cases} (A - \lambda I)^p \mathbf{x} = \mathbf{0} \\ (A - \lambda I)^{p-1}\mathbf{x} \neq \mathbf{0} \end{cases} \Leftrightarrow \begin{cases} (A^{-1} - \lambda^{-1}I)^p \mathbf{x} = \mathbf{0} \\ (A^{-1} - \lambda^{-1}I)^{p-1}\mathbf{x} \neq \mathbf{0}. \end{cases}$$

We then generalize the above spectral property of the regular inverse to the group inverse and Drazin inverse.

Lemma 2.2.1 *If* $\mathrm{Ind}(A) = 1$, *then the* 0*-vectors of* A *are all of grade* 1.

Proof Since $\mathrm{Ind}(A) = 1$, we have $\mathrm{rank}(A^2) = \mathrm{rank}(A)$. Thus $\mathcal{N}(A^2) = \mathcal{N}(A)$. Let **x** be a 0-vector of A of grade p, then

$$A^p \mathbf{x} = \mathbf{0}, \text{ i.e., } A^2(A^{p-2}\mathbf{x}) = \mathbf{0}, \quad \text{implying } A^{p-2}\mathbf{x} \in \mathcal{N}(A^2) = \mathcal{N}(A).$$

Thus $A^{p-1}\mathbf{x} = \mathbf{0}$. Continuing the same process, we get $A\mathbf{x} = \mathbf{0}$. □

Lemma 2.2.2 *Let* **x** *be a* λ-*vector of* A *of grade* p, $\lambda \neq 0$, *then* $\mathbf{x} \in \mathcal{R}(A^l)$, *where* l *is a positive integer.*

Proof By the assumption, we have

$$(A - \lambda I)^p \mathbf{x} = \mathbf{0}.$$

Expanding the left side by the Binomial Theorem gives

$$\sum_{i=0}^{p} (-1)^i \lambda^i C_p^i A^{p-i} \mathbf{x} = \mathbf{0}.$$

Moving the last term to the right side and dividing the both sides by its coefficient $(-1)^{p-1}\lambda^p \neq 0$, we get

$$\sum_{i=0}^{p-1} (-1)^{i-p+1} \lambda^{i-p} C_p^i A^{p-i} \mathbf{x} = \mathbf{x}.$$

Set $j = i + 1$, then the above equation becomes

$$\sum_{i=1}^{p} (-1)^{j-p} \lambda^{j-p-1} C_p^{j-1} A^{p-j+1} \mathbf{x} = \mathbf{x}.$$

Set $c_j = (-1)^{j-p} \lambda^{j-p-1} C_p^{j-1}$, then the above equation gives

$$\sum_{i=0}^{p} c_j A^{p-j+1} \mathbf{x} = \mathbf{x}.$$

Hence

$$\mathbf{x} = c_1 A^p \mathbf{x} + c_2 A^{p-1} \mathbf{x} + \cdots + c_p A \mathbf{x}. \tag{2.2.2}$$

Successive premultiplication of (2.2.2) by A gives

$$
\begin{aligned}
A\mathbf{x} &= c_1 A^{p+1}\mathbf{x} \quad +c_2 A^p\mathbf{x} \quad +\cdots + c_p A^2\mathbf{x} \\
A^2\mathbf{x} &= c_1 A^{p+2}\mathbf{x} \quad +c_2 A^{p+1}\mathbf{x} \quad +\cdots + c_p A^3\mathbf{x} \\
&\ \ \vdots \\
A^{l-1}\mathbf{x} &= c_1 A^{p+l-1}\mathbf{x} +c_2 A^{p+l-2}\mathbf{x} +\cdots + c_p A^l\mathbf{x}.
\end{aligned}
\tag{2.2.3}
$$

Successive substitutions of the equations in (2.2.3) into the terms on the right side of (2.2.2) eventually give

$$
\mathbf{x} = A^l q(A)\mathbf{x},
$$

where q is some polynomial.   □

Lemma 2.2.3 *Let $A \in \mathbb{C}^{n \times n}$ and*

$$
X A^{l+1} = A^l
\tag{2.2.4}
$$

for some positive integer l, then every λ-vector of A of grade p for $\lambda \neq 0$ is a λ^{-1}-vector of X of grade p.

Proof The proof will be by induction on the grade p. For $p = 1$, let $\lambda \neq 0$ and $\mathbf{x}$ be a λ-vector of A of grade 1, that is, $A\mathbf{x} = \lambda\mathbf{x}$, then $A^{l+1}\mathbf{x} = \lambda^{l+1}\mathbf{x}$, and therefore $\mathbf{x} = \lambda^{-l-1} A^{l+1}\mathbf{x}$, consequently, by (2.2.4),

$$
X\mathbf{x} = \lambda^{-l-1} X A^{l+1}\mathbf{x} = \lambda^{-l-1} A^l\mathbf{x} = \lambda^{-1}\mathbf{x}.
$$

Thus the lemma is true for $p = 1$.

Now, suppose that it is true for $p = 1, 2, \ldots, r$, and let $\mathbf{x}$ be a λ-vector of A of grade $r + 1$, then, by Lemma 2.2.2, $\mathbf{x} \in \mathcal{R}(A^l)$, i.e., $\mathbf{x} = A^l\mathbf{y}$ for some $\mathbf{y}$. Thus

$$
\begin{aligned}
(X - \lambda^{-1}I)\mathbf{x} &= (X - \lambda^{-1}I)A^l\mathbf{y} \\
&= X(A^l - \lambda^{-1} A^{l+1})\mathbf{y} \\
&= X(I - \lambda^{-1}A)A^l\mathbf{y} \\
&= X(I - \lambda^{-1}A)\mathbf{x} \\
&= -\lambda^{-1}X(A - \lambda I)\mathbf{x}.
\end{aligned}
\tag{2.2.5}
$$

Since $\mathbf{x}$ is a λ-vector of A of grade $r + 1$,

$$
(A - \lambda I)^{r+1}\mathbf{x} = \mathbf{0} \quad \text{and} \quad (A - \lambda I)^r\mathbf{x} \neq \mathbf{0},
$$

that is,

$$
(A - \lambda I)^r((A - \lambda I)\mathbf{x}) = \mathbf{0}, \quad \text{and} \quad (A - \lambda I)^{r-1}((A - \lambda I)\mathbf{x}) \neq \mathbf{0}.
$$

Thus $(A - \lambda I)\mathbf{x}$ is a λ-vector of A of grade r. By the induction hypothesis, $(A - \lambda I)\mathbf{x}$ is a λ^{-1}-vector of X of grade r. Consequently,

$$(X - \lambda^{-1}I)^r(A - \lambda I)\mathbf{x} = \mathbf{0},$$

and

$$\mathbf{z} \equiv (X - \lambda^{-1}I)^{r-1}(A - \lambda I)\mathbf{x} \neq \mathbf{0}. \tag{2.2.6}$$

Thus $(X - \lambda^{-1}I)\mathbf{z} = \mathbf{0}$ and

$$X\mathbf{z} = \lambda^{-1}\mathbf{z}. \tag{2.2.7}$$

It is clear that

$$(X - \lambda^{-1}I)X = X(X - \lambda^{-1}I). \tag{2.2.8}$$

By using (2.2.5)–(2.2.8), we have

$$\begin{aligned}
(X - \lambda^{-1}I)^{r+1}\mathbf{x} &= (X - \lambda^{-1}I)^r(-\lambda^{-1}X(A - \lambda I)\mathbf{x}) \\
&= -\lambda^{-1}X(X - \lambda^{-1}I)^r(A - \lambda I)\mathbf{x} \\
&= \mathbf{0},
\end{aligned}$$

and

$$\begin{aligned}
(X - \lambda^{-1}I)^r\mathbf{x} &= (X - \lambda^{-1}I)^{r-1}(-\lambda^{-1}X(A - \lambda I)\mathbf{x}) \\
&= -\lambda^{-1}X\mathbf{z} \\
&= -\lambda^{-2}\mathbf{z} \\
&\neq \mathbf{0}.
\end{aligned}$$

This completes the induction. $\qquad\qquad\qquad\qquad\qquad\qquad\qquad\qquad\square$

The following theorem shows the spectral property of the group inverse.

Theorem 2.2.4 Let $A \in \mathbb{C}^{n \times n}$ and $\text{Ind}(A) = 1$, then $\mathbf{x}$ is a λ-vector of A of grade p if and only if $\mathbf{x}$ is a $\lambda^\dagger$-vector of A_g of grade p.

Proof Since $X = A_g$, we have $XA^2 = A$ and $AX^2 = X$, it then follows from Lemma 2.2.3 that, for all $\lambda \neq 0$, $\mathbf{x}$ is a λ-vector of A of grade p if and only if $\mathbf{x}$ is a λ^{-1}-vector of A_g of grade p. For $\lambda = 0$, since $\text{Ind}(A) = 1 = \text{Ind}(A_g)$, from Lemma 2.2.1, the 0-vectors of A and A_g are all of grade 1. Let $\mathbf{x}$ be a 0-vector of A, then $A\mathbf{x} = \mathbf{0}$ and $\mathbf{x} \neq \mathbf{0}$. Thus $\mathbf{x} \in \mathcal{N}(A) = \mathcal{N}(A_g)$. Therefore $A_g\mathbf{x} = \mathbf{0}$, i.e., $\mathbf{x}$ is also a zero vector of A_g and vice versa. $\qquad\qquad\qquad\qquad\square$

The following theorem shows that the spectral property of the Drazin inverse is the same as that of the group inverse with regard to nonzero eigenvalues and associated eigenvectors, but weaker than the group inverse for 0-vectors.

Theorem 2.2.5 *Let $A \in \mathbb{C}^{n \times n}$ and $\mathrm{Ind}(A) = k$, then for all $\lambda \neq 0$, $\mathbf{x}$ is a λ-vector of A of grade p if and only if $\mathbf{x}$ is a λ^{-1}-vector of A_d of grade p, and for $\lambda = 0$, $\mathbf{x}$ is a 0-vector of A if and only if $\mathbf{x}$ is a 0-vector of A_d (with no regard to grade).*

Proof By the assumption of $\mathrm{Ind}(A) = k$, A_d satisfies

$$A_d A^{k+1} = A^k \quad \text{and} \quad A(A_d)^2 = A_d.$$

From Lemma 2.2.3, for all $\lambda \neq 0$, $\mathbf{x}$ is a λ-vector of A of grade p if and only if $\mathbf{x}$ is a λ^{-1}-vector of A_d of grade p. For $\lambda = 0$, from Theorem 2.1.9, we have $\mathrm{Ind}(A_d) = 1$. By using Lemma 2.2.1, the 0-vectors of A_d are all of grade 1. Let $\mathbf{x}$ be a 0-vector of A_d, that is, $A_d \mathbf{x} = \mathbf{0}$ and $\mathbf{x} \neq \mathbf{0}$, then

$$\mathbf{x} \in \mathcal{N}(A_d) = \mathcal{N}(A^l), \quad l \geq k.$$

Therefore $A^l \mathbf{x} = \mathbf{0}$. So $\mathbf{x}$ is a 0-vector of A of grade l and vice versa. $\square$

Exercises 2.2

1. Prove that $(A_g)_g = A$.
2. Prove that $(A^*)_g = (A_g)^*$.
3. Prove that $(A^l)_g = (A_g)^l$, $\quad l = 1, 2, \ldots$.
4. Prove that $(A_d)_g = A^2 A_d$.
5. Prove that $A_d(A_d)_g = A A_d$.
6. Prove that if A is nilpotent, then $A_d = O$.
7. Let $A \in \mathbb{C}^{n \times n}$, $\dim(\mathcal{N}(A)) = 1$, and $\mathbf{x} \neq \mathbf{0}$ and $\mathbf{y} \neq \mathbf{0}$ satisfy

$$A\mathbf{x} = \mathbf{0} \quad \text{and} \quad A^*\mathbf{y} = \mathbf{0}.$$

   Prove that
   (1) A_g exists $\Leftrightarrow \mathbf{y}^* \mathbf{x} \neq 0$.
   (2) If $\mathbf{y}^* \mathbf{x} \neq 0$, then $I - AA_g = \dfrac{\mathbf{x}\mathbf{y}^*}{\mathbf{y}^*\mathbf{x}}$.
8. Let $A \in \mathbb{C}^{n \times n}_{n-r}$, $\mathrm{Ind}(A) = 1$, and $U \in \mathbb{C}^{n \times r}_r$ and $V \in \mathbb{C}^{n \times r}_r$ satisfy

$$AU = O, \quad A^*V = O, \quad \text{and} \quad V^*U = I.$$

   Prove that
   (1) $A^\dagger = (I - UU^\dagger)A_g(I - VV^\dagger)$.
   (2) $A_g = (I - UV^*)A^\dagger(I - UV^*)$.

2.3 W-Weighted Drazin Inverse

In 1980, the definition of the Drazin inverse of a square matrix was extended to rectangular matrix by Cline and Greville [1].

Definition 2.3.1 Let $A \in \mathbb{C}^{m \times n}$ and $W \in \mathbb{C}^{n \times m}$, then the matrix $X \in \mathbb{C}^{m \times n}$ satisfying

$$(AW)^{k+1} XW = (AW)^k, \quad \text{for some nonnegative integer } k, \qquad (2.3.1)$$
$$XWAWX = X, \qquad (2.3.2)$$
$$AWX = XWA \qquad (2.3.3)$$

is called the W-weighted Drazin inverse of A and denoted by $X = A_{d,W}$.

It is clear that if $W = I$ and $A \in \mathbb{C}^{n \times n}$, then $X = A_d$.

The existence and uniqueness of the W-weighted Drazin inverse are given in the following theorems.

Theorem 2.3.1 *Let $A \in \mathbb{C}^{m \times n}$. If there exists a matrix $X \in \mathbb{C}^{m \times n}$ satisfying (2.3.1)–(2.3.3) for some $W \in \mathbb{C}^{n \times m}$, then it must be unique.*

Proof Suppose both X_1 and X_2 satisfy (2.3.1)–(2.3.3) for some nonnegative integers k_1 and k_2 respectively. Let $k = \max(k_1, k_2)$, then

$$
\begin{aligned}
X_1 &= X_1 WAWX_1 \\
&= X_1 WAWX_1 WAWX_1 \\
&= AWX_1 WAWX_1 WX_1 \\
&= AWAWX_1 WX_1 WX_1 \\
&= (AW)^2 X_1 (WX_1)^2 \\
&= \cdots \\
&= (AW)^k X_1 (WX_1)^k \\
&= (AW)^{k+1} X_2 WX_1 (WX_1)^k \\
&= X_2 (WA)^{k+1} WX_1 (WX_1)^k \\
&= X_2 (WA)^k WAWX_1 WX_1 (WX_1)^{k-1} \\
&= X_2 (WA)^k WX_1 WAWX_1 (WX_1)^{k-1} \\
&= X_2 (WA)^k WX_1 (WX_1)^{k-1} \\
&= \cdots \\
&= X_2 WAWX_1.
\end{aligned}
$$

Similarly,

$$X_2 = (AW)^{k+1} X_2 (WX_2)^{k+1}.$$

Postmultiplying (2.3.3) with W gives

$$(AW)(X_2 W) = (X_2 W)(AW).$$

Thus

$$X_2 W = (AW)^{k+1}(X_2 W)^{k+2} = (X_2 W)^{k+2}(AW)^{k+1}.$$

Moreover,

$$\begin{aligned}
X_1 &= X_2 W A W X_1 \\
&= (X_2 W)^{k+2}(AW)^{k+1} A W X_1 \\
&= (X_2 W)^{k+2}(AW)^{k+1} X_1 W A \\
&= (X_2 W)^{k+2}(AW)^k A \\
&= (X_2 W)^{k+1} X_2 (W A)^{k+1} \\
&= (X_2 W)^k X_2 W X_2 W A (W A)^k \\
&= (X_2 W)^k X_2 W A W X_2 (W A)^k \\
&= (X_2 W)^k X_2 (W A)^k \\
&= \cdots \\
&= X_2 W X_2 W A \\
&= X_2 W A W X_2 \\
&= X_2,
\end{aligned}$$

which completes the proof.   □

To derive the conditions for the existence of the W-weighted Drazin inverse, we need some results when $\text{Ind}(WA) = k$.

Theorem 2.3.2 *Let* $A \in \mathbb{C}^{m \times n}$, $W \in \mathbb{C}^{n \times m}$, *and* $\text{Ind}(WA) = k$, *then*

$$(AW)_d = A(WA)_d^2 W, \quad and \quad \text{Ind}(AW) \le k + 1. \qquad (2.3.4)$$

Proof By the assumption of $\text{Ind}(WA) = k$, $(WA)_d$ satisfies

$$\begin{aligned}
(WA)_d(WA)^{k+1} &= (WA)^k, \\
(WA)_d^2(WA) &= (WA)_d, \\
(WA)_d(WA) &= (WA)(WA)_d.
\end{aligned} \qquad (2.3.5)$$

Setting $X = A(WA)_d^2 W$, we have

$$(AW)^{k+2}X = (AW)^{k+2}A(WA)_d^2W$$
$$= A(WA)^{k+2}(WA)_d^2W$$
$$= A(WA)^kW$$
$$= (AW)^{k+1},$$
$$X^2(AW) = (A(WA)_d^2W)^2AW$$
$$= A(WA)_d(WA)_dW$$
$$= X,$$

and

$$X(AW) = A(WA)_d^2WAW$$
$$= AWA(WA)_d^2W$$
$$= AWX.$$

Consequently, $X = (AW)_d$ and $\mathrm{Ind}(AW) \le k+1$. $\square$

Corollary 2.3.1 *Under the assumptions in Theorem* 2.3.2, *we have*

$$W(AW)_d^p = (WA)_d^pW \tag{2.3.6}$$

and

$$A(WA)_d^p = (AW)_d^pA, \tag{2.3.7}$$

for any positive integer p.

Proof The proof is by induction on the positive integer p. The assertion (2.3.6) is true for $p = 1$. It follows from (2.3.4) and (2.3.5) that

$$W(AW)_d = WA(WA)_d^2W = (WA)_d^2WAW = (WA)_dW.$$

Suppose that the assertion (2.3.6) is true for all the positive integer less than p, then we have

$$W(AW)_d^{p-1} = (WA)_d^{p-1}W,$$

implying that

$$W(AW)_d^p = W(AW)_d^{p-1}(AW)_d$$
$$= (WA)_d^{p-1}W(AW)_d$$
$$= (WA)_d^{p-1}(WA)_dW$$
$$= (WA)_d^pW.$$

The proof of (2.3.7) is similar and is left to the reader as an exercise. $\square$

By using $(AW)_d$ and $(WA)_d$, the W-weighted Drazin inverse $A_{d,W}$ can be constructed as shown in the following theorem.

Theorem 2.3.3 *Let* $A \in \mathbb{C}^{m \times n}$, $W \in \mathbb{C}^{n \times m}$, *and* $\mathrm{Ind}(AW) = k$, *then*

$$A_{d,W} = A(WA)_d^2 = (AW)_d^2 A.$$

Proof Let $X = A(WA)_d^2$. It can be verified that X satisfies (2.3.1)–(2.3.3):

$$
\begin{aligned}
(AW)^{k+1} X W &= (AW)^{k+1} A(WA)_d^2 W \\
&= A(WA)^{k+1}(WA)_d^2 W \\
&= A(WA)^k (WA)_d W \\
&= A(WA)^k W(AW)_d \\
&= (AW)^{k+1}(AW)_d \\
&= (AW)^k, \\
XWAWX &= A(WA)_d^2 WAWA(WA)_d^2 \\
&= A(WA)_d(WA)_d \\
&= X, \\
AWX &= AWA(WA)_d^2 \\
&= A(WA)_d^2 WA \\
&= XWA.
\end{aligned}
$$

Hence $X = A_{d,W}$. The proof of $A_{d,W} = (AW)_d^2 A$ is similar and is left to the reader as an exercise.   □

In particular, we have the following corollary.

Corollary 2.3.2 $A_{dw} W = (AW)_d$ *and* $W A_{dw} = (WA)_d$.

The basic properties of the W-weighted Drazin inverse are as follows.

Theorem 2.3.4 ([2]) *Let* $A \in \mathbb{C}^{m \times n}$, $W \in \mathbb{C}^{n \times m}$, $\mathrm{Ind}(AW) = k_1$, *and* $\mathrm{Ind}(WA) = k_2$, *then*

(a) $\mathcal{R}(A_{dw}) = \mathcal{R}((AW)_d) = \mathcal{R}((AW)^{k_1})$;
(b) $\mathcal{N}(A_{dw}) = \mathcal{N}((WA)_d) = \mathcal{N}((WA)^{k_2})$;
(c) $WAWA_{dw} = WA(WA)_d = P_{\mathcal{R}((WA)^{k_2}),\mathcal{N}((WA)^{k_2})}$;
$\quad A_{dw}WAW = AW(AW)_d = P_{\mathcal{R}((AW)^{k_1}),\mathcal{N}((AW)^{k_1})}$.

Theorem 2.3.5 *Suppose that* $A \in \mathbb{C}^{m \times n}$ *and* $W \in \mathbb{C}^{n \times m}$, *and let* $k = \max\{\mathrm{Ind}(AW), \mathrm{Ind}(WA)\}$, *then*

$$A_{dw} = (WAW)_{\mathcal{R}(G),\mathcal{N}(G)}^{(2)} = G(GWAWG)^\dagger G,$$

where $G = A(WA)^k$.

Proof The proof is left to the reader as an exercise. □

By using the core-nilpotent decompositions of AW and WA, we can obtain another expression of A_{dw}.

Theorem 2.3.6 ([3]) *Suppose that* $A \in \mathbb{C}^{m \times n}$, $W \in \mathbb{C}^{n \times m}$, *and* $k = \max\{\text{Ind}(AW), \text{Ind}(WA)\}$, *then*

$$A = P \begin{bmatrix} A_{11} & O \\ O & A_{22} \end{bmatrix} Q^{-1}, \quad W = Q \begin{bmatrix} W_{11} & O \\ O & W_{22} \end{bmatrix} P^{-1},$$

and

$$A_{dw} = P \begin{bmatrix} (W_{11} A_{11} W_{11})^{-1} & O \\ O & A_{22} \end{bmatrix} Q^{-1},$$

where A_{11}, W_{11}, P, *and* Q *are nonsingular matrices.*

At last, a characteristic property of the W-weighted Drazin inverse is given as follows.

Theorem 2.3.7 *Let* $A, X \in \mathbb{C}^{m \times n}$, *then for some* $W \in \mathbb{C}^{n \times m}$, $X = A_{d,w}$ *if and only if* X *has a decomposition*

$$X = AYAYA,$$

where $Y \in \mathbb{C}^{n \times m}$ *satisfies* $\text{Ind}(AY) = \text{Ind}(YA) = 1$.

Proof If: Let

$$W = Y(AY)_d^2 = (YA)_d^2 Y,$$

then $WA = (YA)_d$. Since $\text{Ind}(YA) = 1$ and $X = AYAYA$, we have

$$WX = WAYAYA = (YA)_d YAYA = YA. \tag{2.3.8}$$

By Theorem 2.1.9 and $\text{Ind}(YA) = 1$,

$$YA = ((YA)_d)_d = (WA)_d.$$

Substituting YA in (2.3.8) with the above equation, we get

$$WX = (WA)_d.$$

Similarly, since $\text{Ind}(AY) = 1$ and $X = AYAYA$, we have

$$AW = (AY)_d, \quad AY = (AW)_d,$$

and

$$XW = AYAYAW = AYAY(AY)_d = AY = (AW)_d.$$

It then follows that

$$(AW)^{k+1}XW = (AW)^{k+1}(AW)_d$$
$$= (AW)^k,$$
$$XWAWX = (AW)_d A(WA)_d$$
$$= (AW)_d (AW)_d A$$
$$= AYAYA$$
$$= X,$$

and

$$AWX = A(WA)_d$$
$$= (AW)_d A$$
$$= XWA.$$

Therefore $X = A_{d,W}$.

Only if: If $X = A_{d,W}$, for some W, then from Corollary 2.3.1,

$$X = A(WA)_d^2$$
$$= (AW)_d A(WA)_d$$
$$= (AW)(AW)_d^2 A(WA)_d^2 (WA)$$
$$= A(WA)_d^2 WA(WA)_d^2 (WA)$$
$$= AYAYA,$$

where $Y = (WA)_d^2 W$. Thus $AY = (AW)_d$ and $YA = (WA)_d$. By Theorem 2.1.9, we have

$$\mathrm{Ind}(AY) = \mathrm{Ind}((AW)_d) = 1$$

and

$$\mathrm{Ind}(YA) = \mathrm{Ind}((WA)_d) = 1.$$

The proof is completed.   □

Exercises 2.3

1. Prove (2.3.7) in Corollary 2.3.1.
2. Prove that $A_{d,W} = (AW)_d^2 A$.
3. Let $A \in \mathbb{C}^{m \times n}$ and $W \in \mathbb{C}^{n \times m}$ with $\mathrm{Ind}(AW) = k_1$ and $\mathrm{Ind}(WA) = k_2$, show

   (1) $A_{d,W}W = (AW)_d$; $WA_{d,W} = (WA)_d$.
   (2) $\mathcal{R}(A_{d,W}) = \mathcal{R}((AW)_d) = \mathcal{R}((AW)^{k_1})$;
       $\mathcal{N}(A_{d,W}) = \mathcal{N}((WA)_d) = \mathcal{N}((WA)^{k_2})$.
   (3) $WAWA_{d,W} = WA(WA)_d = P_{\mathcal{R}((WA)^{k_2}),\mathcal{N}((WA)^{k_2})}$;
       $A_{d,W}WAW = (AW)_d AW = P_{\mathcal{R}((AW)^{k_1}),\mathcal{N}((AW)^{k_1})}$.

4. Prove that

$$A_{d,W} = (WAW)^{(2)}_{\mathcal{R}((AW)^k), \mathcal{N}((WA)^k)},$$

where $k = \max\{\text{Ind}(AW), \text{Ind}(WA)\}$.

Remarks

The concept of the Drazin inverse is based on the associative ring and the semigroup [4]. Greville further investigated the Drazin inverse of a square matrix in [5]. As for the applications of the Drazin inverse and group inverse, such as, in finite Markov chain, linear differential equations, linear difference equations, the model of population growth and optimal control can be found in [6–10].

A characterization and representation of the Drazin inverse can be found in [11]. A characterization of the Drazin index can be found in [12, 13]. Full-rank and determinantal representations can be found in [14]. The group inverse of a triangular matrix is discussed in [15] and the group inverse of M-matrix is discussed in [16]. The Drazin inverse of a 2×2 block matrix is presented by Hartwig et al. [17] and more results in [12, 18–20]. Representations of the Drazin inverse of a block or modified matrix can be found in [21–24].

References

1. R.E. Cline, T.N.E. Greville, A Drazin inverse for rectangular matrices. Linear Algebra Appl. **29**, 54–62 (1980)
2. Y. Wei, A characterization for the W-weighted Drazin inverse and Cramer rule for W-weighted Drazin inverse solution. Appl. Math. Comput. **125**, 303–310 (2002)
3. Y. Wei, Integral representation of the W-weighted Drazin inverse. Appl. Math. Comput. **144**, 3–10 (2003)
4. M.P. Drazin, Pseudo-inverses in associate rings and semigroups. Amer. Math. Monthly **65**, 506–514 (1958)
5. T.N.E. Greville, Spectral generalized inverses of square matrtces. MRC Technical Science Report 823 (Mathematics Research Center, University of Wisconsin, Madison, 1967)
6. C.D. Meyer, The role of the group generalized inverse in the theory of finite Markov chains. SIAM Rev. **17**, 443–464 (1975)
7. C.D. Meyer, The condition number of a finite Markov chains and perturbation bounds for the limiting probabilities. SIAM J. Alg. Disc. Math. **1**, 273–283 (1980)
8. A. Ben-Israel, T.N.E. Greville, *Generalized Inverses: Theory and Applications*, 2nd edn. (Springer Verlag, New York, 2003)
9. S.L. Campbell, C.D. Meyer Jr., *Generalized Inverses of Linear Transformations* (Pitman, London, 1979)
10. J.H. Wilkinson, Note on the practical significance of the Drazin inverse, in *Recent Applications of Generalized Inverses*, ed. by S.L. Campbell (Pitman, London, 1982)
11. Y. Wei, A characterization and representation of the Drazin inverse. SIAM J. Matrix Anal. Appl. **17**, 744–747 (1996)
12. Q. Xu, Y. Wei, C. Song, Explicit characterization of the Drazin index. Linear Algebra Appl. **436**, 2273–2298 (2012)
13. C. Zhu, G. Chen, Index splitting for the Drazin inverse of linear operator in Banach space. Appl. Math. Comput. **135**, 201–209 (2003)

14. P.S. Stanimirović, D. Djordjević, Full-rank and determinantal representation of the Drazin inverse. Linear Algebra Appl. **311**, 131–151 (2000)
15. R.E. Hartwig, The group inverse of a triangular matrix. Linear Algebra Appl. **237–238**, 97–108 (1996)
16. S. Kirkland, M. Neumann, *Group Inverses of M-Matrices and Their Applications* (CRC Press, Boca Raton, FL, 2013)
17. R.E. Hartwig, X. Li, Y. Wei, Representations for the Drazin inverse of a 2 × 2 block matrix. SIAM J. Matrix Anal. Appl **27**, 767–771 (2006)
18. C. Deng, Y. Wei, Representations for the Drazin inverse of 2 × 2 block matrix with singular Schur complement. Linear Algebra Appl. **435**, 2766–2783 (2011)
19. Y. Wei, Expression for the Drazin inverse of a 2 × 2 block matrix. Linear Multilinear Algebra **45**, 131–146 (1998)
20. Y. Wei, H. Diao, On group inverse of singular Toeplitz matrices. Linear Algebra Appl. **399**, 109–123 (2005)
21. J. Chen, Z. Xu, Representations for the weighted Drazin inverse of a modified matrix. Appl. Math. Comput. **203**(1), 202–209 (2008)
22. D.S. Cvetković-Ilić, J. Chen, Z. Xu, Explicit representations of the Drazin inverse of block matrix and modified matrix. Linear Multilinear Algebra **57**(4), 355–364 (2009)
23. J.M. Shoaf, The Drazin inverse of a rank-one modification of a square matrix. Ph.D. thesis, North Carolina State University, 1975
24. Y. Wei, The Drazin inverse of a modified matrix. Appl. Math. Comput. **125**, 295–301 (2002)

Chapter 3
Generalization of the Cramer's Rule and the Minors of the Generalized Inverses

It is well known that the Cramer's rule for the solution $\mathbf{x}$ of a nonsingular equation

$$A\mathbf{x} = \mathbf{b} \quad (A \in \mathbb{C}^{n \times n}, \ \mathbf{b} \in \mathbb{C}^n, \ \mathbf{x} = [x_1, x_2, \ldots, x_n]^T)$$

is

$$x_i = \frac{\det(A(i \to \mathbf{b}))}{\det(A)}, \quad i = 1, 2, \ldots, n,$$

where $A(i \to \mathbf{b})$ denotes the matrix obtained by replacing the ith column of A with $\mathbf{b}$.

In 1970, Steve Robinson [1] gave an elegant proof of the Cramer's rule by rewriting $A\mathbf{x} = \mathbf{b}$ as

$$A I(i \to \mathbf{x}) = A(i \to \mathbf{b}),$$

where I is the identity matrix of order n, and taking determinants

$$\det(A) \det(I(i \to \mathbf{x})) = \det(A(i \to \mathbf{b})).$$

The Cramer's rule then follows from $\det(I(i \to \mathbf{x})) = x_i, i = 1, 2, \ldots, n$.

Since 1982, the Robinson's trick has been used to derive a series of the Cramer's rules for the minimum-norm solution and the minimum-norm (N) solution of consistent linear equations; for the unique solutions of special consistent restricted linear equations; for the minimum-norm least-squares solution and the minimum-norm (N) least-squares (M) solution of inconsistent linear equations; for the unique solutions of a class of singular equations; and for the best approximate solution of a matrix equation $AXH = K$ [2–12].

The basic idea of these Cramer's rules is to construct a nonsingular bordered matrix by adjoining certain matrices to the original matrix. The solution of the original system is then obtained from the new nonsingular system.

© Springer Nature Singapore Pte Ltd. and Science Press 2018
G. Wang et al., *Generalized Inverses: Theory and Computations*,
Developments in Mathematics 53, https://doi.org/10.1007/978-981-13-0146-9_3

As we know, the jth column of the inverse of a nonsingular matrix can be computed by solving a linear system with the jth unit vector $\mathbf{e}_j$ as the right-hand-side. Thus, by applying the Cramer's rule, the inverse of a nonsingular matrix A can be expressed in terms of the determinants of A and modified A. This chapter presents determinantal expressions of the generalized inverses.

3.1   Nonsingularity of Bordered Matrices

Given a matrix A, an associated bordered matrix

$$\begin{bmatrix} A & B \\ C & D \end{bmatrix}$$

is an expanded matrix that contains A as its leading principal submatrix as shown above. This section establishes relations between the generalized inverses and the nonsingularity of bordered matrices.

3.1.1   Relations with $A^{\dagger}_{MN}$ and $A^{\dagger}$

In 1986, Wang [8] showed the following results on a relation between $A^{\dagger}$ and the inverse of a bordered matrix.

Theorem 3.1.1 *Let $A \in \mathbb{C}_r^{m \times n}$, M and N be Hermitian positive definite matrices of orders m and n respectively, and the columns of $U \in \mathbb{C}_{m-r}^{m \times (m-r)}$ and $V^* \in \mathbb{C}_{n-r}^{n \times (n-r)}$ form bases for $\mathcal{N}(A^*)$ and $\mathcal{N}(A)$ respectively, then the bordered matrix*

$$A_2 = \begin{bmatrix} A & M^{-1}U \\ VN & O \end{bmatrix}$$

is nonsingular and its inverse

$$A_2^{-1} = \begin{bmatrix} A^{\dagger}_{MN} & V^*(VNV^*)^{-1} \\ (U^*M^{-1}U)^{-1}U^* & O \end{bmatrix}. \tag{3.1.1}$$

Proof Obviously, $(VNV^*)^{-1}$ and $(U^*M^{-1}U)^{-1}$ exist and

$$VNV^*(VNV^*)^{-1} = I_{n-r}. \tag{3.1.2}$$

It follows from $AV^* = O$ that

$$AV^*(VNV^*)^{-1} = O \tag{3.1.3}$$

and from (1.4.15) that

$$
\begin{aligned}
VNA_{MN}^{\dagger} &= VNA_{MN}^{\dagger}AA_{MN}^{\dagger} \\
&= V(NA_{MN}^{\dagger}A)^{*}A_{MN}^{\dagger} \\
&= VA^{*}(NA_{MN}^{\dagger})^{*}A_{MN}^{\dagger} \\
&= O.
\end{aligned}
\tag{3.1.4}
$$

Finally, let

$$
F = M^{-1}U(U^{*}M^{-1}U)^{-1}U^{*} \quad \text{and} \quad E = AA_{MN}^{\dagger},
$$

then E and F are idempotent, and therefore they are the projectors

$$
\begin{aligned}
E &= P_{\mathcal{R}(E),\mathcal{N}(E)} = P_{\mathcal{R}(U)^{\perp},M^{-1}\mathcal{R}(U)}, \\
F &= P_{\mathcal{R}(F),\mathcal{N}(F)} = P_{M^{-1}\mathcal{R}(U),\mathcal{R}(U)^{\perp}}.
\end{aligned}
$$

Since

$$
P_{\mathcal{R}(U)^{\perp},M^{-1}\mathcal{R}(U)} + P_{M^{-1}\mathcal{R}(U),\mathcal{R}(U)^{\perp}} = I_{m},
$$

we have

$$
AA_{MN}^{\dagger} + M^{-1}U(U^{*}M^{-1}U)^{-1}U^{*} = I_{m}.
\tag{3.1.5}
$$

From (3.1.2)–(3.1.5),

$$
\begin{bmatrix} A & M^{-1}U \\ VN & O \end{bmatrix}
\begin{bmatrix} A_{MN}^{\dagger} & V^{*}(VNV^{*})^{-1} \\ (U^{*}M^{-1}U)^{-1}U^{*} & O \end{bmatrix}
= \begin{bmatrix} I_{m} & O \\ O & I_{n-r} \end{bmatrix},
$$

which is the desired result (3.1.1). $\qquad \square$

Corollary 3.1.1 *Let $A \in \mathbb{C}_{r}^{m \times n}$, and $U \in \mathbb{C}_{m-r}^{m \times (m-r)}$ and $V^{*} \in \mathbb{C}_{n-r}^{n \times (n-r)}$ be matrices whose columns form the bases for $\mathcal{N}(A^{*})$ and $\mathcal{N}(A)$ respectively, then*

$$
A_{1} = \begin{bmatrix} A & U \\ V & O \end{bmatrix}
$$

is nonsingular and

$$
A_{1}^{-1} = \begin{bmatrix} A^{\dagger} & V^{\dagger} \\ U^{\dagger} & O \end{bmatrix}.
\tag{3.1.6}
$$

Proof Applying Theorem 3.1.1 and

$$
V^{\dagger} = V^{*}(VV^{*})^{-1}, \quad U^{\dagger} = (U^{*}U)^{-1}U^{*},
$$

we obtain (3.1.6) immediately. $\qquad \square$

Corollary 3.1.2 *Let $A \in \mathbb{C}_r^{m \times n}$, M and N be Hermitian positive definite matrices of orders m and n respectively, and $U \in \mathbb{C}_{m-r}^{m \times (m-r)}$ and $V^* \in \mathbb{C}_{n-r}^{n \times (n-r)}$ satisfy*

$$AV^* = O, \ VNV^* = I_{n-r}; \quad A^*U = O, \ U^*M^{-1}U = I_{m-r},$$

then

$$A_2 = \begin{bmatrix} A & M^{-1}U \\ VN & O \end{bmatrix}$$

is nonsingular and its inverse

$$A_2^{-1} = \begin{bmatrix} A_{MN}^{\dagger} & V^* \\ U^* & O \end{bmatrix}.$$

Corollary 3.1.3 *([2]) Suppose that $A \in \mathbb{C}_r^{m \times n}$, $U \in \mathbb{C}_{m-r}^{m \times (m-r)}$, and $V^* \in \mathbb{C}_{n-r}^{n \times (n-r)}$ satisfy*

$$AV^* = O, \ VV^* = I_{n-r}, \quad A^*U = O, \ U^*U = I_{m-r},$$

then

$$A_1 = \begin{bmatrix} A & U \\ V & O \end{bmatrix}$$

is nonsingular and its inverse

$$A_1^{-1} = \begin{bmatrix} A^{\dagger} & V^* \\ U^* & O \end{bmatrix}.$$

3.1.2 Relations Between the Nonsingularity of Bordered Matrices and A_d and A_g

In 1989, Wang [9, 13] showed relations between the Drazin inverse and group inverse and the nonsingularity of bordered matrices.

Lemma 3.1.1 *Let $U \in \mathbb{C}_p^{n \times p}$, $V^* \in \mathbb{C}_p^{n \times p}$, and $\mathcal{R}(U) \oplus \mathcal{N}(V) = \mathbb{C}^n$, then VU is nonsingular.*

Proof If $(VU)\mathbf{x} = \mathbf{0}$, then $U\mathbf{x} \in \mathcal{N}(V)$ and $U\mathbf{x} \in \mathcal{R}(U)$. By the assumption,

$$U\mathbf{x} \in \mathcal{R}(U) \cap \mathcal{N}(V) = \{\mathbf{0}\}.$$

Thus $U\mathbf{x} = \mathbf{0}$. Since U is of full column rank, $\mathbf{x} = \mathbf{0}$. This shows that VU has linearly independent columns and $VU \in \mathbb{C}^{p \times p}$ is nonsingular.   □

The following results can be found in [9, 13].

Theorem 3.1.2 *Let* $A \in \mathbb{C}^{n \times n}$, $\text{Ind}(A) = k$, $\text{rank}(A^k) = r < n$, *and* $U, V^* \in \mathbb{C}_{n-r}^{n \times (n-r)}$ *be matrices whose columns form the bases for* $\mathcal{N}(A^k)$ *and* $\mathcal{N}(A^{k*})$ *respectively, then*

$$A_4 = \begin{bmatrix} A & U \\ V & O \end{bmatrix}$$

is nonsingular and its inverse

$$A_4^{-1} = \begin{bmatrix} A_d & U(VU)^{-1} \\ (VU)^{-1}V & -(VU)^{-1}VAU(VU)^{-1} \end{bmatrix}. \tag{3.1.7}$$

Proof By the assumptions on U and V, we have

$$\mathcal{R}(U) = \mathcal{N}(A^k) \quad \text{and} \quad \mathcal{N}(V) = \mathcal{R}(A^k).$$

From Theorem 3.1.1 and Lemma 3.1.1, VU is nonsingular, thus its inverse $(VU)^{-1}$ exists. Setting

$$X = \begin{bmatrix} A_d & U(VU)^{-1} \\ (VU)^{-1}V & -(VU)^{-1}VAU(VU)^{-1} \end{bmatrix},$$

we have

$$A_4X = \begin{bmatrix} AA_d + U(VU)^{-1}V & (I - U(VU)^{-1}V)AU(VU)^{-1} \\ VA_d & VU(VU)^{-1} \end{bmatrix}.$$

Obviously,

$$VU(VU)^{-1} = I_{n-r}. \tag{3.1.8}$$

Denoting $G = (VU)^{-1}V$, we have $UGU = U$, $G \in U\{1\}$, and

$$G^* = V^*((VU)^{-1})^*, \quad \mathcal{R}(G^*) = \mathcal{R}(V^*) = \mathcal{N}(A^{k*}), \quad \mathcal{N}(G) = \mathcal{R}(A^k).$$

Since $UG = U(VU)^{-1}V$ is idempotent, it is a projector and

$$U(VU)^{-1}V = UG = P_{\mathcal{R}(UG),\mathcal{N}(UG)} = P_{\mathcal{R}(U),\mathcal{N}(G)} = P_{\mathcal{N}(A^k),\mathcal{R}(A^k)}.$$

Thus

$$AA_d + U(VU)^{-1}V = P_{\mathcal{R}(A^k),\mathcal{N}(A^k)} + P_{\mathcal{N}(A^k),\mathcal{R}(A^k)} = I_n. \tag{3.1.9}$$

Since $\mathcal{N}(V) = \mathcal{R}(A^k)$, we have $VA^k = O$. It then follows from (2.1.17) that,

$$VA_d = VA^k(A^{2k+1})^{(1)}A^k = O. \tag{3.1.10}$$

Finally, setting $F = U(VU)^{-1}$, we get $A^kF = O$, since $\mathcal{R}(F) = \mathcal{R}(U) = \mathcal{N}(A^k)$. It follows from (3.1.9) and (2.1.17) that

$$(I - U(VU)^{-1}V)AU(VU)^{-1} = AA_dAF$$
$$= A^{k+1}(A^{2k+1})^{(1)}A^{k+1}F$$
$$= O. \tag{3.1.11}$$

From (3.1.8)–(3.1.11), we have $A_4X = I_{2n-r}$, which proves (3.1.7).   □

Corollary 3.1.4 *Let* $A \in \mathbb{C}^{n \times n}$, $\mathrm{Ind}(A) = 1$, $\mathrm{rank}(A) = r < n$, *and* $U, V^* \in \mathbb{C}_{n-r}^{n \times (n-r)}$ *be matrices whose columns form the bases for* $\mathcal{N}(A)$ *and* $\mathcal{N}(A^*)$ *respectively, then*

$$A_3 = \begin{bmatrix} A & U \\ V & O \end{bmatrix}$$

is nonsingular and its inverse

$$A_3^{-1} = \begin{bmatrix} A_g & U(VU)^{-1} \\ (VU)^{-1}V & -(VU)^{-1}VAU(VU)^{-1} \end{bmatrix}.$$

3.1.3   Relations Between the Nonsingularity of Bordered Matrices and $A_{T,S}^{(2)}$, $A_{T,S}^{(1,2)}$, and $A_{(L)}^{(-1)}$

Now, we investigate the relations between the nonsingularity of bordered matrices and the generalized inverses $A_{T,S}^{(2)}$, $A_{T,S}^{(1,2)}$, and $A_{(L)}^{(-1)}$.

Theorem 3.1.3 ([3, 4]) *Suppose* $A \in \mathbb{C}_r^{m \times n}$, $T \subset \mathbb{C}^n$, $S \subset \mathbb{C}^m$, $\dim(T) = \dim(S^{\perp}) = t \leq r$ *and* $AT \oplus S = \mathbb{C}^m$. *Let* B *and* C^* *be of full column rank such that*

$$S = \mathcal{R}(B) \quad and \quad T = \mathcal{N}(C),$$

then the bordered matrix

$$A_6 = \begin{bmatrix} A & B \\ C & O \end{bmatrix}$$

is nonsingular and

$$A_6^{-1} = \begin{bmatrix} A_{T,S}^{(2)} & (I - A_{T,S}^{(2)}A)C^{\dagger} \\ B^{\dagger}(I - AA_{T,S}^{(2)}) & B^{\dagger}(AA_{T,S}^{(2)}A - A)C^{\dagger} \end{bmatrix}. \tag{3.1.12}$$

Proof By Theorem 1.3.8 and the assumptions, we have

$$\mathcal{R}(U) = T = \mathcal{N}(C) \quad and \quad \mathcal{N}(V) = S = \mathcal{R}(B),$$

consequently,

$$CU = O \quad and \quad VB = O. \tag{3.1.13}$$

Set

$$X = \begin{bmatrix} A_{T,S}^{(2)} & (I - A_{T,S}^{(2)}A)C^\dagger \\ B^\dagger(I - AA_{T,S}^{(2)}) & B^\dagger(AA_{T,S}^{(2)}A - A)C^\dagger \end{bmatrix},$$

then

$$A_6 X = \begin{bmatrix} P_{AT,S} + BB^\dagger P_{S,AT} & A(I - A_{T,S}^{(2)}A)C^\dagger - BB^\dagger P_{S,AT}AC^\dagger \\ CA_{T,S}^{(2)} & C(I - A_{T,S}^{(2)}A)C^\dagger \end{bmatrix},$$

where

$$P_{AT,S} = AA_{T,S}^{(2)} \quad \text{and} \quad P_{S,AT} = I - AA_{T,S}^{(2)}.$$

Using (3.1.13) and

$$A_{T,S}^{(2)} = U(VAU)^{-1}V,$$

we have

$$CA_{T,S}^{(2)} = CU(VAU)^{-1}V = O. \tag{3.1.14}$$

Noting that $\mathcal{R}(C^\dagger) = \mathcal{R}(C^*) = T^\perp$ and C is of full row rank, we get

$$C(I - A_{T,S}^{(2)}A)C^\dagger = CC^\dagger = I. \tag{3.1.15}$$

Moreover, since

$$BB^\dagger = P_{\mathcal{R}(B)} = P_S$$

and

$$BB^\dagger P_{S,AT} = P_S P_{S,AT} = P_{S,AT},$$

we have

$$P_{AT,S} + BB^\dagger P_{S,AT} = P_{AT,S} + P_{S,AT} = I. \tag{3.1.16}$$

Finally,

$$\begin{aligned} & A(I - A_{T,S}^{(2)}A)C^\dagger - BB^\dagger P_{S,AT}AC^\dagger \\ &= (I - AA_{T,S}^{(2)})AC^\dagger - P_{S,AT}AC^\dagger \\ &= O. \end{aligned} \tag{3.1.17}$$

From (3.1.14)–(3.1.17), we have $A_6 X = I_{m+n-t}$, hence (3.1.12) holds. $\qquad\square$

When $t = r$ in the above theorem, by Corollary 1.3.4, $AT \oplus S = \mathbb{C}^m$ is equivalent to $\mathcal{R}(A) \oplus S = \mathbb{C}^m$ and $\mathcal{N}(A) \oplus T = \mathbb{C}^n$. In this case, $A_{T,S}^{(2)}$ becomes $A_{T,S}^{(1,2)}$, and we have the following theorem.

Theorem 3.1.4 ([14]) *Let $A \in \mathbb{C}_r^{m \times n}$, $T \subset \mathbb{C}^n$, $S \subset \mathbb{C}^m$, $\dim(T) = \dim(S^{\perp}) = r$ and $\mathcal{R}(A) \oplus S = \mathbb{C}^m$, $\mathcal{N}(A) \oplus T = \mathbb{C}^n$, and B and C^* be of full column rank such that*

$$S = \mathcal{R}(B) \quad and \quad T = \mathcal{N}(C),$$

then the bordered matrix

$$A_5 = \begin{bmatrix} A & B \\ C & O \end{bmatrix}$$

is nonsingular and

$$A_5^{-1} = \begin{bmatrix} A_{T,S}^{(1,2)} & (I - A_{T,S}^{(1,2)} A) C^{\dagger} \\ B^{\dagger}(I - A A_{T,S}^{(1,2)}) & O \end{bmatrix}.$$

By Theorem 1.5.1, the Bott-Duffin inverse $A_L^{(-1)} = A_{L,L^{\perp}}^{(2)}$, thus we have the following theorem.

Theorem 3.1.5 ([15]) *Let $A \in \mathbb{C}_r^{n \times n}$, $U \in \mathbb{C}_p^{n \times p}$, and $L = \mathcal{N}(U^*)$, then*

$$A_7 = \begin{bmatrix} A & U \\ U^* & O \end{bmatrix}$$

is nonsingular if and only if

$$A\mathcal{N}(U^*) \oplus \mathcal{R}(U) = \mathbb{C}^n.$$

In this case the inverse of A_7 is

$$A_7^{-1} = \begin{bmatrix} A_L^{(-1)} & (I - A_L^{(-1)} A) U^{*\dagger} \\ U^{\dagger}(I - A A_L^{(-1)}) & U^{\dagger}(A A_L^{(-1)} A - A) U^{*\dagger} \end{bmatrix}.$$

Exercises 3.1

1. Prove $I - A A_{T,S}^{(2)} = P_{S,AT}$.
2. Show the relation between $A_{(L)}^{(\dagger)}$ and the nonsingularity of a bordered matrix.

3.2 Cramer's Rule for Solutions of Linear Systems

Using the relations between the generalized inverses and nonsingular bordered matrices discussed in the previous sections, in this section, we give the Cramer's rules for the solutions of systems of linear equations and matrix equations.

We adopt the following notations and definitions.

- Let $A \in \mathbb{C}^{m \times n}$, $\mathbf{x} \in \mathbb{C}^m$, and $\mathbf{y} \in \mathbb{C}^n$, $A(j \to \mathbf{x})$ denotes the matrix obtained by replacing the jth column of A with $\mathbf{x}$; $A(\mathbf{y}^* \leftarrow i)$ denotes the matrix obtained by replacing the ith row of A with $\mathbf{y}^*$.
- Let $A \in \mathbb{C}^{n \times n}$, $\det(A)$ denotes the determinant of A.
- $\mathcal{R}_c(A) = \{S : S \oplus \mathcal{R}(A) = \mathbb{C}^m\}$ and $\mathcal{N}_c(A) = \{T : T \oplus \mathcal{N}(A) = \mathbb{C}^n\}$ denote the complements of $\mathcal{R}(A)$ and of $\mathcal{N}(A)$ respectively.

3.2.1 Cramer's Rule for the Minimum-Norm (N) Least-Squares (M) Solution of an Inconsistent System of Linear Equations

Let $A \in \mathbb{C}^{m \times n}$, $\mathbf{b} \in \mathbb{C}^m$, M and N are Hermitian positive definite matrices of orders m and n respectively. The vector $\mathbf{u} \in \mathbb{C}^n$ is called the least-squares (M) solution of the inconsistent system of linear equations

$$Ax = b \quad (A \in \mathbb{C}^{m \times n}, \ \mathbf{b} \notin \mathcal{R}(A), \ \mathbf{x} = [x_1, x_2, \ldots, x_n]^T), \tag{3.2.1}$$

if

$$\|A\mathbf{u} - \mathbf{b}\|_M \le \|A\mathbf{v} - \mathbf{b}\|_M, \quad \text{for all } \mathbf{v} \in \mathbb{C}^n.$$

Thus the least-squares (M) solution of (3.2.1) is not unique. If $\mathbf{u}$ is the least-squares (M) solution of (3.2.1) and

$$\|\mathbf{u}\|_N < \|\mathbf{w}\|_N,$$

for all the least-squares (M) solution $\mathbf{w} \ne \mathbf{u}$ of (3.2.1), then $\mathbf{u}$ is called the minimum-norm (N) least-squares (M) solution of (3.2.1).

We know that the minimum-norm (N) least-squares (M) solution of (3.2.1) is $\mathbf{x} = A_{MN}^{\dagger}\mathbf{b}$. The following theorem about the Cramer's rule for finding the solution $\mathbf{x} = A_{MN}^{\dagger}\mathbf{b}$ is given by Wang [8].

Theorem 3.2.1 *Let $A \in \mathbb{C}_r^{m \times n}$, and $M \in \mathbb{C}^{m \times m}$, $N \in \mathbb{C}^{n \times n}$ be Hermitian positive definite matrices, and $U \in \mathbb{C}_{m-r}^{m \times (m-r)}$ and $V^* \in \mathbb{C}_{n-r}^{n \times (n-r)}$ be matrices whose columns form the bases for $\mathcal{N}(A^*)$ and $\mathcal{N}(A)$ respectively, then the minimum-norm (N) least-squares (M) solution $\mathbf{x}$ of (3.2.1) satisfies*

$$\mathbf{x} \in N^{-1}\mathcal{R}(A^*), \quad \mathbf{b} - A\mathbf{x} \in M^{-1}\mathcal{N}(A^*), \tag{3.2.2}$$

and its components are given by

$$x_j = \frac{\det \begin{bmatrix} A(j \to \mathbf{b}) & M^{-1}U \\ VN(j \to \mathbf{0}) & O \end{bmatrix}}{\det \begin{bmatrix} A & M^{-1}U \\ VN & O \end{bmatrix}}, \quad j = 1, 2, \ldots, n. \tag{3.2.3}$$

Proof Let $A = FG$ be a full rank factorization. By Theorem 1.4.4,

$$A_{MN}^{\dagger} = N^{-1}G^*(F^*MAN^{-1}G^*)^{-1}F^*M.$$

Thus

$$VN\mathbf{x} = VNA_{MN}^{\dagger}\mathbf{b} \equiv V\mathbf{h},$$

where

$$\mathbf{h} = G^*(F^*MAN^{-1}G^*)^{-1}F^*M\mathbf{b} \in \mathcal{R}(G^*).$$

Since $G^*F^* = A^*$, we have $\mathcal{R}(A^*) \subset \mathcal{R}(G^*)$. Because F^* is of full row rank, $F^*F^{*(1)} = I$. Thus

$$G^* = A^*F^{*(1)}, \quad \mathcal{R}(G^*) \subset \mathcal{R}(A^*), \quad \mathbf{h} \in \mathcal{R}(G^*) = \mathcal{R}(A^*).$$

By the assumption $\mathcal{R}(V^*) = \mathcal{N}(A)$, we have

$$\mathcal{N}(V) = \mathcal{R}(A^*) \quad \text{and} \quad V\mathbf{h} = \mathbf{0}.$$

Therefore

$$VN\mathbf{x} = \mathbf{0}. \tag{3.2.4}$$

It then follows that

$$N\mathbf{x} \in \mathcal{N}(V) = \mathcal{R}(A^*), \quad \mathbf{x} \in N^{-1}\mathcal{R}(A^*),$$

which proves the first statement in (3.2.2). Since

$$A^*MAA_{MN}^{\dagger} = A^*(MAA_{MN}^{\dagger})^* = (MAA_{MN}^{\dagger}A)^* = (MA)^* = A^*M,$$

we have

$$A^*M\mathbf{b} = A^*MAA_{MN}^{\dagger}\mathbf{b} = A^*MA\mathbf{x} \quad \text{and} \quad A^*M(\mathbf{b} - A\mathbf{x}) = \mathbf{0}.$$

It then follows that

$$M(\mathbf{b} - A\mathbf{x}) \in \mathcal{N}(A^*) \quad \text{and} \quad \mathbf{b} - A\mathbf{x} \in M^{-1}\mathcal{N}(A^*),$$

which proves the second statement in (3.2.2).

From $M(\mathbf{b} - A\mathbf{x}) \in \mathcal{N}(A^*)$, we get

$$M(\mathbf{b} - A\mathbf{x}) = U\mathbf{z}, \quad \mathbf{z} \in \mathbb{C}^{m-r}.$$

Thus

$$\mathbf{b} = A\mathbf{x} + M^{-1}U\mathbf{z}.$$

It then follows from (3.2.4) and the above equation that the minimum-norm (N) least-squares (M) solution $\mathbf{x}$ of (3.2.1) satisfies

$$
\begin{bmatrix} A & M^{-1}U \\ VN & O \end{bmatrix} \begin{bmatrix} \mathbf{x} \\ \mathbf{z} \end{bmatrix} = \begin{bmatrix} \mathbf{b} \\ \mathbf{0} \end{bmatrix}. \tag{3.2.5}
$$

By Theorem 3.1.1, the coefficient matrix of (3.2.5) is nonsingular, and (3.2.3) follows from the standard Cramer's rule.   □

The following corollary is a result in [2].

Corollary 3.2.1 *Let* $A \in \mathbb{C}_r^{m \times n}$, $\mathbf{b} \in \mathbb{C}^m$, $\mathbf{b} \notin \mathcal{R}(A)$, *and* $U \in \mathbb{C}_{m-r}^{m \times (m-r)}$, $V^* \in \mathbb{C}_{n-r}^{n \times (n-r)}$ *be matrices whose columns form the bases for* $\mathcal{N}(A^*)$ *and* $\mathcal{N}(A)$ *respectively, then components of the minimum-norm least-squares solution* $\mathbf{x} = A^\dagger \mathbf{b}$ *of* (3.2.1) *are given by*

$$
x_j = \frac{\det \begin{bmatrix} A(j \to \mathbf{b}) & U \\ V(j \to \mathbf{0}) & O \end{bmatrix}}{\det \begin{bmatrix} A & U \\ V & O \end{bmatrix}}, \quad j = 1, 2, \ldots, n.
$$

3.2.2 Cramer's Rule for the Solution of a Class of Singular Linear Equations

Let $A \in \mathbb{C}_r^{n \times n}$, $r < n$, and $\mathrm{Ind}(A) = k$. We consider the following problem: For a given $\mathbf{b} \in \mathcal{R}(A^k)$ find a vector $\mathbf{x} \in \mathcal{R}(A^k)$ such that

$$
A\mathbf{x} = \mathbf{b}. \tag{3.2.6}
$$

From (2.1.6), $\mathcal{R}(A^k) \oplus \mathcal{N}(A^k) = \mathbb{C}^n$, thus, for any $\mathbf{x} \in \mathbb{C}^n$,

$$
\mathbf{x} = \mathbf{u} + \mathbf{v}, \quad \mathbf{u} \in \mathcal{R}(A^k) \text{ and } \mathbf{v} \in \mathcal{N}(A^k).
$$

It follows from (2.1.15) that

$$
A A_d \mathbf{u} = P_{\mathcal{R}(A^k), \mathcal{N}(A^k)} \mathbf{u} = \mathbf{u},
$$
$$
A A_d \mathbf{v} = P_{\mathcal{R}(A^k), \mathcal{N}(A^k)} \mathbf{v} = \mathbf{0}.
$$

Therefore

$$
A A_d \mathbf{x} = \mathbf{u}, \quad \mathbf{x} \in \mathbb{C}^n, \ \mathbf{u} \in \mathcal{R}(A^k). \tag{3.2.7}
$$

Set $A_I = A \big|_{\mathcal{R}(A^k)}$, that is, A is restricted to $\mathcal{R}(A^k)$. If $\mathbf{u} \in \mathcal{R}(A^k)$, then $\mathbf{u} = A^k \mathbf{z}$, $\mathbf{z} \in \mathbb{C}^n$, and

$$A_I \mathbf{u} = A_I A^k \mathbf{z}$$
$$= A^{k+1} \mathbf{z} \in \mathcal{R}(A^{k+1}) = \mathcal{R}(A^k).$$

Clearly, the linear transformation $A_I: \mathcal{R}(A^k) \to \mathcal{R}(A^k)$ is 1-1 onto and invertible. Thus there exists A_I^{-1} such that premultiplying (3.2.7) with A_I^{-1} gives

$$A_d \mathbf{x} = A_I^{-1} \mathbf{u}, \quad \mathbf{x} = \mathbf{u} + \mathbf{v} \in \mathbb{C}^n, \quad \mathbf{u} \in \mathcal{R}(A^k), \ \mathbf{v} \in \mathcal{N}(A^k).$$

The linear transformation A_d defined by the above equation is called the Drazin inverse of A. The proof of the equivalence between this definition and Definition 2.1.2 is given in the following theorem.

Theorem 3.2.2 *Let $A \in \mathbb{C}^{n \times n}$, $\mathrm{Ind}(A) = k$, then A_d is the Drazin inverse of A if and only if*

$$A_d \mathbf{x} = A_I^{-1} \mathbf{u}, \quad \text{for all } \mathbf{x} = \mathbf{u} + \mathbf{v} \in \mathbb{C}^n, \tag{3.2.8}$$

where $\mathbf{u} \in \mathcal{R}(A^k)$ and $\mathbf{v} \in \mathcal{N}(A^k)$.

Proof ONLY IF: It has been shown above.
IF: Firstly, from (3.2.8), we have $A_d \mathbf{v} = \mathbf{0}$, $\mathbf{v} \in \mathcal{N}(A^k)$ and $A_d \mathbf{u} = A_I^{-1} \mathbf{u}$. Thus

$$AA_d \mathbf{v} = \mathbf{0}, \qquad \qquad \mathbf{v} \in \mathcal{N}(A^k),$$
$$AA_d \mathbf{u} = AA_I^{-1} \mathbf{u} = \mathbf{u}, \ \mathbf{u} \in \mathcal{R}(A^k). \tag{3.2.9}$$

Obviously, $\mathbf{v} \in \mathcal{N}(A^k)$, $A^k \mathbf{v} = \mathbf{0}$, $A^{k+1} \mathbf{v} = \mathbf{0}$, and $A\mathbf{v} \in \mathcal{N}(A^k) = \mathcal{N}(A_d)$, thus

$$A_d A \mathbf{v} = \mathbf{0}, \quad \mathbf{v} \in \mathcal{N}(A^k). \tag{3.2.10}$$

On the other hand, $\mathbf{u} \in \mathcal{R}(A^k)$ and $A\mathbf{u} \in \mathcal{R}(A^{k+1}) = \mathcal{R}(A^k)$, thus

$$A_d A \mathbf{u} = A_I^{-1} A \mathbf{u} = \mathbf{u}, \quad \mathbf{u} \in \mathcal{R}(A^k). \tag{3.2.11}$$

It follows from (3.2.9)–(3.2.11) that $AA_d \mathbf{x} = A_d A \mathbf{x}$, for all $\mathbf{x} \in \mathbb{C}^n$, thus

$$AA_d = A_d A. \tag{3.2.12}$$

Secondly,

$$A_d A A_d \mathbf{x} = A_d A A_I^{-1} \mathbf{u} = A_d \mathbf{u} = A_d(\mathbf{u} + \mathbf{v}) = A_d \mathbf{x}, \quad \text{for all } \mathbf{x} \in \mathbb{C}^n, \text{ thus}$$

$$A_d A A_d = A_d. \tag{3.2.13}$$

Finally,

$$A^{k+1} A_d \mathbf{x} = A^k A A_I^{-1} \mathbf{u} = A^k \mathbf{u} = A^k(\mathbf{u} + \mathbf{v}) = A^k \mathbf{x}, \quad \text{for all } \mathbf{x} \in \mathbb{C}^n, \text{ thus}$$

$$A^{k+1} A_d = A^k. \tag{3.2.14}$$

Therefore, the sufficiency is proved by (3.2.12)–(3.2.14). □

It follows that the unique solution of (3.2.6) is

$$\mathbf{x} = A_I^{-1} \mathbf{b} = A_d \mathbf{b}.$$

The Cramer's rule for the unique solution $\mathbf{x} = A_d \mathbf{b}$ of (3.2.6) is given in the following theorem.

Theorem 3.2.3 ([9]) *Suppose that $A \in \mathbb{C}^{n \times n}$, $\mathrm{Ind}(A) = k$, $\mathrm{rank}(A^k) = r < n$, and $U, V^* \in \mathbb{C}_{n-r}^{n \times (n-r)}$ be matrices whose columns form the bases for $\mathcal{N}(A^k)$ and $\mathcal{N}(A^{k^*})$ respectively. Let $\mathbf{b} \in \mathcal{R}(A^k)$, then the components of the unique solution $\mathbf{x} = A_d \mathbf{b}$ of (3.2.6) are given by*

$$x_j = \frac{\det \begin{bmatrix} A(j \rightarrow \mathbf{b}) \ U \\ V(j \rightarrow \mathbf{0}) \ O \end{bmatrix}}{\det \begin{bmatrix} A \ U \\ V \ O \end{bmatrix}}, \quad j = 1, 2, \ldots, n. \tag{3.2.15}$$

Proof Since $\mathbf{x} = A_d \mathbf{b} \in \mathcal{R}(A^k)$ and $\mathcal{N}(V) = \mathcal{R}(A^k)$, we have

$$V\mathbf{x} = \mathbf{0}.$$

It follows from (3.2.6) and the above equation that the solution of (3.2.6) satisfies

$$\begin{bmatrix} A \ U \\ V \ O \end{bmatrix} \begin{bmatrix} \mathbf{x} \\ \mathbf{0} \end{bmatrix} = \begin{bmatrix} \mathbf{b} \\ \mathbf{0} \end{bmatrix}.$$

By Theorem 3.1.2, the coefficient matrix in the above equation is nonsingular. Using (3.1.7), we have $\mathbf{x} = A_d \mathbf{b}$. Consequently, (3.2.15) follows from the standard Cramer's rule. □

3.2.3 Cramer's Rule for the Solution of a Class of Restricted Linear Equations

Let $A \in \mathbb{C}_r^{m \times n}$, $\mathbf{b} \in \mathcal{R}(A)$ and $T \subset \mathbb{C}^n$. The Cramer's rule for the unique solution of a class of restricted linear equations:

$$A\mathbf{x} = \mathbf{b}, \quad \mathbf{x} \in T \tag{3.2.16}$$

is given by Chen [3].

Theorem 3.2.4 *The Eq.* (3.2.16) *have a unique solution if and only if*

$$\mathbf{b} \in AT \quad and \quad T \cap \mathcal{N}(A) = \{\mathbf{0}\}.$$

Proof If $\mathbf{b} \in AT$, then it is obvious that (3.2.16) has a solution $\mathbf{x}_0 \in T$. Let the general solution of (3.2.16) be $\mathbf{x} = \mathbf{x}_0 + \mathbf{y} \in T$, where $\mathbf{y} \in \mathcal{N}(A)$, then

$$\mathbf{y} = \mathbf{x} - \mathbf{x}_0 \in T.$$

Since $T \cap \mathcal{N}(A) = \{\mathbf{0}\}$, we get $\mathbf{y} = \mathbf{0}$. Therefore (3.2.16) has a unique solution $\mathbf{x} = \mathbf{x}_0$.

Conversely, let the general solution of (3.2.16) be $\mathbf{x} = \mathbf{x}_0 + \mathbf{y} \in T$, where $\mathbf{y} \in \mathcal{N}(A)$ and $\mathbf{x}_0 \in T$ is a particular solution of (3.2.16). Since (3.2.16) has a unique solution, we have $\mathbf{y} = \mathbf{0}$. Moreover,

$$\mathbf{x} = \mathbf{x}_0 \quad and \quad \mathbf{b} = A\mathbf{x}_0 \in AT.$$

It follows from $\mathbf{y} \in \mathcal{N}(A)$, $\mathbf{y} = \mathbf{x} - \mathbf{x}_0 \in T$, and $\mathbf{y} = \mathbf{0}$ that $T \cap \mathcal{N}(A) = \{\mathbf{0}\}$.   □

Lemma 3.2.1 *Let $A \in \mathbb{C}_r^{m \times n}$ and T be a subspace of $\mathbb{C}^n$, then the following conditions are equivalent:*

(1) $T \cap \mathcal{N}(A) = \{\mathbf{0}\}$;
(2) $\dim(AT) = \dim(T) = s \leq r$;
(3) *There exists a subspace S of $\mathbb{C}^m$ of dimension $m - \dim(T)$ such that*

$$AT \oplus S = \mathbb{C}^m$$

or equivalently

$$A^* S^\perp \oplus T^\perp = \mathbb{C}^n.$$

Proof (1)$\Leftrightarrow$(2): It follows from the equation:

$$\dim(AT) = \dim(T) - \dim(T \cap \mathcal{N}(A)).$$

(2)$\Rightarrow$(3): If $\dim(AT) = \dim(T) = s < r$, then there exists $S \subset \mathbb{C}^m$ with $\dim(S) = m - s$ such that $AT \oplus S = \mathbb{C}^m$. If $\dim(AT) = \dim(T) = s = r$, by the equation $\dim(\mathcal{N}(A)) = n - r$, then, from the equivalence of (1) and (2) and $T \cap \mathcal{N}(A) = \{\mathbf{0}\}$, $T \oplus \mathcal{N}(A) = \mathbb{C}^n$. Since $\dim(\mathcal{R}(A)) = r$, there exists an $S \subset \mathbb{C}^m$ such that $\mathcal{R}(A) \oplus S = \mathbb{C}^m$, which is equivalent to $AT \oplus S = \mathbb{C}^m$ by Corollary 1.3.4.
(3)$\Rightarrow$(2): From Exercise 1.3.9, $AT \oplus S = \mathbb{C}^m$ is equivalent to $A^* S^\perp \oplus T^\perp = \mathbb{C}^n$. Thus

$$T \oplus (A^* S^\perp)^\perp = \mathbb{C}^n.$$

However,

$$\mathcal{N}(A) = \mathcal{R}(A^*)^{\perp} \subset (A^* S^{\perp})^{\perp},$$

therefore

$$T \cap \mathcal{N}(A) = \{0\},$$

which is equivalent to

$$\dim(AT) = \dim(T) = s \le r$$

by the equivalence of (1) and (2). $\qquad\square$

Theorem 3.2.5 Let $A \in \mathbb{C}_r^{m \times n}$, $T \subset \mathbb{C}^n$, and the condition in Lemma 3.2.1 be satisfied, then the unique solution of the restricted linear equations

$$A\mathbf{x} = \mathbf{b}, \quad \mathbf{x} \in T$$

is given by

$$\mathbf{x} = A_{T,S}^{(2)} \mathbf{b},$$

for any subspace S of $\mathbb{C}^m$ satisfying $AT \oplus S = \mathbb{C}^m$.

Proof Obviously, $\mathbf{x} = A_{T,S}^{(2)} \mathbf{b} \in T$. Since $AA_{T,S}^{(2)}$ is the projector $P_{AT,S}$ and $\mathbf{b} \in AT$, we have $A\mathbf{x} = AA_{T,S}^{(2)} \mathbf{b} = \mathbf{b}$. Thus $\mathbf{x} = A_{T,S}^{(2)} \mathbf{b}$ is a solution of (3.2.16).

By the condition in Lemma 3.2.1 and Theorem 3.2.4, the solution of (3.2.16) is unique, independent of the choice of the subspace S. $\qquad\square$

Corollary 3.2.2 Let $A \in \mathbb{C}_r^{m \times n}$, $\mathbf{b} \in AT$, $\dim(AT) = \dim(T) = r$, $T \oplus \mathcal{N}(A) = \mathbb{C}^n$, and $S \subset \mathbb{C}^m$ satisfy

$$\mathcal{R}(A) \oplus S = \mathbb{C}^m,$$

then the unique solution of the restricted linear Eq. (3.2.16) is given by

$$\mathbf{x} = A_{T,S}^{(1,2)} \mathbf{b} \tag{3.2.17}$$

for any subspace S of $\mathbb{C}^m$ satisfying $\mathcal{R}(A) \oplus S = \mathbb{C}^m$.

Proof From Theorem 3.2.5 and the note after Theorem 3.1.3, we obtain (3.2.17) immediately. $\qquad\square$

Now, we have the Cramer's rule for the solution $\mathbf{x} = A_{T,S}^{(2)} \mathbf{b}$ or $\mathbf{x} = A_{T,S}^{(1,2)} \mathbf{b}$.

Theorem 3.2.6 Let $A \in \mathbb{C}_r^{m \times n}$, $T \subset \mathbb{C}^n$, the condition in Lemma 3.2.1 be satisfied, and both B and C^* be of full column rank and satisfy

$$S = \mathcal{R}(B) \quad and \quad T = \mathcal{N}(C), \tag{3.2.18}$$

then the components x_j of the unique solution $\mathbf{x} = A_{T,S}^{(2)} \mathbf{b}$ of (3.2.16) are given by

$$x_j = \frac{\det \begin{bmatrix} A(j \to \mathbf{b}) & B \\ C(j \to \mathbf{0}) & O \end{bmatrix}}{\det \begin{bmatrix} A & B \\ C & O \end{bmatrix}}, \quad j = 1, 2, \ldots, n. \tag{3.2.19}$$

Proof From (3.2.18),

$$\mathbf{x} = A_{T,S}^{(2)}\mathbf{b} \in T = \mathcal{N}(C) \iff C\mathbf{x} = \mathbf{0}.$$

Thus (3.2.16) can be rewritten as

$$\begin{bmatrix} A & B \\ C & O \end{bmatrix} \begin{bmatrix} \mathbf{x} \\ \mathbf{0} \end{bmatrix} = \begin{bmatrix} \mathbf{b} \\ \mathbf{0} \end{bmatrix}.$$

By Theorem 3.1.3, the coefficient matrix of the above equation is nonsingular. Thus $\mathbf{x} = A_{T,S}^{(2)}\mathbf{b}$ and $\mathbf{y} = \mathbf{0}$ is the unique solution of the nonsingular linear equations

$$\begin{pmatrix} A & B \\ C & O \end{pmatrix} \begin{pmatrix} \mathbf{x} \\ \mathbf{y} \end{pmatrix} = \begin{pmatrix} \mathbf{b} \\ \mathbf{0} \end{pmatrix}.$$

Consequently, (3.2.19) follows from the standard Cramer's rule for the above equation. □

Theorem 3.2.7 *Let $A \in \mathbb{C}_r^{m \times n}$, $T \subset \mathbb{C}^n$ and $S \subset \mathbb{C}^m$ satisfy*

$$\mathcal{R}(A) \oplus S = \mathbb{C}^m \quad and \quad T \oplus \mathcal{N}(A) = \mathbb{C}^n, \tag{3.2.20}$$

and both B and C^* be of full column rank and satisfy

$$S = \mathcal{R}(B) \quad and \quad T = \mathcal{N}(C),$$

then the components x_j of the unique solution $\mathbf{x} = A_{T,S}^{(1,2)}\mathbf{b}$ of (3.2.16) are given by

$$x_j = \frac{\det \begin{bmatrix} A(j \to \mathbf{b}) & B \\ C(j \to \mathbf{0}) & O \end{bmatrix}}{\det \begin{bmatrix} A & B \\ C & O \end{bmatrix}}, \quad j = 1, 2, \ldots, n. \tag{3.2.21}$$

Proof Since $\dim(\mathcal{N}(A)) = n - r$ and $\dim(\mathcal{R}(A)) = r$, we have

$$\dim(T) = r \quad and \quad \dim(S) = m - r$$

by (3.2.20). Since (3.2.20) is equivalent to $AT \oplus S = \mathbb{C}^m$, we have $AT = \mathcal{R}(A)$, $\mathbf{b} \in \mathcal{R}(A) = AT$ and $\dim(T) = \dim(AT) = r$ are satisfied. It follows from

Corollary 3.2.2 that the unique solution of (3.2.16) is $\mathbf{x} = A_{T,S}^{(1,2)}\mathbf{b}$. Consequently, (3.2.21) follows from (3.2.19), which is the unique solution of (3.2.16). □

Corollary 3.2.3 *Let $A \in \mathbb{C}_r^{n \times n}$ and $L \subset \mathbb{C}^n$ satisfy*

$$AL \oplus L^\perp = \mathbb{C}^n,$$

then the restricted equations

$$A\mathbf{x} + \mathbf{y} = \mathbf{b}, \quad \mathbf{x} \in L, \quad \mathbf{y} \in L^\perp, \tag{3.2.22}$$

have a unique pair of solutions $\mathbf{x}$ and $\mathbf{y}$. Let U be of full column rank and satisfy

$$L = \mathcal{N}(U^*),$$

then the components of the solution $\mathbf{x} = A_{(L)}^{(-1)}\mathbf{b}$ of (3.2.22) are given by

$$x_j = \frac{\det \begin{bmatrix} A(j \to \mathbf{b}) & U \\ U^*(j \to \mathbf{0}) & O \end{bmatrix}}{\det \begin{bmatrix} A & U \\ U^* & O \end{bmatrix}}, \quad j = 1, 2, \ldots, n \tag{3.2.23}$$

Proof From Theorem 1.5.1, the unique solution of (3.2.22) is $\mathbf{x} = A_{(L)}^{(-1)}\mathbf{b}$. Since $A_{(L)}^{(-1)} = A_{L,L^\perp}^{(2)} = A_{\mathcal{N}(U^*),\mathcal{R}(U)}^{(2)}$, (3.2.23) follows from setting $B = U$ and $C = U^*$ in (3.2.19). □

3.2.4 An Alternative and Condensed Cramer's Rule for the Restricted Linear Equations

In this section, we consider again the restricted linear Eq. (3.2.16):

$$A\mathbf{x} = \mathbf{b}, \quad \mathbf{x} \in T,$$

where $A \in \mathbb{C}^{m \times n}$ and $T \subset \mathbb{C}^n$, and assume that the conditions in Lemma 3.2.1 are satisfied.

Recall that a component of the unique solution of (3.2.16) is expressed by (3.2.19) as the quotient of the determinants of two square matrices both of order $m + n - r$. Because these matrices are possibly considerably larger than A, the aim of this section is to derive a condensed Cramer's rule for the solution of (3.2.16).

First we give an explicit expression of the generalized inverse $A_{T,S}^{(2)}$ in terms of the group inverse.

Lemma 3.2.2 ([16]) *Let $A \in \mathbb{C}_r^{m \times n}$, $T \subset \mathbb{C}^n$, $S \subset \mathbb{C}^m$, and $\dim(T) = \dim(S^{\perp}) = t \leq r$. In addition, suppose that $G \in \mathbb{C}^{n \times m}$ satisfies*

$$\mathcal{R}(G) = T \quad and \quad \mathcal{N}(G) = S.$$

If A has a $\{2\}$-inverse $A_{T,S}^{(2)}$, then

$$\mathrm{Ind}(AG) = \mathrm{Ind}(GA) = 1. \tag{3.2.24}$$

Furthermore, we have

$$A_{T,S}^{(2)} = G(AG)_g = (GA)_g G.$$

Proof It is easy to verify

$$\mathcal{R}(AG) = A\mathcal{R}(G) = AT \quad and \quad S = \mathcal{N}(G) \subset \mathcal{N}(AG).$$

By Theorem 1.3.8, we have

$$\begin{aligned}
\dim(AT) &= m - \dim(S) \\
&= m - (m - t) \\
&= t.
\end{aligned}$$

Now

$$\dim(\mathcal{R}(AG)) + \dim(\mathcal{N}(AG)) = m,$$

hence

$$\begin{aligned}
\dim(\mathcal{N}(AG)) &= m - \dim(\mathcal{R}(AG)) \\
&= m - \dim(AT) \\
&= m - t \\
&= \dim(S).
\end{aligned}$$

Thus $\mathcal{N}(AG) = S$, implying that

$$\mathcal{R}(AG) \oplus \mathcal{N}(AG) = AT \oplus S = \mathbb{C}^m.$$

By Theorem 2.1.1, we have
$$\mathrm{Ind}(AG) = 1.$$

Let $X = G(AG)_g$. We can verify

$$XAX = G(AG)_g AG(AG)_g$$
$$= G(AG)_g$$
$$= X,$$

and

$$\mathcal{R}(X) = \mathcal{R}(G(AG)_g)$$
$$\subset \mathcal{R}(G)$$
$$= T;$$
$$\mathcal{N}(X) = \mathcal{N}(G(AG)_g)$$
$$\supset \mathcal{N}((AG)_g)$$
$$= \mathcal{N}(AG)$$
$$\supset \mathcal{N}(G)$$
$$= S.$$

Obviously, $\mathrm{rank}(X) \leq \dim(T)$. On the other hand,

$$\mathrm{rank}(X) = \mathrm{rank}(G(AG)_g)$$
$$\geq \mathrm{rank}(AG(AG)_g)$$
$$= \mathrm{rank}(AG)$$
$$= s$$
$$= \dim(T).$$

Thus, $\mathcal{R}(X) = T$.

Similarly, we can show that $\mathcal{N}(X) = S$, which is the desired result $A_{T,S}^{(2)} = G(AG)_g$. Similarly, it follows that $\mathrm{Ind}(GA) = 1$ and $A_{T,S}^{(2)} = (GA)_g G$. □

Theorem 3.2.8 *Given A, T, S, and G as in Lemma 3.2.2, and*

$$AT \oplus S = \mathbb{C}^m.$$

Suppose that the columns of V and U^* form the bases for $\mathcal{N}(GA)$ and $\mathcal{N}((GA)^*)$ respectively. We define

$$E = V(UV)^{-1}U.$$

Then E satisfies

$$\mathcal{R}(E) = \mathcal{R}(V) = \mathcal{N}(GA) \quad \text{and} \quad \mathcal{N}(E) = \mathcal{N}(U) = \mathcal{R}(GA), \qquad (3.2.25)$$

and $GA + E$ is nonsingular and its inverse

$$(GA + E)^{-1} = (GA)_g + E_g.$$

Proof From the assumptions on V and U, we have

$$\mathcal{R}(V) = \mathcal{N}(GA), \quad \mathcal{N}(U) = \mathcal{R}(GA).$$

By Lemmas 3.2.2 and 3.1.1, we have

$$\mathrm{Ind}(GA) = 1 \Leftrightarrow \mathcal{R}(GA) \oplus \mathcal{N}(GA) = \mathbb{C}^n$$
$$\Leftrightarrow \mathcal{R}(V) \oplus \mathcal{N}(U) = \mathbb{C}^n$$

Thus UV is nonsingular and $E = V(UV)^{-1}U$ exists. It follows from $E^j = E, j = 1, 2, \ldots,$ that

$$E_d = E_g = E.$$

From

$$\mathcal{R}(E) = \mathcal{R}(V) = \mathcal{N}(GA)$$

and

$$\mathcal{N}(E) = \mathcal{N}(U) = \mathcal{R}(GA) = \mathcal{R}((GA)_g),$$

we have

$$(GA)E = O \quad \text{and} \quad E(GA)_g = O,$$

and

$$(GA + E)((GA)_g + E_g) = (GA)(GA)_g + E E_g$$
$$= P_{\mathcal{R}(GA),\mathcal{N}(GA)} + P_{\mathcal{R}(E),\mathcal{N}(E)}$$
$$= P_{\mathcal{N}(E),\mathcal{R}(E)} + P_{\mathcal{R}(E),\mathcal{N}(E)}$$
$$= I.$$

This completes the proof.   □

Theorem 3.2.9 *Given* $A, T, S, G, V, U^*,$ *and* E *as above, and* $\mathbf{b} \in AT,$ *the restricted linear Eq.* (3.2.16)

$$A\mathbf{x} = \mathbf{b}, \quad \mathbf{x} \in T$$

is equivalent to the nonsingular linear equations

$$(GA + E)\mathbf{x} = G\mathbf{b}, \quad \mathbf{x} \in T. \tag{3.2.26}$$

The components of the unique solution of (3.2.16) *are given by*

$$x_i = \frac{\det((GA + E)(i \to G\mathbf{b}))}{\det(GA + E)}, \quad i = 1, 2, \ldots, n. \tag{3.2.27}$$

Proof From the assumptions, $\mathbf{b} \in AT = \mathcal{R}(GA)$, then $\mathbf{b} = AG\mathbf{y}$ for some $\mathbf{y}$. By (3.2.25), $EGA = O$, so

$$EG\mathbf{b} = EGAG\mathbf{y} = \mathbf{0}.$$

It follows from Theorem 3.2.8 and Lemma 3.2.2 that the unique solution of the nonsingular linear Eq. (3.2.26) is

$$\begin{aligned}
\mathbf{x} &= (GA + E)^{-1}G\mathbf{b} \\
&= ((GA)_g + E_g)G\mathbf{b} \\
&= (GA)_g G\mathbf{b} + EG\mathbf{b} \\
&= A_{T,S}^{(2)}\mathbf{b} \in \mathcal{R}(A_{T,S}^{(2)}) = T.
\end{aligned}$$

From Theorem 3.2.5, the unique solution of the restricted linear Eq. (3.2.16) is also $\mathbf{x} = A_{T,S}^{(2)}\mathbf{b}$. This completes the proof of the equivalence between (3.2.16) and (3.2.26). Consequently, (3.2.27) follows from the standard Cramer's rule for (3.2.26). $\qquad\square$

Corollary 3.2.4 *Let* $A \in \mathbb{C}_r^{m \times n}$, $T = \mathcal{R}(A^*)$, $S = \mathcal{N}(A^*)$, $\mathbf{b} \in A\mathcal{R}(A^*) = \mathcal{R}(A)$, *and the columns of* $V \in \mathbb{C}_{n-r}^{n \times (n-r)}$ *form an orthonormal basis for* $\mathcal{N}(A^*A)$. *We define*

$$E = VV^*.$$

Then E satisfies

$$\mathcal{R}(E) = \mathcal{R}(V) = \mathcal{N}(A^*A), \quad \mathcal{N}(E) = \mathcal{N}(V^*) = \mathcal{R}(A^*A),$$

and $A^*A + E$ *is nonsingular and*

$$\begin{aligned}
(A^*A + E)^{-1} &= (A^*A)_g + E_g \\
&= (A^*A)^\dagger + E^\dagger.
\end{aligned}$$

The consistent restricted linear equations

$$A\mathbf{x} = \mathbf{b}, \quad \mathbf{x} \in \mathcal{R}(A^*) \tag{3.2.28}$$

are equivalent to the nonsingular linear equations

$$(A^*A + E)\mathbf{x} = A^*\mathbf{b}, \quad \mathbf{x} \in \mathcal{R}(A^*).$$

The components of the unique solution $\mathbf{x} = A^\dagger\mathbf{b}$ *of* (3.2.28) *are given by*

$$x_i = \frac{\det((A^*A + E)(i \to A^*\mathbf{b}))}{\det(A^*A + E)}, \quad i = 1, 2, \ldots, n.$$

Corollary 3.2.5 *Let $A \in \mathbb{C}_r^{m \times n}$, M and N be Hermitian positive definite matrices of orders m and n respectively, $T = \mathcal{R}(A^\#)$, $S = \mathcal{N}(A^\#)$, $A^\# = N^{-1}A^*M$, $\mathbf{b} \in A\mathcal{R}(A^\#) = \mathcal{R}(A)$, and the columns of $V, U^* \in \mathbb{C}_{n-r}^{n \times (n-r)}$ form the bases for $\mathcal{N}(A^\#A)$ and $\mathcal{N}((A^\#A)^*)$ respectively. We define*

$$E = V(UV)^{-1}U.$$

Then E satisfies

$$\mathcal{R}(E) = \mathcal{R}(V) = \mathcal{N}(A^\#A) = \mathcal{N}(A^*MA),$$
$$\mathcal{N}(E) = \mathcal{N}(U) = \mathcal{R}(A^\#A) = N^{-1}(\mathcal{R}(A^*MA)),$$

and $A^\#A + E$ is nonsingular and

$$(A^\#A + E)^{-1} = (A^\#A)_g + E_g.$$

The consistent restricted linear equations

$$A\mathbf{x} = \mathbf{b}, \quad \mathbf{x} \in \mathcal{R}(A^\#) \tag{3.2.29}$$

is equivalent to the nonsingular linear equations

$$(A^\#A + E)\mathbf{x} = A^\#\mathbf{b}, \quad \mathbf{x} \in \mathcal{R}(A^\#).$$

The components of the unique solution $\mathbf{x} = A_{MN}^\dagger \mathbf{b}$ of (3.2.29) are given by

$$x_i = \frac{\det((A^\#A + E)(i \to A^\#\mathbf{b}))}{\det(A^\#A + E)}, \quad i = 1, 2, \ldots, n.$$

Corollary 3.2.6 *Let $A \in \mathbb{C}^{n \times n}$, $\text{Ind}(A) = k$, $\text{rank}(A^k) = r$, $T = \mathcal{R}(A^k)$, $S = \mathcal{N}(A^k)$, $\mathbf{b} \in \mathcal{R}(A^k)$, and the columns of $V, U^* \in \mathbb{C}_{n-r}^{n \times (n-r)}$ form the bases for $\mathcal{N}(A^k)$ and $\mathcal{N}(A^{k^*})$ respectively. We define*

$$E = V(UV)^{-1}U.$$

Then E satisfies

$$\mathcal{R}(E) = \mathcal{R}(V) = \mathcal{N}(A^k), \quad \mathcal{N}(E) = \mathcal{N}(U) = \mathcal{R}(A^k),$$

and $A^k + E$ is nonsingular and

$$(A^k + E)^{-1} = (A^k)_g + E_g.$$

The restricted linear equations

$$Ax = b, \quad x \in \mathcal{R}(A^k) \tag{3.2.30}$$

is equivalent to the nonsingular linear equations

$$(A^k + E)x = A^{k-1}b, \quad x \in \mathcal{R}(A^k).$$

The components of the unique solution $x = A_d b$ *of (3.2.30) are given by*

$$x_i = \frac{\det((A^k + E)(i \to A^{k-1}b))}{\det(A^k + E)}, \quad i = 1, 2, \ldots, n.$$

Corollary 3.2.7 *Let* $A \in \mathbb{C}^{n \times n}$, $\mathrm{Ind}(A) = 1$, $\mathrm{rank}(A) = r$, $T = \mathcal{R}(A)$, $S = \mathcal{N}(A)$, $b \in \mathcal{R}(A)$, *and the columns of* $V, U^* \in \mathbb{C}^{n \times (n-r)}_{n-r}$ *form the bases for* $\mathcal{N}(A)$ *and* $\mathcal{R}(A)$ *respectively. We define*

$$E = V(UV)^{-1}U.$$

Then E satisfies

$$\mathcal{R}(E) = \mathcal{R}(V) = \mathcal{N}(A), \quad \mathcal{N}(E) = \mathcal{N}(U) = \mathcal{R}(A),$$

and $A + E$ *is nonsingular and*

$$(A + E)^{-1} = A_g + E_g.$$

The restricted linear equations

$$Ax = b, \quad x \in \mathcal{R}(A) \tag{3.2.31}$$

is equivalent to the nonsingular linear equations

$$(A + E)x = b, \quad x \in \mathcal{R}(A).$$

The components of the unique solution $x = A_g b$ *of (3.2.31) are given by*

$$x_i = \frac{\det((A + E)(i \to b))}{\det(A + E)}, \quad i = 1, 2, \ldots, n.$$

Exercises 3.2

1. Prove Corollary 3.2.2.
2. Prove Corollary 3.2.3.
3. Let A, T and S be the same as in Lemma 3.2.1, and both $B \in \mathbb{C}^{s \times m}$ and $C \in \mathbb{C}^{(n-s) \times n}$ be of full row rank and satisfy

$$\mathcal{R}(B^*) = S^\perp \quad \text{and} \quad \mathcal{R}(C^*) = T^\perp.$$

Prove that the matrix

$$\begin{bmatrix} BA \\ C \end{bmatrix}$$

is nonsingular and the components of the unique solution of (3.2.16) are given by

$$x_j = \frac{\det \begin{bmatrix} (BA)(j \to B\mathbf{b}) \\ C(j \to \mathbf{0}) \end{bmatrix}}{\det \begin{bmatrix} BA \\ C \end{bmatrix}}, \quad j = 1, 2, \ldots, n.$$

3.3   Cramer's Rule for Solution of a Matrix Equation

We consider the problem of solving the matrix equation $AXB = D$ using the Cramer's rule. We start with the nonsingular case where an exact solution X can be found, then a general case where a best approximation solution can be found. Analogous to the linear systems in the previous section, we also study restricted matrix equations and a condensed form of the Cramer rule for solving restricted matrix equations.

3.3.1   Cramer's Rule for the Solution of a Nonsingular Matrix Equation

First, we discuss the Cramer's rule for the unique solution of the matrix equation $AXB = D$.

Lemma 3.3.1 (1) *Let $A \in \mathbb{C}_n^{n \times n}$, and $D = [\mathbf{d}_1, \mathbf{d}_2, \ldots, \mathbf{d}_p] \in \mathbb{C}^{n \times p}$, then the unique solution of the matrix equation*

$$AY = D$$

is $Y = A^{-1}D = [\mathbf{y}_1, \mathbf{y}_2, \ldots, \mathbf{y}_p] \in \mathbb{C}^{n \times p}$, whose elements are given by

$$y_{ik} = \frac{\det(A(i \to \mathbf{d}_k))}{\det(A)}, \quad i = 1, 2, \ldots, n; \quad k = 1, 2, \ldots, p. \tag{3.3.1}$$

(2) *Let $B \in \mathbb{C}_p^{p \times p}$ and I be the identity matrix of order p, then the unique solution of the matrix equation*

$$ZB = I \tag{3.3.2}$$

is $Z = B^{-1} = [\mathbf{z}_1, \mathbf{z}_2, \ldots, \mathbf{z}_p]^T \in \mathbb{C}^{p \times p}$, whose elements are given by

$$z_{kj} = \frac{\det(B(\mathbf{e}_k^T \leftarrow j))}{\det(B)}, \quad k = 1, 2, \ldots, p; \quad j = 1, 2, \ldots, p.$$

Proof (1) It is easy to verify (3.3.1).

(2) It follows from (3.3.2) that

$$B^* Z^* = I.$$

Let $Z^* = [\widetilde{z}_{jk}] \in \mathbb{C}^{p \times p}$, then $\widetilde{z}_{jk} = \overline{z}_{kj}$. By part(1),

$$\widetilde{z}_{jk} = \frac{\det(B^*(j \to \mathbf{e}_k))}{\det(B^*)}.$$

Thus

$$z_{kj} = \frac{\det((B^*(j \to \mathbf{e}_k))^*)}{\det(B)}$$

$$= \frac{\det(B(\mathbf{e}_k^T \leftarrow j))}{\det(B)},$$

which completes the proof.   □

Now we have the Cramer's rule for the matrix equation $AXB = D$ when both A and B are nonsingular.

Theorem 3.3.1 *Let* $A \in \mathbb{C}_n^{n \times n}$, $B \in \mathbb{C}_p^{p \times p}$, *and* $D = [\mathbf{d}_1, \mathbf{d}_2, \ldots, \mathbf{d}_p] \in \mathbb{C}^{n \times p}$, *then the unique solution of the matrix equation*

$$AXB = D \tag{3.3.3}$$

is $X = A^{-1}DB^{-1} = (x_{ij}) \in \mathbb{C}^{n \times p}$, *whose elements are given by*

$$x_{ij} = \frac{\sum_{k=1}^{p} \det(A(i \to \mathbf{d}_k)) \det(B(\mathbf{e}_k^T \leftarrow j))}{\det(A) \det(B)}, \tag{3.3.4}$$

where $\mathbf{d}_k$ *is the k-th column of D and* $\mathbf{e}_k$ *denotes the p-vector whose k-th component is unity and other components are zero.*

Proof Let $A^{-1}D = Y$ and $B^{-1} = Z$, then

$$X = YZ.$$

It follows from Lemma 3.3.1 that the unique solution of (3.3.3) is $X = [x_{ij}] \in \mathbb{C}^{n \times p}$, whose elements are given by

$$
\begin{aligned}
x_{ij} &= \sum_{k=1}^{p} y_{ik} z_{kj} \\
&= \frac{\sum_{k=1}^{p} \det(A(i \to \mathbf{d}_k)) \det(B(\mathbf{e}_k^T \leftarrow j))}{\det(A) \det(B)},
\end{aligned}
$$

which is (3.3.4).   □

3.3.2   Cramer's Rule for the Best-Approximate Solution of a Matrix Equation

The matrix equation $AXB = D$ may not have exact solution when A and B are rectangle matrices. In this general case, we find the best-approximate solution in the following sense.

If X_0 satisfies

(1) $\|AX_0B - D\|_F \le \|AXB - D\|_F, \quad \forall X$;
(2) $\|X_0\|_F \le \|X\|_F, \forall X \ne X_0$, where X satisfies $\|AXB - D\|_F = \|AX_0B - D\|_F$,

then X_0 is called the best-approximate solution of the matrix equation

$$
AXB = D, \tag{3.3.5}
$$

where $\|P\|_F = (\mathrm{tr}(P^*P))^{\frac{1}{2}}$ is the Frobenius-norm of the matrix P.

The best-approximate solution of (3.3.5) is discussed in [17], and the Cramer's rule for it is given in [18].

Definition 3.3.1 Let $A = [a_{ij}] \in \mathbb{C}^{m \times n}$ and $B \in \mathbb{C}^{p \times q}$, then the Kronecker product $A \otimes B$ of A and B is the $mp \times nq$ matrix expressible in partitioned form:

$$
A \otimes B =
\begin{bmatrix}
a_{11}B & a_{12}B & \dots & a_{1n}B \\
a_{21}B & a_{22}B & \dots & a_{2n}B \\
\vdots & \vdots & & \vdots \\
a_{m1}B & a_{m2}B & \dots & a_{mn}B
\end{bmatrix}.
$$

The properties of the Kronecker product can be found in [19–21].

Lemma 3.3.2 *Let* A, B, A_1, A_2, B_1, *and* B_2 *be matrices whose orders are suitable for the following operations, then*

(1) $O \otimes A = A \otimes O = O$, *where* O *is a zero matrix;*
(2) $(A_1 + A_2) \otimes B = A_1 \otimes B + A_2 \otimes B$;
(3) $A \otimes (B_1 + B_2) = A \otimes B_1 + A \otimes B_2$;
(4) $\alpha A \otimes \beta B = (\alpha\beta)(A \otimes B)$;
(5) $(A_1 A_2) \otimes (B_1 B_2) = (A_1 \otimes B_1)(A_2 \otimes B_2)$;
(6) $(A \otimes B)^{-1} = A^{-1} \otimes B^{-1}$;
(7) $(A \otimes B)^* = A^* \otimes B^*$,
    $(A \otimes B)^T = A^T \otimes B^T$;
(8) $(A \otimes B)^\dagger = A^\dagger \otimes B^\dagger$;
(9) $(A \otimes B)^\dagger_{C,D} = A^\dagger_{MN} \otimes B^\dagger_{PQ}$, *where* $C = M \otimes P$, $D = N \otimes Q$, *and* M, N, P *and* Q *are Hermitian positive definite matrices of orders suitable for the operations;*
(10) $(A \otimes B)_d = A_d \otimes B_d$, *where* A *and* B *are square matrices;*
(11) $(A \otimes B)_{d,W} = A_{d,W_1} \otimes B_{d,W_2}$, *where* $W = W_1 \otimes W_2$.

An important application of the Kronecker product is that we can rewrite the matrix Eq. (3.3.5) as a vector equation. For any $X = [x_{ij}] \in \mathbb{C}^{n \times r}$, let the vector $\mathbf{v}(X) = [v_k] \in \mathbb{C}^{nr}$ be the transpose of the row vector obtained by lining the rows of X end to end with the first row on the left and the last row on the right. In other words,

$$v_{r(i-1)+j} = x_{ij}, \quad i = 1, 2, \ldots, n, \quad j = 1, 2, \ldots, r.$$

It is easy to verify that

$$\mathbf{v}(AXB) = (A \otimes B^T)\mathbf{v}(X).$$

By using the above equation, the matrix Eq. (3.3.5) can be rewritten as the vector equation:

$$(A \otimes B^T)\mathbf{v}(X) = \mathbf{v}(D). \tag{3.3.6}$$

It is clear that the best-approximate solution of (3.3.5) is equivalent to the minimum-norm least-squares solution of (3.3.6). It follows from Theorem 1.1.6 and Lemma 3.3.2 that

$$\begin{aligned}
\mathbf{v}(X) &= (A \otimes B^T)^\dagger \mathbf{v}(D) \\
&= (A^\dagger \otimes (B^T)^\dagger)\mathbf{v}(D) \\
&= \mathbf{v}(A^\dagger D B^\dagger).
\end{aligned}$$

Thus, the best-approximate solution of (3.3.5) is simply

$$X = A^\dagger D B^\dagger. \tag{3.3.7}$$

The Cramer's rules for the best-approximate solutions of two special matrix equations $AY = D$ and $ZB = I$ are given in the following two lemmas.

Lemma 3.3.3 *Let* $A \in \mathbb{C}_r^{m \times n}$, $D = [\mathbf{d}_1, \mathbf{d}_2, \ldots, \mathbf{d}_w] \in \mathbb{C}^{m \times w}$, *and* $U \in \mathbb{C}_{m-r}^{m \times (m-r)}$ *and* $V^* \in \mathbb{C}_{n-r}^{n \times (n-r)}$ *be matrices whose columns form the bases for* $\mathcal{N}(A^*)$ *and* $\mathcal{N}(A)$ *respectively, then the best-approximate solution of the matrix equation*

$$AY = D$$

is $Y = A^\dagger D = [y_{ik}] \in \mathbb{C}^{n \times w}$, *whose elements are given by*

$$y_{ik} = \frac{\det \begin{bmatrix} A(i \to \mathbf{d}_k) \ U \\ V(i \to \mathbf{0}) \ O \end{bmatrix}}{\det \begin{bmatrix} A \ U \\ V \ O \end{bmatrix}}, \quad i = 1, 2, \ldots, n; \ k = 1, 2, \ldots, w. \quad (3.3.8)$$

Proof Let $Y = [\mathbf{y}_1, \mathbf{y}_2, \ldots, \mathbf{y}_w]$ be the column partition of Y. Since

$$\|D - AY\|_F^2 = \sum_{k=1}^{w} \|\mathbf{d}_k - A\mathbf{y}_k\|_2^2$$

and

$$\|Y\|_F^2 = \sum_{k=1}^{w} \|\mathbf{y}_k\|_2^2,$$

the matrix Y is the best-approximate solution of $AY = D$ if and only if $\mathbf{y}_k$, $k = 1, 2, \ldots, w$, is the minimum-norm least-squares solution of $A\mathbf{y}_k = \mathbf{d}_k$. Then (3.3.8) follows from Corollary 3.2.1. □

Lemma 3.3.4 *Let* $B \in \mathbb{C}_p^{q \times w}$, I_w *be the identity matrix of order* w, *and* $P \in \mathbb{C}_{q-p}^{q \times (q-p)}$ *and* $Q^* \in \mathbb{C}_{w-p}^{w \times (w-p)}$ *matrices whose columns form the bases for* $\mathcal{N}(B^*)$ *and* $\mathcal{N}(B)$ *respectively, then the best-approximate solution of the matrix equation*

$$ZB = I_w \quad (3.3.9)$$

is $Z = B^\dagger = [\mathbf{z}_1, \mathbf{z}_2, \ldots, \mathbf{z}_w]^T \in \mathbb{C}^{w \times q}$, *whose elements are given by*

$$z_{kj} = \frac{\det \begin{bmatrix} B(\mathbf{e}_k^T \leftarrow j) \ P(\mathbf{0} \leftarrow j) \\ Q \qquad\qquad O \end{bmatrix}}{\det \begin{bmatrix} B \ P \\ Q \ O \end{bmatrix}}, \quad k = 1, 2, \ldots, w; \ j = 1, 2, \ldots, q.$$

$$(3.3.10)$$

Proof It follows from (3.3.9) that

$$B^* Z^* = I.$$

Let $Z^* = (\tilde{z}_{jk}) \in \mathbb{C}^{q \times w}$, then $\tilde{z}_{jk} = \overline{z}_{kj}$. Using Lemma 3.3.3, we have

$$
\tilde{z}_{jk} = \frac{\det \begin{bmatrix} B^*(j \to e_k) & Q^* \\ P^*(j \to 0) & O \end{bmatrix}}{\det \begin{bmatrix} B^* & Q^* \\ P^* & O \end{bmatrix}},
$$

which implies (3.3.10).   □

Putting Lemmas 3.3.3 and 3.3.4 together, we have the following Cramer's rule for the best-approximate solution of (3.3.5).

Theorem 3.3.2 *Let $A \in \mathbb{C}_r^{m \times n}$, $B \in \mathbb{C}_p^{q \times w}$, $D = [d_1, d_2, \ldots, d_w] \in \mathbb{C}^{m \times w}$, and $U \in \mathbb{C}_{m-r}^{m \times (m-r)}$, $V^* \in \mathbb{C}_{n-r}^{n \times (n-r)}$, $P \in \mathbb{C}_{q-p}^{q \times (q-p)}$, and $Q^* \in \mathbb{C}_{w-p}^{w \times (w-p)}$ be matrices whose columns form the bases for $\mathcal{N}(A^*)$, $\mathcal{N}(A)$, $\mathcal{N}(B^*)$, and $\mathcal{N}(B)$ respectively, then the best-approximate solution of the matrix Eq. (3.3.5):*

$$
AXB = D
$$

is $X = [x_{ij}] \in \mathbb{C}^{n \times q}$, whose elements are given by

$$
x_{ij} =
$$

$$
\frac{\displaystyle\sum_{k=1}^{w} \det \begin{bmatrix} A(i \to d_k) & U \\ V(i \to 0) & O \end{bmatrix} \det \begin{bmatrix} B(e_k^T \leftarrow j) & P(0 \leftarrow j) \\ Q & O \end{bmatrix}}{\det \begin{bmatrix} A & U \\ V & O \end{bmatrix} \det \begin{bmatrix} B & P \\ Q & O \end{bmatrix}},
$$

$$
i = 1, 2, \ldots, n; \quad j = 1, 2, \ldots, q, \tag{3.3.11}
$$

where e_k is the kth unit vector of dimension w.

Proof By (3.3.7), the best-approximate solution of (3.3.5) is $X = A^\dagger D B^\dagger$. Setting

$$
A^\dagger D = Y \quad \text{and} \quad B^\dagger = Z,
$$

we get

$$
X = YZ.
$$

Thus the best-approximate solution of (3.3.5) is $X = [x_{ij}] \in \mathbb{C}^{n \times q}$, whose elements are given by

$$
x_{ij} = \sum_{k=0}^{w} y_{ik} z_{kj}
$$

and (3.3.11) follows from Lemmas 3.3.3 and 3.3.4.   □

3.3.3   Cramer's Rule for the Unique Solution of a Restricted Matrix Equation

The Cramer's rule for the unique solution of the restricted matrix equation

$$AXB = D, \quad \mathcal{R}(X) \subset T, \quad \mathcal{N}(X) \supset \tilde{S}, \tag{3.3.12}$$

is given as follows, where $A \in \mathbb{C}_r^{m \times n}$, $B \in \mathbb{C}_{\tilde{r}}^{p \times q}$, $D \in \mathbb{C}^{m \times q}$, $T \subset \mathbb{C}^n$, $S \subset \mathbb{C}^m$, $\tilde{T} \subset \mathbb{C}^q$, and $\tilde{S} \subset \mathbb{C}^p$ satisfy

$$\dim(T) = \dim(S^\perp) = t \le r, \quad \dim(\tilde{T}) = \dim(\tilde{S}^\perp) = \tilde{t} \le \tilde{r} \tag{3.3.13}$$

and

$$AT \oplus S = \mathbb{C}^m \text{ equivalent to } T \oplus (A^*S^\perp)^\perp = \mathbb{C}^n$$
$$B\tilde{T} \oplus \tilde{S} = \mathbb{C}^p \text{ equivalent to } \tilde{T} \oplus (B^*\tilde{S}^\perp)^\perp = \mathbb{C}^q. \tag{3.3.14}$$

If we define the range and null space of a pair of matrices A and B as

$$\mathcal{R}(A, B) = \{Y = AXB : X \in \mathbb{C}^{n \times p}\}$$

and

$$\mathcal{N}(A, B) = \{X \in \mathbb{C}^{n \times p} : AXB = O\}$$

respectively, then the unrestricted matrix equation

$$AXB = D$$

has a solution if $D \in \mathcal{R}(A, B)$.

Now, we consider the solution of the restricted matrix Eq. (3.3.12).

Theorem 3.3.3 *Given the matrices A, B and D, and the subspaces T, S, $\tilde{T}$, and $\tilde{S}$ as above. Suppose that matrices $G \in \mathbb{C}^{n \times m}$ and $\tilde{G} \in \mathbb{C}^{q \times p}$ satisfy*

$$\mathcal{R}(G) = T, \ \mathcal{N}(G) = S, \ \mathcal{R}(\tilde{G}) = \tilde{T}, \ and \ \mathcal{N}(\tilde{G}) = \tilde{S}. \tag{3.3.15}$$

If

$$D \in \mathcal{R}(AG, \tilde{G}B),$$

then the restricted matrix Eq. (3.3.12) has the unique solution

$$X = A_{T,S}^{(2)} D B_{\tilde{T},\tilde{S}}^{(2)}. \tag{3.3.16}$$

Proof From the definitions of the range and null space of a pair of matrices, $D = AGY\tilde{G}B$ for some Y. Consequently,

$$\mathcal{R}(D) \subset \mathcal{R}(AG) \quad \text{and} \quad \mathcal{N}(D) \supset \mathcal{N}(\tilde{G}B),$$

equivalently,

$$\mathcal{R}(D) \subset AT \quad \text{and} \quad \mathcal{N}(D) \supset (B^*\tilde{S}^\perp)^\perp,$$

since $\mathcal{R}(AG) = A\mathcal{R}(G) = AT$ and $\mathcal{R}(D^*) \subset \mathcal{R}(B^*\tilde{G}^*) = B^*\mathcal{R}(\tilde{G}^*) = B^*\tilde{S}^\perp$. Thus we can verify that

$$AA_{T,S}^{(2)}DB_{\tilde{T},\tilde{S}}^{(2)}B = P_{AT,S}DP_{\tilde{T},(B^*\tilde{S}^\perp)^\perp} = D,$$

that is, X in (3.3.16) is a solution of the matrix Eq. (3.3.12). The solution $A_{T,S}^{(2)}DB_{\tilde{T},\tilde{S}}^{(2)}$ also satisfies the restricted conditions because

$$\mathcal{R}(X) \subset \mathcal{R}(A_{T,S}^{(2)}) = T \quad \text{and} \quad \mathcal{N}(X) \supset \mathcal{N}(B_{\tilde{T},\tilde{S}}^{(2)}) = \tilde{S}.$$

Finally, we prove the uniqueness. If X_0 is also a solution of the restricted matrix Eq. (3.3.12), then

$$\begin{aligned}
X &= A_{T,S}^{(2)}DB_{\tilde{T},\tilde{S}}^{(2)} \\
&= A_{T,S}^{(2)}AX_0BB_{\tilde{T},\tilde{S}}^{(2)} \\
&= P_{T,(A^*S^\perp)^\perp}X_0P_{B\tilde{T},\tilde{S}} \\
&= X_0,
\end{aligned}$$

since $\mathcal{R}(X_0) \subset T$ and $\mathcal{N}(X_0) \supset \tilde{S}$. $\qquad\square$

Next, we show a Cramer's rule for solving the restricted matrix Eq. (3.3.12).

Theorem 3.3.4 *Given the matrices A, B and D, and the subspaces T, S, $\tilde{T}$ and $\tilde{S}$ as above. Let*

$$L \in \mathbb{C}_{m-t}^{m\times(m-t)}, \ M^* \in \mathbb{C}_{n-t}^{n\times(n-t)}, \ \tilde{L} \in \mathbb{C}_{p-\tilde{t}}^{p\times(p-\tilde{t})}, \ \text{and} \ \tilde{M}^* \in \mathbb{C}_{q-\tilde{t}}^{q\times(q-\tilde{t})}$$

be matrices such that

$$\mathcal{R}(L) = S, \ \mathcal{N}(M) = T, \ \mathcal{R}(\tilde{L}) = \tilde{S}, \ \text{and} \ \mathcal{N}(\tilde{M}) = \tilde{T},$$

then the elements of the unique solution $X = [x_{ij}]$ of the restricted matrix Eq. (3.3.12) are given by

$$x_{ij} = \frac{\sum_{k=1}^{q}\det\begin{bmatrix} A(i \to \mathbf{d}_k) \ L \\ M(i \to \mathbf{0}) \ O \end{bmatrix}\det\begin{bmatrix} B(\mathbf{e}_k^T \leftarrow j) \ \tilde{L}(\mathbf{0}^T \leftarrow j) \\ \tilde{M} \qquad O \end{bmatrix}}{\det\begin{bmatrix} A \ L \\ M \ O \end{bmatrix}\det\begin{bmatrix} B \ \tilde{L} \\ \tilde{M} \ O \end{bmatrix}},$$

$$i = 1, 2, \ldots, n; \ j = 1, 2, \ldots, p, \qquad (3.3.17)$$

where $\mathbf{d}_k$ is the kth column of D and $\mathbf{e}_k$ is the kth column of the $q \times q$ identity matrix.

Proof Since X is the solution of the restricted matrix Eq. (3.3.12), we have

$$\mathcal{R}(X) \subset T = \mathcal{N}(M) \quad \text{and} \quad \mathcal{N}(X) \supset \widetilde{S} = \mathcal{R}(\widetilde{L}).$$

It follows that

$$MX = O \quad \text{and} \quad X\widetilde{L} = O$$

and

$$\begin{bmatrix} A & L \\ M & O \end{bmatrix} \begin{bmatrix} X & O \\ O & O \end{bmatrix} \begin{bmatrix} B & \widetilde{L} \\ \widetilde{M} & O \end{bmatrix} = \begin{bmatrix} AXB & O \\ O & O \end{bmatrix} = \begin{bmatrix} D & O \\ O & O \end{bmatrix}. \tag{3.3.18}$$

From Theorem 3.1.3, the two coefficient matrices on the left of the above equation are nonsingular and

$$\begin{bmatrix} A & L \\ M & O \end{bmatrix}^{-1} = \begin{bmatrix} A_{T,S}^{(2)} & (I - A_{T,S}^{(2)}A)M^\dagger \\ L^\dagger(I - AA_{T,S}^{(2)}) & L^\dagger(AA_{T,S}^{(2)}A - A)M^\dagger \end{bmatrix}$$

and

$$\begin{bmatrix} B & \widetilde{L} \\ \widetilde{M} & O \end{bmatrix}^{-1} = \begin{bmatrix} B_{\widetilde{T},\widetilde{S}}^{(2)} & (I - B_{\widetilde{T},\widetilde{S}}^{(2)}B)\widetilde{M}^\dagger \\ \widetilde{L}^\dagger(I - BB_{\widetilde{T},\widetilde{S}}^{(2)}) & \widetilde{L}^\dagger(BB_{\widetilde{T},\widetilde{S}}^{(2)}B - B)\widetilde{M}^\dagger \end{bmatrix}.$$

By using $\mathcal{R}(X) \subset T$ and $\mathcal{N}(X) \supset \widetilde{S}$, we have

$$(I - A_{T,S}^{(2)}A)X = P_{(A^*S^\perp)^\perp,T}X = O$$

and

$$XB(I - B_{\widetilde{T},\widetilde{S}}^{(2)}B) = X(I - BB_{\widetilde{T},\widetilde{S}}^{(2)})B = XP_{\widetilde{S},B\widetilde{T}}B = O.$$

Therefore

$$\begin{bmatrix} X & O \\ O & O \end{bmatrix} = \begin{bmatrix} A & L \\ M & O \end{bmatrix}^{-1} \begin{bmatrix} D & O \\ O & O \end{bmatrix} \begin{bmatrix} B & \widetilde{L} \\ \widetilde{M} & O \end{bmatrix}^{-1}$$

$$= \begin{bmatrix} A_{T,S}^{(2)}DB_{\widetilde{T},\widetilde{S}}^{(2)} & O \\ O & O \end{bmatrix}.$$

From the above equation, the unique solution of (3.3.12) is the same as that of (3.3.18). Applying Theorem 3.3.1 to (3.3.18), we obtain (3.3.17) immediately. $\square$

Corollary 3.3.1 *Let* $A \in \mathbb{C}_r^{m \times n}$, $B \in \mathbb{C}_{\widetilde{r}}^{p \times q}$, *M, N, P, and Q be Hermitian positive define matrices of orders m, n, p, and q respectively, and* $U \in \mathbb{C}_{m-r}^{m \times (m-r)}$, $V^* \in \mathbb{C}_{n-r}^{n \times (n-r)}$, $\widetilde{U} \in \mathbb{C}_{p-\widetilde{r}}^{p \times (p-\widetilde{r})}$ *and* $\widetilde{V}^* \in \mathbb{C}_{q-\widetilde{r}}^{q \times (q-\widetilde{r})}$ *be matrices whose columns form* the bases for $\mathcal{N}(A^*)$, $\mathcal{N}(A)$, $\mathcal{N}(B^*)$, and $\mathcal{N}(B)$ respectively. Suppose that $A^\# = N^{-1}A^*M$, $B^\# = Q^{-1}B^*P$ and $D \in \mathbb{C}^{m\times q}$ satisfy

$$D \in \mathcal{R}(AA^\#, B^\# B). \tag{3.3.19}$$

Then the restricted matrix equation

$$AXB = D, \quad \mathcal{R}(X) \subset N^{-1}\mathcal{R}(A^*), \quad \mathcal{N}(X) \supset P^{-1}\mathcal{N}(B^*),$$

has a unique solution

$$X = A_{MN}^\dagger D B_{PQ}^\dagger = [x_{ij}],$$

whose elements are given by

$$x_{ij} = \frac{\sum_{k=1}^{q} \det\begin{bmatrix} A(i \to \mathbf{d}_k) & M^{-1}U \\ VN(i \to \mathbf{0}) & O \end{bmatrix} \det\begin{bmatrix} B(\mathbf{e}_k^T \leftarrow j) & P^{-1}\tilde{U}(\mathbf{0}^T \leftarrow j) \\ \tilde{V}Q & O \end{bmatrix}}{\det\begin{bmatrix} A & M^{-1}U \\ VN & O \end{bmatrix} \det\begin{bmatrix} B & P^{-1}\tilde{U} \\ \tilde{V}Q & O \end{bmatrix}},$$

for $i = 1, 2, \ldots, n$ and $j = 1, 2, \ldots, p$, where $\mathbf{d}_k$ is the kth column of D and $\mathbf{e}_k$ is the kth column of the $q \times q$ identity matrix.

Corollary 3.3.2 Given the matrices A, B, U, V, $\tilde{U}$, and $\tilde{V}$ as above. Let $D \in \mathbb{C}^{m\times q}$ satisfy

$$D \in \mathcal{R}(AA^*, B^*B), \tag{3.3.20}$$

then the restricted matrix equation

$$AXB = D, \quad \mathcal{R}(X) \subset \mathcal{R}(A^*), \quad \mathcal{N}(X) \supset \mathcal{N}(B^*)$$

has a unique solution

$$X = A^\dagger D B^\dagger = [x_{ij}],$$

whose elements are given by

$$x_{ij} = \frac{\sum_{k=1}^{q} \det\begin{bmatrix} A(i \to \mathbf{d}_k) & U \\ V(i \to \mathbf{0}) & O \end{bmatrix} \det\begin{bmatrix} B(\mathbf{e}_k^T \leftarrow j) & \tilde{U}(\mathbf{0}^T \leftarrow j) \\ \tilde{V} & O \end{bmatrix}}{\det\begin{bmatrix} A & U \\ V & O \end{bmatrix} \det\begin{bmatrix} B & \tilde{U} \\ \tilde{V} & O \end{bmatrix}},$$

$$i = 1, 2, \ldots, n; \quad j = 1, 2, \ldots, p,$$

where $\mathbf{d}_k$ is the kth column of D and $\mathbf{e}_k$ is the kth column of the $q \times q$ identity matrix.

Corollary 3.3.3 Let $A \in \mathbb{C}^{n\times n}$ with $\text{Ind}(A) = k$, $B \in \mathbb{C}^{p\times p}$ with $\text{Ind}(B) = \tilde{k}$, rank $(A^k) = r < n$, rank$(B^{\tilde{k}}) = \tilde{r} < p$, and $U, V^* \in \mathbb{C}^{n\times(n-r)}_{n-r}$ and $\tilde{U}, \tilde{V}^* \in \mathbb{C}^{p\times(p-\tilde{r})}_{p-\tilde{r}}$ be *matrices whose columns form the bases for* $\mathcal{N}(A^k)$, $\mathcal{N}(A^{k*})$, $\mathcal{N}(B^{\widetilde{k}})$, *and* $\mathcal{N}(B^{\widetilde{k}*})$ *respectively. Suppose that* $D \in \mathbb{C}^{n \times p}$ *satisfies*

$$D \in \mathcal{R}(A^{k+1}, B^{\widetilde{k}+1}). \tag{3.3.21}$$

Then the restricted matrix equation

$$AXB = D, \quad \mathcal{R}(X) \subset \mathcal{R}(A^k), \quad \mathcal{N}(X) \supset \mathcal{N}(B^{\widetilde{k}})$$

has a unique solution

$$X = A_d D B_d = [x_{ij}],$$

whose elements are given by

$$x_{ij} = \frac{\sum\limits_{k=1}^{q} \det \begin{bmatrix} A(i \rightarrow \mathbf{d}_k) & U \\ V(i \rightarrow \mathbf{0}) & O \end{bmatrix} \det \begin{bmatrix} B(\mathbf{e}_k^T \leftarrow j) & \widetilde{U}(\mathbf{0}^T \leftarrow j) \\ \widetilde{V} & O \end{bmatrix}}{\det \begin{bmatrix} A & U \\ V & O \end{bmatrix} \det \begin{bmatrix} B & \widetilde{U} \\ \widetilde{V} & O \end{bmatrix}},$$

$$i = 1, 2, \ldots, n; \quad j = 1, 2, \ldots, p,$$

where $\mathbf{d}_k$ *is the kth column of D and* $\mathbf{e}_k$ *is the kth column of the* $q \times q$ *identity matrix.*

Corollary 3.3.4 *Let* $A \in \mathbb{C}^{n \times n}$ *with* $\text{Ind}(A) = 1$, $B \in \mathbb{C}^{p \times p}$ *with* $\text{Ind}(B) = 1$, *rank* $(A) = r < n$, $\text{rank}(B) = \widetilde{r} < p$, *and* $U, V^* \in \mathbb{C}^{n \times (n-r)}_{n-r}$ *and* $\widetilde{U}, \widetilde{V}^* \in \mathbb{C}^{p \times (p-\widetilde{r})}_{p-\widetilde{r}}$ *be matrices whose columns form the bases for* $\mathcal{N}(A)$, $\mathcal{N}(A^*)$, $\mathcal{N}(B)$ *and* $\mathcal{N}(B^*)$ *respectively. Suppose that* $D \in \mathbb{C}^{n \times p}$ *satisfies*

$$D \in \mathcal{R}(A^2, B^2). \tag{3.3.22}$$

Then the restricted matrix equation

$$AXB = D, \quad \mathcal{R}(X) \subset \mathcal{R}(A), \quad \mathcal{N}(X) \supset \mathcal{N}(B)$$

has a unique solution

$$X = A_g D B_g = [x_{ij}],$$

whose elements are given by

$$x_{ij} = \frac{\sum\limits_{k=1}^{q} \det \begin{bmatrix} A(i \rightarrow \mathbf{d}_k) & U \\ V(i \rightarrow \mathbf{0}) & O \end{bmatrix} \det \begin{bmatrix} B(\mathbf{e}_k^T \leftarrow j) & \widetilde{U}(\mathbf{0}^T \leftarrow j) \\ \widetilde{V} & O \end{bmatrix}}{\det \begin{bmatrix} A & U \\ V & O \end{bmatrix} \det \begin{bmatrix} B & \widetilde{U} \\ \widetilde{V} & O \end{bmatrix}},$$

$$i = 1, 2, \ldots, n; \quad j = 1, 2, \ldots, p,$$

where $\mathbf{d}_k$ is the kth column of D and $\mathbf{e}_k$ is the kth column of the $q \times q$ identity matrix.

Let $A \in \mathbb{C}^{m \times n}$, $B \in \mathbb{C}^{p \times q}$, $T \subset \mathbb{C}^n$, and $S \subset \mathbb{C}^p$, then the Bott-Duffin inverses

$$(A^*A)_{(T)}^{(-1)} = (A^*A)_{T,T^\perp}^{(2)} \quad \text{and} \quad (BB^*)_{(S^\perp)}^{(-1)} = (BB^*)_{S^\perp,S}^{(2)}.$$

It is obviously that the orthogonal projectors P_T and $P_{S^\perp}$ satisfy

$$T = \mathcal{R}(P_T), \quad T^\perp = \mathcal{N}(P_T), \quad S^\perp = \mathcal{R}(P_{S^\perp}), \quad S = \mathcal{N}(P_{S^\perp}).$$

Thus, when setting $G = P_T$ and $\widetilde{G} = P_{S^\perp}$ in Theorem 3.3.3, we have the following result immediately.

Corollary 3.3.5 *Let $A \in \mathbb{C}^{m \times n}$, $B \in \mathbb{C}^{p \times q}$, $D \in \mathbb{C}^{m \times q}$, $T \subset \mathbb{C}^n$ and $S \subset \mathbb{C}^p$. If*

$$A^*DB^* \in \mathcal{R}(A^*AP_T, P_{S^\perp}BB^*),$$

then the restricted matrix equation

$$A^*AXBB^* = A^*DB^*, \quad \mathcal{R}(X) \subset T, \quad \mathcal{N}(X) \supset S \qquad (3.3.23)$$

has the unique solution

$$X = (A^*A)_{(T)}^{(-1)} A^*DB^* (BB^*)_{(S^\perp)}^{(-1)}.$$

By using Theorem 3.3.4, we have the Cramer's rule for solving the restricted matrix Eq. (3.3.23).

Corollary 3.3.6 *Given A, B, D and the subspaces T and S as in Corollary 3.3.5. Let*

$$T = \mathcal{N}(E), \quad S = \mathcal{R}(F)$$

and

$$T^\perp = \mathcal{R}(E^*), \quad S^\perp = \mathcal{R}(F^*),$$

then the elements of the unique solution $X = [x_{ij}]$ of the restricted matrix Eq. (3.3.23) are given by

$$x_{ij} = \frac{\sum\limits_{k=1}^{q} \det \begin{bmatrix} A^*A(i \to \mathbf{d}_k) & E^* \\ E(i \to \mathbf{0}) & O \end{bmatrix} \det \begin{bmatrix} BB^*(\mathbf{e}_k^T \leftarrow j) & F(\mathbf{0}^T \leftarrow j) \\ F^* & O \end{bmatrix}}{\det \begin{bmatrix} A^*A & E^* \\ E & O \end{bmatrix} \det \begin{bmatrix} BB^* & F \\ F^* & O \end{bmatrix}},$$
$$i = 1, \ldots, n, \ j = 1, \ldots, p,$$

where $\mathbf{d}_k$ is the kth column of A^*DB^* and $\mathbf{e}_k$ is the kth column of the $q \times q$ identity matrix.

3.3.4   An Alternative Condensed Cramer's Rule for a Restricted Matrix Equation

In this section, we revisit the restricted matrix Eq. (3.3.12):

$$AXB = D, \quad \mathcal{R}(X) \subset T, \quad \mathcal{N}(X) \supset \widetilde{S},$$

where $A \in \mathbb{C}_r^{m \times n}$, $B \in \mathbb{C}_{\widetilde{r}}^{p \times q}$, $D \in \mathbb{C}^{m \times q}$, $T \subset \mathbb{C}^n$, $S \subset \mathbb{C}^m$, $\widetilde{T} \subset \mathbb{C}^q$, $\widetilde{S} \subset \mathbb{C}^p$ and the conditions (3.3.13) and (3.3.14) are satisfied.

It follows from Lemma 3.2.2 and Theorem 3.2.8, we have the following results.

Lemma 3.3.5 *Given B, $\widetilde{T}$, $\widetilde{S}$ as above. In addition, suppose $\widetilde{G} \in \mathbb{C}^{q \times p}$ such that*

$$\mathcal{R}(\widetilde{G}) = \widetilde{T} \quad and \quad \mathcal{N}(\widetilde{G}) = \widetilde{S}.$$

If B has a $\{2\}$-inverse $B_{\widetilde{T}, \widetilde{S}}^{(2)}$, then

$$\mathrm{Ind}(B\widetilde{G}) = \mathrm{Ind}(\widetilde{G}B) = 1.$$

Furthermore, we have

$$B_{\widetilde{T}, \widetilde{S}}^{(2)} = \widetilde{G}(B\widetilde{G})_g = (\widetilde{G}B)_g \widetilde{G}. \tag{3.3.24}$$

Theorem 3.3.5 *Given B, $\widetilde{T}$, $\widetilde{S}$, and $\widetilde{G}$ as in Lemma 3.3.5, and*

$$B\widetilde{T} \oplus \widetilde{S} = \mathbb{C}^p.$$

Let $\widetilde{V}$ and $\widetilde{U}^*$ be matrices whose columns form the bases for $\mathcal{N}(B\widetilde{G})$ and $\mathcal{N}((B\widetilde{G})^*)$ respectively. We define

$$F = \widetilde{V}(\widetilde{U}\widetilde{V})^{-1}\widetilde{U}.$$

Then F satisfies

$$\mathcal{R}(F) = \mathcal{R}(\widetilde{V}) = \mathcal{N}(B\widetilde{G}), \quad \mathcal{N}(F) = \mathcal{N}(\widetilde{U}) = \mathcal{R}(B\widetilde{G}),$$

and $B\widetilde{G} + F$ is nonsingular and

$$(B\widetilde{G} + F)^{-1} = (B\widetilde{G})_g + F_g.$$

Next, we show a condensed Cramer's rule for solving the restricted matrix Eq. (3.3.12).

Theorem 3.3.6 *Given A, B, T, S, $\widetilde{T}$, $\widetilde{S}$, G, $\widetilde{G}$, V, U^*, $\widetilde{V}$, $\widetilde{U}^*$, E, and F as in Theorems 3.2.8 and 3.3.5. If*

$$D \in \mathcal{R}(AG, \widetilde{G}B),$$

then the restricted matrix Eq. (3.3.12):

$$AXB = D, \quad \mathcal{R}(X) \subset T, \ \mathcal{N}(X) \supset \widetilde{S}$$

is equivalent to the nonsingular matrix equation

$$(GA + E)X(B\widetilde{G} + F) = GD\widetilde{G} \tag{3.3.25}$$

and the entries of the unique solution X of (3.3.12) are given by

$$x_{ij} = \frac{\sum\limits_{k=1}^{p} \det((GA + E)(i \to \widetilde{\mathbf{d}}_k))\det((B\widetilde{G} + F)(\mathbf{e}_k^T \leftarrow j))}{\det(GA + E)\det(B\widetilde{G} + F)},$$
$$i = 1, 2, \ldots, n, \ j = 1, 2, \ldots, p, \tag{3.3.26}$$

where $\widetilde{\mathbf{d}}_k$ is the kth column of $GD\widetilde{G}$ and $\mathbf{e}_k$ is the kth column of the identity matrix of order p.

Proof It follows from Theorems 3.2.8 and 3.3.5 that $GA + E$ and $B\widetilde{G} + F$ are nonsingular, and that the unique solution of (3.3.25) is

$$\begin{aligned} X &= (GA + E)^{-1}GD\widetilde{G}(B\widetilde{G} + F)^{-1} \\ &= (GA)_g GD\widetilde{G}(B\widetilde{G})_g + (GA)_g GD\widetilde{G}F_g \\ &\quad + E_g GD\widetilde{G}(B\widetilde{G})_g + E_g GD\widetilde{G}F_g. \end{aligned} \tag{3.3.27}$$

From assumptions, we have

$$D = AGY\widetilde{G}B \quad \text{for some } Y,$$

and

$$\mathcal{N}(E) = \mathcal{N}(U) = \mathcal{R}(GA), \quad \mathcal{R}(F) = \mathcal{R}(\widetilde{V}) = \mathcal{N}(B\widetilde{G}).$$

Thus

$$EGA = O \quad \text{and} \quad B\widetilde{G}F = O,$$

implying

$$E_g GD = EGAGY\widetilde{G}B = O \quad \text{and} \quad D\widetilde{G}F_g = AGY\widetilde{G}B\widetilde{G}F = O. \tag{3.3.28}$$

It follows from (3.3.27) and (3.3.28), (3.2.24) and (3.3.24) that

$$\begin{aligned} X &= (GA)_g GD\widetilde{G}(B\widetilde{G})_g \\ &= A_{T,S}^{(2)} DB_{\widetilde{T},\widetilde{S}}^{(2)}. \end{aligned}$$

The restricted matrix Eq. (3.3.12) has the unique solution

$$X = A_{T,S}^{(2)} D B_{\tilde{T},\tilde{S}}^{(2)}$$

by Theorem 3.3.3, hence the restricted matrix Eq. (3.3.12) is equivalent to the nonsingular matrix Eq. (3.3.25). Applying the Cramer rule (3.3.4)–(3.3.25), we can obtain the expression (3.3.26) immediately.   □

Similar to Corollaries 3.3.1–3.3.5, we have a series of condensed Cramer rules for those corresponding restricted matrix equations, see [22]. It is omitted here and left as an exercise.

Exercises 3.3

1. Prove (9), (10) and (11) of Lemma 3.3.2. (cf. [4])
2. Can we use the conditions

$$\mathcal{R}(D) \subset \mathcal{R}(A) \quad \text{and} \quad \mathcal{N}(D) \supset \mathcal{N}(B)$$

instead of the conditions (3.3.19) and (3.3.20) in Corollaries 3.3.1 and 3.3.2?
3. Can we use the conditions

$$\mathcal{R}(D) \subset \mathcal{R}(A^k) \quad \text{and} \quad \mathcal{N}(D) \supset \mathcal{N}(B^{\tilde{k}})$$

instead of the condition (3.3.21) in Corollary 3.3.3?
4. Can we use the conditions

$$\mathcal{R}(D) \subset \mathcal{R}(A) \quad \text{and} \quad \mathcal{N}(D) \supset \mathcal{N}(B)$$

instead of the condition (3.3.22) in Corollary 3.3.4?

3.4   Determinantal Expressions of the Generalized Inverses and Projectors

It is well known that if A is nonsingular, then the inverse of A is given by

$$A^{-1} = \frac{\text{adj}(A)}{\det(A)},$$

where

$$(\text{adj}(A))_{ij} = \det(A(i \to \mathbf{e}_j)), \quad i, j = 1, 2, \ldots, n.$$

Let $A^{-1} = [\alpha_{ij}]$, then

$$\alpha_{ij} = \frac{\det(A(i \to \mathbf{e}_j))}{\det(A)} \quad i, j = 1, 2, \ldots, n.$$

The determinantal expression of an ordinary inverse can be extended to the generalized inverses. These results offer a useful tool for the theory and computations of the generalized inverses.

By using the results in Sect. 3.1, the determinantal expressions of the generalized inverses $A_{MN}^{\dagger}$, $A^{\dagger}$, A_d, A_g, $A_{T,S}^{(2)}$, $A_{T,S}^{(1,2)}$, and $A_{(L)}^{(-1)}$ and the projectors $A^{\dagger}A$ and $AA^{\dagger}$ are given in the following theorems.

Theorem 3.4.1 *Let $A \in \mathbb{C}_r^{m \times n}$, M and N be Hermitian positive definite matrices of orders m and n respectively, and $U \in \mathbb{C}_{m-r}^{m \times (m-r)}$ and $V^* \in \mathbb{C}_{n-r}^{n \times (n-r)}$ matrices whose columns form bases for $\mathcal{N}(A^*)$ and $\mathcal{N}(A)$ respectively. Also, let*

$$A_2 = \begin{bmatrix} A & M^{-1}U \\ VN & O \end{bmatrix} \quad \text{and} \quad A_1 = \begin{bmatrix} A & U \\ V & O \end{bmatrix},$$

and $A_{MN}^{\dagger} = [\alpha_{ij}^{(2)}]$ and $A^{\dagger} = [\alpha_{ij}^{(1)}]$, then

$$\alpha_{ij}^{(l)} = \frac{\det(A_l(i \to \mathbf{e}_j))}{\det(A_l)}, \quad l = 1, 2. \tag{3.4.1}$$

Proof It follows from Theorem 3.1.1 and Corollary 3.1.1 that A_2 and A_1 are nonsingular and the $n \times m$ submatrices in the upper-left corners of A_2^{-1} and A_1^{-1} are $A_{MN}^{\dagger}$ and $A^{\dagger}$ respectively. Using $A_l A_l^{-1} = I, l = 1, 2$, we obtain (3.4.1) immediately. $\square$

Theorem 3.4.2 *Let $A \in \mathbb{C}^{n \times n}$, $\mathrm{Ind}(A) = k$ and $\mathrm{rank}(A^k) = r < n$, and $U, V^* \in \mathbb{C}_{n-r}^{n \times (n-r)}$ be matrices whose columns form bases for $\mathcal{N}(A^k)$ and $\mathcal{N}(A^{k*})$ respectively. Let*

$$A_4 = \begin{bmatrix} A & U \\ V & O \end{bmatrix}, \quad k > 1, \quad \text{and} \quad A_3 = \begin{bmatrix} A & U \\ V & O \end{bmatrix}, \quad k = 1,$$

and $A_d = [\alpha_{ij}^{(4)}]$ and $A_g = [\alpha_{ij}^{(3)}]$, then

$$\alpha_{ij}^{(l)} = \frac{\det(A_l(i \to \mathbf{e}_j))}{\det(A_l)}, \quad l = 3, 4. \tag{3.4.2}$$

Proof It follows from Theorem 3.1.2 and Corollary 3.1.4 that A_4 and A_3 are nonsingular and the $n \times m$ submatrices in the upper-left corners of A_4^{-1} and A_3^{-1} are A_d and A_g respectively. Using $A_l A_l^{-1} = I, l = 3, 4$, we obtain (3.4.2) immediately. $\square$

Theorem 3.4.3 *Let $A \in \mathbb{C}_r^{m \times n}$, $T \subset \mathbb{C}^n$, $S \subset \mathbb{C}^m$, $\dim(T) = \dim(S^{\perp}) = t \leq r$ and $AT \oplus S = \mathbb{C}^m$, and both B and C^* be of full column rank and satisfy*

$$S = \mathcal{R}(B) \quad and \quad T = \mathcal{N}(C).$$

Let

$$A_6 = \begin{bmatrix} A & B \\ C & O \end{bmatrix}, \ t < r, \quad and \quad A_5 = \begin{bmatrix} A & B \\ C & O \end{bmatrix}, \ t = r,$$

and $A_{T,S}^{(2)} = [\alpha_{ij}^{(6)}]$ *and* $A_{T,S}^{(1,2)} = [\alpha_{ij}^{(5)}]$, *then*

$$\alpha_{ij}^{(l)} = \frac{\det(A_l(i \to \mathbf{e}_j))}{\det(A_l)}, \quad l = 5, 6. \tag{3.4.3}$$

Proof It follows from Theorems 3.1.3 and 3.1.4 that A_6 and A_5 are nonsingular and the $n \times m$ submatrices in the upper-left corners of A_6^{-1} and A_5^{-1} are $A_{T,S}^{(2)}$ and $A_{T,S}^{(1,2)}$ respectively. Using $A_l A_l^{-1} = I, l = 5, 6$, we obtain (3.4.3) immediately.   □

Theorem 3.4.4 *Let* $A \in \mathbb{C}_r^{n \times n}$, $U \in \mathbb{C}_p^{n \times p}$ *satisfy* $\mathcal{N}(U^*) = L$ *and* $AL \oplus L^\perp = \mathbb{C}^n$, *and*

$$A_7 = \begin{bmatrix} A & U \\ U^* & O \end{bmatrix},$$

and $A_{(L)}^{(-1)} = [\alpha_{ij}^{(7)}]$, *then*

$$\alpha_{ij}^{(7)} = \frac{\det(A_7(i \to \mathbf{e}_j))}{\det(A_7)}. \tag{3.4.4}$$

Proof It follows from Theorem 3.1.5 that A_7 is nonsingular and the $n \times n$ submatrix in the upper-left corner of A_7 is $A_{(L)}^{(-1)}$. Using $A_7 A_7^{-1} = I$, we obtain (3.4.4) immediately.   □

Finally, since the projectors $A^\dagger A$ and $AA^\dagger$ are the best-approximate solutions of $AY = A$ and $YA = A$ respectively, using Lemmas 3.3.3 and 3.3.4, we have the following theorem.

Theorem 3.4.5 *Let* $A = [\mathbf{a}_1, \mathbf{a}_2, \ldots, \mathbf{a}_n] \in \mathbb{C}_r^{m \times n}$, $A^* = [\tilde{\mathbf{a}}_1, \tilde{\mathbf{a}}_2, \ldots, \tilde{\mathbf{a}}_m] \in \mathbb{C}_r^{n \times m}$, *and* $U \in \mathbb{C}_{m-r}^{m \times (m-r)}$ *and* $V^* \in \mathbb{C}_{n-r}^{n \times (n-r)}$ *be matrices whose columns form the bases for* $\mathcal{N}(A^*)$ *and* $\mathcal{N}(A)$ *respectively. Also, let*

$$A^\dagger A = [\varphi_{ij}] \in \mathbb{C}^{n \times n} \quad and \quad AA^\dagger = [\psi_{ij}] \in \mathbb{C}^{m \times m},$$

then

$$\varphi_{ij} = \frac{\det \begin{bmatrix} A(i \to \mathbf{a}_j) & U \\ V(i \to \mathbf{0}) & O \end{bmatrix}}{\det \begin{bmatrix} A & U \\ V & O \end{bmatrix}}, \quad i, j = 1, 2, \ldots, n.$$

and

$$\psi_{ij} = \frac{\det \begin{bmatrix} A(\widehat{\mathbf{a}}_i \leftarrow j) & U(\mathbf{0} \leftarrow j) \\ V & O \end{bmatrix}}{\det \begin{bmatrix} A & U \\ V & O \end{bmatrix}}, \quad i, j = 1, 2, \ldots, m.$$

The determinantal expressions of the projectors $A_{MN}^\dagger A$, $A A_{MN}^\dagger$, $A A_d$, and $A A_g$ are given in [21]. The determinantal expressions of the projectors $A_{T,S}^{(1,2)} A$, $A A_{T,S}^{(1,2)}$, $A_{T,S}^{(2)} A$ and $A A_{T,S}^{(2)}$ are left to the reader.

Exercises 3.4

1. Show the determinantal expressions of the generalized Bott-Duffin inverses $A_{(L)}^{(\dagger)}$ and the W-weighted Drazin inverse $A_{d,W}$ [23].
2. Show the determinantal expressions of the projectors $A_{T,S}^{(1,2)} A$, $A A_{T,S}^{(1,2)}$, $A_{T,S}^{(2)} A$, $A A_{T,S}^{(2)}$, $W A W A_{d,W}$ and $A_{d,W} W A W$.
3. By using the Werner's method [14], show the condensed Cramer rule for the solutions of linear equations and matrix equations in Sects. 3.2 and 3.3.

3.5 The Determinantal Expressions of the Minors of the Generalized Inverses

In this section, we study the minors, submatrices, of inverses. First, we introduce notations and review the expressions of the minors of the regular inverse of a non-singular matrix. Then we present the minors of various generalized inverses in the following subsections.

If a matrix $A \in \mathbb{R}^{n \times n}$ is nonsingular, then the adjoint formula for its inverse

$$A^{-1} = \frac{\text{adj}(A)}{\det(A)}$$

has a well-known generalization, the Jacobi identity, which relates the minors of A^{-1} to those of A.

Denote the set of strictly increasing sequences of k integers chosen from $\{1, 2, \ldots, n\}$ by

$$Q_{k,n} = \{\alpha : \alpha = (\alpha_1, \alpha_2, \cdots, \alpha_k), \ 1 \leq \alpha_1 < \alpha_2 < \cdots < \alpha_k \leq n\}.$$

For $\alpha, \beta \in Q_{k,n}$, we denote

$A[\alpha, \beta]$ the submatrix of A having row indices α and column indices β,
$A[\alpha', \beta']$ the submatrix obtained from A by deleting rows indexed α and columns indexed β.

Then the Jacobi identity [24] is: for any $\alpha, \beta \in Q_{k,n}$,

$$\det(A^{-1}[\beta, \alpha]) = (-1)^{S(\alpha)+S(\beta)} \frac{\det(A[\alpha', \beta'])}{\det(A)},$$

where $S(\alpha)$ is the sum of the integers in α. By convention,

$$\det(A[\emptyset, \emptyset]) = 1.$$

We adopt the following notations from [25]. For any index sets I and J, let A_{I*}, A_{*J}, and A_{IJ} denote the submatrices of A lying in the rows indexed by I, in the columns indexed by J, and in their intersection, respectively. The principal submatrix A_{JJ} is denoted by A_J. For $A \in \mathbb{R}_r^{m \times n}$, let

$$\begin{aligned}
\mathcal{I}(A) &= \{I \in Q_{r,m} : \text{rank}(A_{I*}) = r\}, \\
\mathcal{J}(A) &= \{J \in Q_{r,n} : \text{rank}(A_{*J}) = r\}, \\
\mathcal{B}(A) &= \{(I, J) \in Q_{r,m} \times Q_{r,n} : \text{rank}(A_{IJ}) = r\},
\end{aligned}$$ (3.5.1)

that is,

$$\mathcal{B}(A) = \mathcal{I}(A) \times \mathcal{J}(A),$$

be the index sets of maximal sets of linearly independent rows, columns and maximal nonsingular submatrices, respectively.

For $\alpha \in Q_{k,m}, \beta \in Q_{k,n}$, let

$$\begin{aligned}
\mathcal{I}(\alpha) &= \{I \in \mathcal{I}(A) : \alpha \subset I\}, \\
\mathcal{J}(\beta) &= \{J \in \mathcal{J}(A) : \beta \subset J\}, \\
\mathcal{B}(\alpha, \beta) &= \{(I, J) \in \mathcal{B}(A) : \alpha \subset I, \beta \subset J\},
\end{aligned}$$ (3.5.2)

that is,

$$\mathcal{B}(\alpha, \beta) = \mathcal{I}(\alpha) \times \mathcal{J}(\beta).$$

For $\alpha = (\alpha_1, \alpha_2, \cdots, \alpha_k)$ and $\beta = (\beta_1, \beta_2, \cdots, \beta_k)$, we denote

$$A(\beta \rightarrow I_\alpha)$$

the matrix obtained from A by replacing the β_ith column with the unit vector $\mathbf{e}_{\alpha_i}$, $i = 1, 2, \ldots, k$, and denote

$$A(\beta \rightarrow O)$$

the matrix obtained from A by replacing the β_ith column with the zero vector, $i = 1, 2, \ldots, k$.

Finally, the coefficient $(-1)^{S(\alpha)+S(\beta)} \det(A[\alpha', \beta'])$ of $\det(A[\alpha, \beta])$ in the Laplace expansion of $\det(A)$ is denoted by

$$\frac{\partial}{\partial|A_{\alpha\beta}|}|A|. \tag{3.5.3}$$

Using the above notations, (3.5.3) can be rewritten as

$$\frac{\partial}{\partial|A_{\alpha\beta}|}|A| = (-1)^{S(\alpha)+S(\beta)} \det(A[\alpha', \beta']) = \det(A(\beta \to I_\alpha)) \tag{3.5.4}$$

and the Jacobi identity as

$$\det(A^{-1}[\beta, \alpha]) = \frac{\det(A(\beta \to I_\alpha))}{\det(A)} \tag{3.5.5}$$

$$= \frac{\det(A^T A(\beta \to I_\alpha))}{\det(A^T A)}. \tag{3.5.6}$$

As in [26], we define the volume of an $m \times n$ matrix A by

$$\text{Vol}(A) = \sqrt{\sum_{(I,J)\in B(A)} \det^2(A_{IJ})}$$

and in particular,

$$\text{Vol}(A) = \sqrt{\det(A^T A)} \quad \text{if } A \text{ has full column rank} \tag{3.5.7}$$

and

$$\text{Vol}(A) = \sqrt{\det(AA^T)} \quad \text{if } A \text{ has full row rank.} \tag{3.5.8}$$

If $A = FG$ is a full rank factorization of A, then

$$\text{Vol}(A) = \text{Vol}(F)\text{Vol}(G). \tag{3.5.9}$$

Let $A \in \mathbb{R}_r^{m\times n}$, and $U \in \mathbb{R}_{m-r}^{m\times(m-r)}$ and $V^T \in \mathbb{R}_{n-r}^{n\times(n-r)}$ be matrices whose columns form the orthonormal bases for $\mathcal{N}(A^T)$ and $\mathcal{N}(A)$ respectively, then

$$B = \begin{bmatrix} A & U \\ V & O \end{bmatrix}$$

is nonsingular by Corollary 3.1.3 and its inverse is

$$B^{-1} = \begin{bmatrix} A^\dagger & V^T \\ U^T & O \end{bmatrix}. \tag{3.5.10}$$

If A is of full column (row) rank, then V (U) is empty. Moreover, by [26],

$$\det(B^T B) = \text{Vol}^2(A). \tag{3.5.11}$$

In the following subsections, we study the determinantal expressions of the minors of the various generalized inverses.

3.5.1 Minors of the Moore-Penrose Inverse

Theorem 3.5.1 ([25]) *Let $A \in \mathbb{R}_r^{m \times n}$ and $1 \le k \le r$, then for any $\alpha \in Q_{k,m}$ and $\beta \in Q_{k,n}$,*

$$\det(A^\dagger[\beta, \alpha]) = \tag{3.5.12}$$

$$\begin{cases} 0, & \text{if } \mathcal{B}(\alpha, \beta) = \emptyset; \\ \text{Vol}^{-2}(A) \displaystyle\sum_{(I,J) \in \mathcal{B}(\alpha,\beta)} \det(A_{IJ}) \dfrac{\partial}{\partial |A_{\alpha\beta}|} |A_{IJ}|, & \text{otherwise.} \end{cases}$$

Proof From (3.5.10) and (3.5.6), we have

$$\det(A^\dagger[\beta, \alpha]) = \det(B^{-1}[\beta, \alpha])$$

$$= \frac{1}{\det(B^T B)} \det(B^T B(\beta \to I_\alpha)). \tag{3.5.13}$$

Now

$$\det(B^T B(\beta \to I_\alpha))$$

$$= \det\left(\begin{bmatrix} A^T & V^T \\ U^T & O \end{bmatrix} \begin{bmatrix} A(\beta \to I_\alpha) & U \\ V(\beta \to O) & O \end{bmatrix} \right)$$

$$= \det\left(\begin{bmatrix} A^T & V^T \end{bmatrix} \begin{bmatrix} A(\beta \to I_\alpha) \\ V(\beta \to O) \end{bmatrix} \right)$$

$$= \sum_{I \in \mathcal{I}(A)} \det\begin{bmatrix} (A^T)_{*I} & V^T \end{bmatrix} \det\begin{bmatrix} (A(\beta \to I_\alpha))_{I*} \\ V(\beta \to O) \end{bmatrix}$$

$$= \sum_{I \in \mathcal{I}(\alpha)} \det\begin{bmatrix} (A_{I*})^T & V^T \end{bmatrix} \det\begin{bmatrix} A_{I*}(\beta \to I_\alpha) \\ V(\beta \to O) \end{bmatrix}. \tag{3.5.14}$$

In the above equation, the equality next to the last is by the Cauchy-Binet formula, noting that the determinant of any $n \times n$ submatrix of $[A^T \ V^T] \in \mathbb{R}^{n \times (m+n-r)}$

consisting of more than r columns of A^T is zero. The last equality holds because the matrix

$$\begin{bmatrix} A(\beta \to I_\alpha)_{I*} \\ V(\beta \to O) \end{bmatrix}$$

has at least one column of zeros if $I \notin \mathcal{I}(\alpha)$.

We claim (and prove later) that for any fixed $I \in \mathcal{I}(\alpha)$,

$$\det[(A_{I*})^T \; V^T]\det\begin{bmatrix} A_{I*}(\beta \to I_\alpha) \\ V(\beta \to O) \end{bmatrix}$$

$$= \sum_{J \in \mathcal{J}(\beta)} \det(A_{IJ})\det(A_{IJ}(\beta \to I_\alpha)). \qquad (3.5.15)$$

Then, using (3.5.11), (3.5.14) and (3.5.4), the Eq. (3.5.13) becomes

$$\det(A^\dagger[\beta, \alpha])$$

$$= \frac{1}{\mathrm{Vol}^2(A)} \sum_{I \in \mathcal{I}(\alpha)} \sum_{J \in \mathcal{J}(\beta)} \det(A_{IJ})\det(A_{IJ}(\beta \to I_\alpha))$$

$$= \frac{1}{\mathrm{Vol}^2(A)} \sum_{(I,J) \in \mathcal{B}(\alpha,\beta)} \det(A_{IJ})\frac{\partial}{\partial |A_{\alpha\beta}|}|A_{IJ}|.$$

Finally, we prove (3.5.15). For any fixed $I \in \mathcal{I}(\alpha)$, the columns of V^T form an orthonormal basis for $\mathcal{N}(A_{I*})$. Let

$$L = \begin{bmatrix} A_{I*} \\ V \end{bmatrix},$$

then, by Corollary 3.1.3, L is nonsingular and its inverse is

$$L^{-1} = [A^\dagger_{I*} \; V^T],$$

and

$$\det((A_{I*})^\dagger[\beta, \alpha])$$
$$= \det(L^{-1}[\beta, \alpha])$$
$$= \frac{1}{\det(L^T L)} \det(L^T L(\beta \to I_\alpha))$$
$$= \frac{1}{\mathrm{Vol}^2(A_{I*})} \det[(A_{I*})^T \; V^T]\det\begin{bmatrix} A_{I*}(\beta \to I_\alpha) \\ V(\beta \to O) \end{bmatrix}. \qquad (3.5.16)$$

Writing $(A_{I*})^T = C$, we have

$$\det((A_{I*})^\dagger[\beta, \alpha]) = \det((C^\dagger)^T[\beta, \alpha]) = \det(C^\dagger[\alpha, \beta]).$$

Let W be a matrix whose columns form an orthonormal basis for $\mathcal{N}(C^T)$ and

$$M = [C \ \ W],$$

then, by Corollary 3.1.3, M is nonsingular and its inverse is

$$M^{-1} = \begin{bmatrix} C^\dagger \\ W^T \end{bmatrix}.$$

Thus

$$(M^T)^{-1} = [(C^T)^\dagger \ \ W]$$

and

$$
\begin{aligned}
&\det((A_{I*}^\dagger)[\beta, \alpha]) \\
&= \det((C^T)^\dagger[\beta, \alpha]) \\
&= \det((M^T)^{-1}[\beta, \alpha]) \\
&= \det((M)^{-1}[\alpha, \beta]) \\
&= \frac{1}{\det(M^T M)} \det(M^T M(\alpha \to I_\beta)) \\
&= \frac{1}{\mathrm{Vol}^2(A_{I*})} \det(A_{I*}(A_{I*})^T(\alpha \to I_\beta)) \\
&= \frac{1}{\mathrm{Vol}^2(A_{I*})} \sum_{J \in \mathcal{J}(\beta)} \det(A_{IJ}) \det((A_{IJ})^T(\alpha \to I_\beta)) \\
&= \frac{1}{\mathrm{Vol}^2(A_{I*})} \sum_{J \in \mathcal{J}(\beta)} \det(A_{IJ}) \det(A_{IJ}(\beta \to I_\alpha)).
\end{aligned}
\tag{3.5.17}
$$

The equality next to the last is by the Cauchy-Binet formula, noting that, if $J \notin \mathcal{J}(\beta)$, then the submatrix of $(A_{I*})^T(\alpha \to I_\beta)$ whose rows are indexed by J has at least one column of zeros.

Finally, (3.5.15) follows by comparing (3.5.16) and (3.5.17).   □

Note that $\mathcal{B}(\alpha, \beta) = \emptyset$ is equivalent to the linear dependence of either the columns of $A_{*\beta}$ or the rows of $A_{\alpha*}$.

As a special case, if $\alpha = I \in \mathcal{I}(A)$ and $\beta = J \in \mathcal{J}(A)$, then $\mathcal{B}(\alpha, \beta)$ contains only one element, i.e., (I, J). Now Theorem 3.5.1 gives the identity:

$$\det(A^\dagger)_{IJ} = \frac{1}{\mathrm{Vol}^2(A)} \det(A_{IJ}), \quad \forall (I, J) \in \mathcal{B}(\alpha, \beta).$$

3.5.2 Minors of the Weighted Moore-Penrose Inverse

Let $A \in \mathbb{R}_r^{m \times n}$, $U \in \mathbb{R}_{m-r}^{m \times (m-r)}$ and $V^T \in \mathbb{R}_{n-r}^{n \times (n-r)}$ be matrices whose columns form the bases for $\mathcal{N}(A^T)$ and $\mathcal{N}(A)$ respectively, and M and N be symmetric positive definite matrices of orders m and n respectively, then

$$\begin{bmatrix} A & M^{-1}U \\ VN & O \end{bmatrix}$$

is nonsingular by Theorem 3.1.1 and its inverse is

$$\begin{bmatrix} A & M^{-1}U \\ VN & O \end{bmatrix}^{-1} = \begin{bmatrix} A_{MN}^\dagger & V^T(VNV^T)^{-1} \\ (U^T M^{-1}U)^{-1}U^T & O \end{bmatrix}.$$

Lemma 3.5.1 Let $B \in \mathbb{R}_r^{r \times n}$, $V^T \in \mathbb{R}_{n-r}^{n \times (n-r)}$ and $BV^T = O$, then

$$\begin{bmatrix} B \\ VN \end{bmatrix}$$

is nonsingular and its inverse is

$$\begin{bmatrix} B \\ VN \end{bmatrix}^{-1} = [B_{IN}^\dagger \quad V^T(VNV^T)^{-1}]. \tag{3.5.18}$$

Proof Since
$$B_{IN}^\dagger = N^{-1/2}(BN^{-1/2})^\dagger,$$

we have

$$BB_{IN}^\dagger = BN^{-1/2}(BN^{-1/2})^\dagger = I,$$
$$VN(V^T(VNV^T)^{-1}) = I,$$
$$VNB_{IN}^\dagger = VNB_{IN}^\dagger BB_{IN}^\dagger$$
$$= V(NB_{IN}^\dagger B)^T B_{IN}^\dagger$$
$$= VB^T(NB_{IN}^\dagger)^T B_{IN}^\dagger$$
$$= O,$$

and
$$BV^T(VNV^T)^{-1} = O,$$

which imply (3.5.18). $\qquad\qquad\square$

Lemma 3.5.2 *Let $C \in \mathbb{R}_r^{n \times r}$, $W \in \mathbb{R}_{n-r}^{n \times (n-r)}$ and $C^T W = O$, then*

$$[C \ NW]$$

is nonsingular and its inverse is

$$[C \ NW]^{-1} = \begin{bmatrix} C_{N^{-1},I}^{\dagger} \\ (W^T N W)^{-1} W^T \end{bmatrix}. \tag{3.5.19}$$

Proof Since

$$[C \ NW]^T = \begin{bmatrix} C^T \\ W^T N \end{bmatrix},$$

by using (3.5.18), we have

$$\begin{bmatrix} C^T \\ W^T N \end{bmatrix}^{-1} = \left[(C^T)_{I,N}^{\dagger} \ \ W(W^T N W)^{-1} \right]$$

$$= \left[(C_{N^{-1},I}^{\dagger})^T \ \ W(W^T N W)^{-1} \right].$$

Thus (3.5.19) holds.   □

Lemma 3.5.3 *Let $A \in \mathbb{R}_r^{m \times n}$ and $A = FG$ be a full rank factorization of A, and M and N be symmetric positive definite matrices of orders m and n respectively. Then*

$$A_{MN}^{\dagger} = G_{IN}^{\dagger} F_{MI}^{\dagger} \tag{3.5.20}$$

and

$$\det(F^T M F) \det(G N^{-1} G^T) = \text{Vol}^2(M^{1/2} A N^{-1/2}). \tag{3.5.21}$$

Proof By using Theorem 1.4.4,

$$\begin{aligned} A_{MN}^{\dagger} &= N^{-1} G^T (F^T M A N^{-1} G^T)^{-1} F^T M \\ &= N^{-1} G^T (G N^{-1} G^T)^{-1} (F^T M F)^{-1} F^T M \\ &= G_{IN}^{\dagger} F_{MI}^{\dagger}. \end{aligned}$$

Since

$$G N^{-1} G^T = (G N^{-1/2})(G N^{-1/2})^T$$

and

$$F^T M F = (M^{1/2} F)^T (M^{1/2} F),$$

where $M^{1/2} F$ is of full column rank and $G N^{-1/2}$ is of full row rank, using (3.5.7)–(3.5.9), we have

$$\det(F^T M F) \det(G N^{-1} G^T)$$
$$= \det((M^{1/2} F)^T (M^{1/2} F)) \det((G N^{-1/2})(G N^{-1/2})^T)$$
$$= \mathrm{Vol}^2(M^{1/2} F) \, \mathrm{Vol}^2(G N^{-1/2})$$
$$= \mathrm{Vol}^2(M^{1/2} A N^{-1/2}),$$

which completes the proof. $\qquad\qquad\qquad\qquad\qquad\qquad\qquad\qquad\qquad$ □

Lemma 3.5.4 *Let $A \in \mathbb{R}_r^{m \times n}$ and $A = FG$ be a full rank factorization of A, $1 \le k \le r$, and $\forall \alpha \in Q_{k,m}$, $\forall \beta \in Q_{k,n}$ and $\forall (I, J) \in \mathcal{B}(\alpha, \beta)$, then*

$$\det(A_{IJ}(\beta \to I_\alpha))$$
$$= \sum_{\omega \in Q_{k,r}} \det((F(\omega \to I_\alpha))_{I*}) \det((G(\beta \to I_\omega))_{*J}). \qquad (3.5.22)$$

Proof From (3.5.5),

$$\det(A_{IJ}(\beta \to I_\alpha)) = \det(A_{IJ}) \det(A_{IJ}^{-1}[\beta, \alpha]). \qquad (3.5.23)$$

Since F is of full column rank and G is of full row rank, we have

$$\det((F(\omega \to I_\alpha))_{I*}) = \det(F_{I*}(\omega \to I_\alpha))$$
$$= \det(F_{I*}) \det(F_{I*}^{-1}[\omega, \alpha]) \qquad (3.5.24)$$

and

$$\det((G(\beta \to I_\omega))_{*J}) = \det(G_{*J}(\beta \to I_\omega))$$
$$= \det(G_{*J}) \det(G_{*J}^{-1}[\beta, \omega]). \qquad (3.5.25)$$

By using $A_{IJ} = F_{I*} G_{*J}$, we have

$$\det(A_{IJ}) = \det(F_{I*}) \det(G_{*J}). \qquad (3.5.26)$$

It follows from $A_{IJ}^{-1} = G_{*J}^{-1} F_{I*}^{-1}$ and the Cauchy-Binet formula that

$$\det(A_{IJ}^{-1}[\beta, \alpha]) = \sum_{\omega \in Q_{k,r}} \det(G_{*J}^{-1}[\beta, \omega]) \det(F_{I*}^{-1}[\omega, \alpha]). \qquad (3.5.27)$$

Thus, (3.5.22) follows by (3.5.23)–(3.5.27). $\qquad\qquad\qquad\qquad\qquad\qquad$ □

Theorem 3.5.2 ([27]) *Let $A \in \mathbb{R}_r^{m \times n}$, and $U \in \mathbb{R}_{m-r}^{m \times (m-r)}$ and $V^T \in \mathbb{R}_{n-r}^{n \times (n-r)}$ be matrices whose columns form the bases for $\mathcal{N}(A^T)$ and $\mathcal{N}(A)$ respectively, and M and N symmetric positive definite matrices of orders m and n respectively, $1 \le k \le r$, then for any $\alpha \in Q_{k,m}$ and $\beta \in Q_{k,n}$,*

$$\det(A^{\dagger}_{MN}[\beta, \alpha]) \tag{3.5.28}$$

$$= \begin{cases} 0, & \text{if } \mathcal{B}(\alpha, \beta) = \emptyset; \\ \text{Vol}^{-2}(\widetilde{A}) \sum_{(I,J)\in\mathcal{B}(\alpha,\beta)} \det((MAN^{-1})_{IJ}) \dfrac{\partial|A_{IJ}|}{\partial|A_{\alpha\beta}|} & \text{otherwise,} \end{cases}$$

where $\widetilde{A} = M^{1/2}AN^{-1/2}$.

Proof Let $A = FG$ be a full rank factorization. It follows from (3.5.20) and the Cauchy-Binet formula that

$$\det(A^{\dagger}_{MN}[\beta, \alpha]) = \sum_{\omega \in Q_{k,r}} \det(G^{\dagger}_{IN}[\beta, \omega]) \, \det(F^{\dagger}_{MI}[\omega, \alpha]).$$

Setting $G^T = D$, by the assumption $AV^T = O$, we have $GV^T = O$, i.e., $D^T V^T = O$. From Lemma 3.5.2,

$$H = [D \ NV^T]$$

is nonsingular and its inverse

$$H^{-1} = \begin{bmatrix} D^{\dagger}_{N^{-1},I} \\ (VNV^T)^{-1}V \end{bmatrix}.$$

Since

$$(D^{\dagger}_{N^{-1},I})^T = (D^T)^{\dagger}_{I,N} = G^{\dagger}_{IN},$$

we have

$$\det(G^{\dagger}_{IN}[\beta, \omega])$$
$$= \det((D^{\dagger}_{N^{-1},I})^T[\beta, \omega])$$
$$= \det(D^{\dagger}_{N^{-1},I}[\omega, \beta])$$
$$= \det(H^{-1}[\omega, \beta])$$
$$= \frac{1}{\det(H^T N^{-1} H)} \det(H^T N^{-1} H(\omega \to I_{\beta}))$$
$$= \frac{1}{\det(GN^{-1}G^T)\det(VNV^T)} \det\begin{bmatrix} GN^{-1}G^T(\omega \to I_{\beta}) & GV^T \\ VG^T(\omega \to I_{\beta}) & VNV^T \end{bmatrix}$$
$$= \frac{1}{\det(GN^{-1}G^T)} \det(GN^{-1}G^T(\omega \to I_{\beta}))$$
$$= \frac{1}{\det(GN^{-1}G^T)} \sum_{J\in\mathcal{J}(\beta)} \det((GN^{-1})_{*J}) \det((G^T(\omega \to I_{\beta}))_{J*})$$
$$= \frac{1}{\det(GN^{-1}G^T)} \sum_{J\in\mathcal{J}(\beta)} \det((GN^{-1})_{*J}) \det((G(\beta \to I_{\omega}))_{*J}). \tag{3.5.29}$$

The equality next to the last is by the Cauchy-Binet formula. Noting that, if $J \notin \mathcal{J}(\beta)$, then the submatrix of $G^T(\omega \leftarrow I_\alpha)$ whose rows are indexed by J has at least one column of zeros.

From the assumption $A^T U = O$, we have $F^T U = O$. By Lemma 3.5.2,

$$P = [F \quad M^{-1}U]$$

is nonsingular and its inverse is

$$P^{-1} = \begin{bmatrix} F^\dagger_{MI} \\ (U^T M^{-1}U)U^T \end{bmatrix}.$$

Thus

$$
\begin{aligned}
&\det(F^\dagger_{MI}[\omega, \alpha]) \\
&= \det(P^{-1}[\omega, \alpha]) \\
&= \frac{1}{\det(P^T M P)} \det(P^T M P(\omega \to I_\alpha)) \\
&= \frac{1}{\det(F^T M F)} \det(F^T M F(\omega \to I_\alpha)) \\
&= \frac{1}{\det(F^T M F)} \sum_{I \in \mathcal{I}(\alpha)} \det((F^T M)_{*I}) \det((F(\omega \to I_\alpha))_{I*}) \\
&= \frac{1}{\det(F^T M F)} \sum_{I \in \mathcal{I}(\alpha)} \det((MF)_{I*}) \det((F(\omega \to I_\alpha))_{I*}). \qquad (3.5.30)
\end{aligned}
$$

If $I \notin \mathcal{I}(\alpha)$, then the submatrix of $F(\omega \leftarrow I_\alpha)$ whose rows are indexed by I has at least one column of zeros.

It follows from (3.5.2), (3.5.4), (3.5.21), (3.5.22), (3.5.29) and (3.5.30) that the second part in (3.5.28) holds.

Note that $\mathcal{B}(\alpha, \beta) = \emptyset$ is equivalent to linear dependence of either the columns of $A_{*\beta}$ or the rows of $A_{\alpha*}$. $\qquad\square$

As a special case, if $\alpha = I \in \mathcal{I}(A)$ and $\beta = J \in \mathcal{J}(A)$, then $\mathcal{B}(\alpha, \beta)$ contains only one element, which is (I, J). Now Theorem 3.5.2 gives the identity:

$$\det((A^\dagger_{MN})_{JI}) = \frac{1}{\mathrm{Vol}^2(\widetilde{A})} \det((MAN^{-1})_{IJ}).$$

3.5.3   Minors of the Group Inverse and Drazin Inverse

Definition 3.5.1 Let $A \in \mathbb{R}_r^{m \times n}$, $0 < k \leq r$, the kth compound matrix of A denoted by $C_k(A)$ is the $\binom{m}{k} \times \binom{n}{k}$ matrix whose elements are determinants of all $k \times k$ submatrices of A in lexicographic order.

Some well known properties of compound matrices are listed below (see for example [28]).

Lemma 3.5.5

(1) $C_k(A^T) = C_k(A)^T$;

(2) If $A \in \mathbb{C}^{m \times p}$, $B \in \mathbb{C}^{p \times n}$ and $\text{rank}(AB) = r$, then

$$C_k(AB) = C_k(A)C_k(B), \quad k \leq \min\{m, r, n\};$$

(3) $\text{rank}(C_k(A)) = \binom{r}{k}$;

(4) $C_k(I) = I$ (appropriate size identity matrix).

Lemma 3.5.6 Let $A \in \mathbb{R}_r^{m \times n}$. If $A = CR$ is a full rank factorization of A, then

$$\frac{\partial}{\partial |A_{\alpha\beta}|}|A_{IJ}| = \sum_{\gamma \in Q_{k,r}} \frac{\partial}{\partial |C_{\alpha\gamma}|}|C_{I*}|\frac{\partial}{\partial |R_{\gamma\beta}|}|R_{*J}|,$$

for any $\alpha \in Q_{k,m}$, $\beta \in Q_{k,n}$, and $(I, J) \in \mathcal{B}(\alpha, \beta)$.

Proof The result follows from the Cauchy-Binet formula

$$\det(A[I \setminus \alpha, \ J \setminus \beta])$$
$$= \sum_{\tau \in Q_{r-k,r}} \det(C[I \setminus \alpha, \ \tau]) \det(R[\tau, \ J \setminus \beta])$$
$$= \sum_{\gamma \in Q_{k,r}} \det(C[I \setminus \alpha, \ \tilde{r} \setminus \gamma]) \det(R[\tilde{r} \setminus \gamma, \ J \setminus \beta]),$$

where $\tilde{r} = \{1, 2, \ldots, r\}$. □

Lemma 3.5.7 Let $R \in \mathbb{R}_r^{r \times n}$ and $C \in \mathbb{R}_r^{m \times r}$. If E and F are a right inverse of R and a left inverse of C respectively, that is, $RE = I$ and $FC = I$, then

$$\det(E[\beta, \gamma]) = \sum_{J \in \mathcal{J}(\beta)} \det(E_{J*})\frac{\partial}{\partial |R_{\gamma\beta}|}|R_{*J}|, \tag{3.5.31}$$

for any $\gamma \in Q_{k,r}$, $\beta \in Q_{k,n}$, and

$$\det(F[\gamma, \alpha]) = \sum_{I \in \mathcal{I}(\alpha)} \det(F_{*I}) \frac{\partial}{\partial |C_{\alpha\gamma}|} |C_{I*}|, \tag{3.5.32}$$

for any $\alpha \in Q_{k,m}$, $\gamma \in Q_{k,r}$.

Proof Let $e_{\beta,\gamma}$ be the right-hand side of (3.5.31), then

$$\begin{aligned} e_{\beta,\gamma} &= \sum_{J \in \mathcal{J}(\beta)} \left(\sum_{\xi \in Q_{k,r}} \det(E[\beta, \xi]) \frac{\partial}{\partial |E_{\beta\xi}|} |E_{J*}| \right) \frac{\partial}{\partial |R_{\gamma\beta}|} |R_{*J}| \\ &= \sum_{J \in \mathcal{J}(\beta)} \sum_{\xi \in Q_{k,r}} \det(E[\beta, \xi])(-1)^{S(\xi)+S(\gamma)} \\ &\qquad \cdot \det(E[J \setminus \beta, \; \widetilde{r} \setminus \xi]) \det(R[\widetilde{r} \setminus \gamma, \; J \setminus \beta]) \\ &= \sum_{\substack{\tau \in Q_{r-k,n} \\ \tau \cap \beta = \emptyset}} \det(E[\beta, \xi])(-1)^{S(\xi)+S(\gamma)} \\ &\qquad \cdot \det(E[\tau, \; \widetilde{r} \setminus \xi]) \det(R[\widetilde{r} \setminus \gamma, \; \tau]), \tag{3.5.33} \end{aligned}$$

noting that if $\tau \in Q_{r-k,n}$ and $\tau \cap \beta \neq \emptyset$, then

$$\sum_{\xi \in Q_{k,r}} (-1)^{S(\xi)} \det(E[\beta, \xi]) \det(E[\tau, \; \widetilde{r} \setminus \xi]) = 0.$$

Thus (3.5.33) becomes

$$e_{\beta,\gamma} = \sum_{\xi \in Q_{k,r}} (-1)^{S(\xi)+S(\gamma)} \det(E[\beta, \xi]) \cdot$$

$$\left(\sum_{\tau \in Q_{r-k,n}} \det(R[\widetilde{r} \setminus \gamma, \; \tau]) \det(E[\tau, \; \widetilde{r} \setminus \xi]) \right). \tag{3.5.34}$$

Now it follows from

$$C_{r-k}(R)C_{r-k}(E) = I,$$

that

$$\sum_{\tau \in Q_{r-k,n}} \det(R[\widetilde{r} \setminus \gamma, \; \tau]) \det(E[\tau, \; \widetilde{r} \setminus \xi]) = \begin{cases} 0, & \text{if } \gamma \neq \xi; \\ 1, & \text{if } \gamma = \xi. \end{cases}$$

Putting the above equation and (3.5.34) together gives

$$e_{\beta,\gamma} = \det(E[\beta, \gamma]).$$

This completes the proof of (3.5.31). The proof of (3.5.32) is similar and left to the reader as an exercise.   □

Theorem 3.5.3 ([29]) *Let* $A \in \mathbb{R}_r^{m \times n}$. *If* $G \in A\{1, 2\}$, *then*

$$\det(G[\beta, \alpha]) = \sum_{(I,J) \in \mathcal{B}(\alpha, \beta)} \det(G_{JI}) \frac{\partial}{\partial |A_{\alpha\beta}|} |A_{IJ}|, \qquad (3.5.35)$$

for any $\alpha \in Q_{k,m}$ *and* $\beta \in Q_{k,n}$, $1 \leq k \leq r$.

Proof Let $A = CR$ be a full rank factorization of A, $E = GC$, and $F = RG$. Since $G \in A\{1\}$, we have $CRGCR = CR$, $RGC = I$. Thus $RE = I$ and

$$FC = RGC = RE = I.$$

It follows from the assumption $G \in A\{2\}$ that

$$EF = GCRG = GAG = G.$$

Denoting the right-hand side of (3.5.35) as $g_{\beta\alpha}$ and using the above equation, we get

$$g_{\beta\alpha} = \sum_{(I,J) \in \mathcal{B}(\alpha, \beta)} \det(E_{J*}) \det(F_{*I}) \frac{\partial}{\partial |A_{\alpha\beta}|} |A_{IJ}|.$$

It then follows from Lemmas 3.5.6 and 3.5.7 that

$$g_{\beta\alpha}$$
$$= \sum_{\gamma \in Q_{k,r}} \left(\sum_{J \in \mathcal{J}(\beta)} \det(E_{J*}) \frac{\partial}{\partial |R_{\gamma\beta}|} |R_{*J}| \right) \left(\sum_{I \in \mathcal{I}(\alpha)} \det(F_{*I}) \frac{\partial}{\partial |C_{\alpha\gamma}|} |C_{I*}| \right)$$
$$= \sum_{\gamma \in Q_{k,r}} \det(E[\beta, \gamma]) \det(F[\gamma, \alpha]),$$

which equals $\det(G[\beta, \alpha])$.   □

The following theorem is a special case of Theorem 3.5.3.

Theorem 3.5.4 *Let* $A \in \mathbb{R}_r^{n \times n}$. *If* $\mathrm{Ind}(A) = 1$, *then* A *has a group inverse* A_g, *and*

$$\det(A_g[\beta, \alpha]) = \nu^{-2} \sum_{(I,J) \in \mathcal{B}(\alpha, \beta)} \det(A_{JI}) \frac{\partial}{\partial |A_{\alpha\beta}|} |A_{IJ}|, \qquad (3.5.36)$$

for any $\alpha, \beta \in Q_{k,n}$, $1 \leq k \leq r$, *where*

$$\nu = \sum_{J \in \mathcal{J}(A)} \det(A_{JJ}).$$

Proof Let $A = CR$ be a full rank factorization of A. By using $A_g = C(RC)^{-2}R$ and the Cauchy-Binet formula, we have

$$\det((A_g)_{JI}) = \det((RC)^{-2})\det(A_{JI})$$

$$= \left(\sum_{J \in \mathcal{J}(A)} \det(A_{JJ})\right)^{-2} \det(A_{JI})$$

$$= \nu^{-2}\det(A_{JI}),$$

which implies (3.5.36) by Theorem 3.5.3. $\square$

Theorem 3.5.5 ([30]) *Let $A \in \mathbb{R}^{n \times n}$, $\mathrm{Ind}(A) = k$ and $\mathrm{rank}(A^k) = r_k$, then A has a Drazin inverse A_d, and*

$$\det(A_d[\beta, \alpha]) = \tag{3.5.37}$$

$$\nu^{-2} \sum_{\omega \in Q_{h,n}} \sum_{(I,J) \in \mathcal{B}(\omega,\beta)} \det((A^k)_{JI})\det(A^{k-1}[\omega, \alpha])\frac{\partial}{\partial |(A^k)_{\omega\beta}|}|(A^k)_{IJ}|,$$

for any $\alpha, \beta, \omega \in Q_{h,n}$ and $I, J \in Q_{r_k,n}$, $1 \le h \le r_k$, where

$$\nu = \sum_{J \in \mathcal{J}(A^k)} \det((A^k)_{JJ}).$$

Proof It is easy to verify that
$$(A_d)^k = (A^k)_g.$$

Applying (3.5.36) and the above equation, we have

$$\det((A_d)^k[\beta, \alpha])$$
$$= \det((A^k)_g[\beta, \alpha])$$
$$= \nu^{-2} \sum_{(I,J) \in \mathcal{B}(\alpha,\beta)} \det((A^k)_{JI})\frac{\partial}{\partial |(A^k)_{\alpha\beta}|}|(A^k)_{IJ}|,$$

where $\nu = \sum_{J \in \mathcal{J}(A^k)} \det((A^k)_{JJ})$. Since $A_d = (A_d)^k A^{k-1}$, from the Cauchy-Binet formula, we have

$$\det(A_d[\beta, \alpha])$$
$$= \sum_{\omega \in Q_{h,n}} \det((A_d)^k[\beta, \omega])\det(A^{k-1}[\omega, \alpha])$$
$$= \nu^{-2} \sum_{\omega \in Q_{h,n}} \sum_{(I,J) \in \mathcal{B}(\omega,\beta)} \det((A^k)_{JI})\det(A^{k-1}[\omega, \alpha])\frac{\partial}{\partial |(A^k)_{\omega\beta}|}|(A^k)_{IJ}|,$$

which completes the proof.   □

As a special case, when $A \in \mathbb{R}^{n \times n}$, $\text{Ind}(A) = k$ and $\text{rank}(A^k) = r_k$, we can show that

$$\det((A_d)_{JI}) = \frac{\det((A^k)_{JI})}{\displaystyle\sum_{(I,J) \in \mathcal{B}(A^k)} \det(A_{JI}) \det((A^k)_{IJ})}, \qquad (3.5.38)$$

for any $(I, J) \in \mathcal{B}(A^k)$. Indeed, it follows from Theorem 2.1.6 that

$$A = B_1 C_1, \quad C_1 B_1 = B_2 C_2, \quad C_2 B_2 = B_3 C_3, \quad \cdots,$$

then

$$A_d = \begin{cases} B_1 B_2 \cdots B_k (C_k B_k)^{-1} C_k C_{k-1} \cdots C_1, & \text{when } (C_k B_k)^{-1} \text{ exists,} \\ O, & \text{when } C_k B_k = O. \end{cases}$$

Set

$$B = B_1 B_2 \cdots B_k, \quad C = C_k C_{k-1} \cdots C_1,$$

then

$$A_d = B(C_k B_k)^{-(k+1)} C = B(CAB)^{-1} C$$

and

$$C_{r_k}(A_d) = C_{r_k}(B(CAB)^{-1})C) = \frac{C_{r_k}(BC)}{\det(CAB)}. \qquad (3.5.39)$$

It is easy to show that $A^k = BC$ is a full rank factorization of A^k, and

$$C_{r_k}(BC) = C_{r_k}(A^k). \qquad (3.5.40)$$

Using the Cauchy-Binet formula, we have

$$\det(CAB) = \sum_{I \in Q_{r_k,n}} \sum_{J \in Q_{r_k,n}} \det(C_{*J}) \det(A_{JI}) \det(B_{I*})$$

$$= \sum_{(I,J) \in \mathcal{B}(A^k)} \det(A_{JI}) \det((A^k)_{IJ}).$$

Applying the above equation and (3.5.40)–(3.5.39), we obtain (3.5.38).

3.5.4 Minors of the Generalized Inverse $A_{T,S}^{(2)}$

In this subsection, we consider the expression of the determinant of a minor of the generalized inverse $A_{T,S}^{(2)}$. In 2001, Wang and Gao [31] gave an expression which includes the group inverse as a special case. Later, in 2006, Yu [32] presented a formula which unifies the expressions (3.5.12) for the Moore-Penrose inverse, (3.5.28) for the weighted Moore-Penrose inverse, (3.5.36) for the group inverse, and (3.5.37) for the Drazin inverse.

Theorem 3.5.6 ([32]) *Let $A \in \mathbb{C}_r^{m \times n}$, T be a subspace of $\mathbb{C}^n$ of dimension $s \leq r$, and S a subspace of $\mathbb{C}^m$ of dimension $m - s$. If A has the generalized inverse $A_{T,S}^{(2)}$ and there exists a matrix $G \in \mathbb{C}^{n \times m}$ such that $\mathcal{R}(G) = T$ and $\mathcal{N}(G) = S$, then*

$$\det(A_{T,S}^{(2)}[\beta, \alpha]) = v^{-1} \sum_{(I,J) \in \mathcal{B}(\alpha, \beta)} \det(G_{JI}) \frac{\partial}{\partial |A_{\alpha, \beta}|} |A_{IJ}|$$

for any $\alpha \in Q_{k,m}$, $\beta \in Q_{k,n}$, $1 \leq k \leq s$, where

$$v = \sum_{J \in \mathcal{J}(AG)} \det((AG)_{JJ}).$$

Proof From Lemma 3.2.2, $A_{T,S}^{(2)} = G(AG)_g$, so

$$\det(A_{T,S}^{(2)}[\beta, \alpha]) = \det(G(AG)_g)[\beta, \alpha].$$

Using the Cauchy-Binet formula and Theorem 3.5.4, we obtain

$$\det(A_{T,S}^{(2)}[\beta, \alpha])$$
$$= \sum_{\varpi} \det(G[\beta, \varpi]) \det((AG)_g[\varpi, \alpha])$$
$$= v^{-2} \sum_{\varpi} \sum_{(I,F) \in \mathcal{B}(\alpha, \varpi)} \det(G[\beta, \varpi]) \det((AG)_{FI}) \frac{\partial}{\partial |(AG)_{\alpha, \varpi}|} |(AG)_{IF}|,$$

where

$$v = \sum_{J \in \mathcal{J}(AG)} \det((AG)_{JJ}).$$

Note that for all I, for which $\alpha \in I$, and all F,

$$\sum_{\varpi} \det(G[\beta, \varpi]) \frac{\partial}{\partial |(AG)_{\alpha, \varpi}|} |(AG)_{IF}| = \det(U_{IF}),$$

where U is the matrix obtained from AG by replacing the β-rows of AG with the α-rows of G. Let $\alpha = \{\alpha_1, \alpha_2, \ldots, \alpha_k\}$ and $\beta = \{\beta_1, \beta_2, \ldots, \beta_k\}$. Then, if we denote B as the matrix obtained from A by replacing the α_ith row of A with the row $[0 \cdots 0\ 1\ 0 \cdots 0]$, where the β_ith entry is 1, $i = 1, 2, \ldots, k$, and all other entries are zero, then we have $U = BG$. Hence

$$\det(U_{IF}) = \sum_J \det(B_{IJ}) \det(G_{JF}).$$

Thus

$$\det(A_{T,S}^{(2)}[\beta, \alpha])$$
$$= v^{-2} \sum_{(I,F)\in\mathcal{B}(\alpha,\beta)} \sum_J \det(AG)_{FI} \det(B_{IJ}) \det(G_{JF})$$
$$= v^{-2} \sum_{(I,F)\in\mathcal{B}(\alpha,\beta)} \sum_J \sum_L \det(A_{FL}) \det(G_{LI}) \det(G_{JF}) \det(B_{IJ}).$$

Since $\mathrm{rank}(G) = s$, $(\mathrm{rank}(C_s(G)) = 1$, where $C_s(G)$ is the sth compound matrix of G. Thus, we can get

$$\det(G_{LI}) \det(G_{JF}) = \det(G_{LF}) \det(G_{JI})$$

and so

$$\det(A_{T,S}^{(2)}[\beta, \alpha])$$
$$= v^{-2} \sum_{(I,F)\in\mathcal{B}(\alpha,\beta)} \sum_J \sum_L \det(A_{FL}) \det(G_{LF}) \det(G_{JI}) \det(B_{IJ})$$
$$= v^{-2} \sum_{(I,F)\in\mathcal{B}(\alpha,\beta)} \sum_J \det((AG)_{FF}) \det(G_{JI}) \det(B_{IJ})$$
$$= v^{-1} \sum_I \sum_J \det(G_{JI}) \det(B_{IJ})$$
$$= v^{-1} \sum_{(I,J)\in\mathcal{B}(\alpha,\beta)} \det(G_{JI}) \frac{\partial}{\partial |A_{\alpha,\beta}|} |A_{IJ}|,$$

which completes the proof. $\square$

Exercises 3.5

1. Prove that if $A = FG$ is a full rank factorization of A, then

$$\mathrm{Vol}(A) = \mathrm{Vol}(F)\, \mathrm{Vol}(G).$$

2. Prove Lemma 3.5.5.

3. Prove (3.5.32).
4. Prove that $(A_d)^k = (A^k)_g$.
5. Prove that $A_d = (A_d)^k A^{k-1}$.
6. Prove that if $A = CR \in \mathbb{C}_r^{n \times n}$ is a full rank factorization of A, then

$$\det(RC) = \sum_{J \in \mathcal{J}(A)} \det(A_{JJ}).$$

Remarks

The trick in Robinson's proof was used by Ben-Israel [2] to derive a Cramer rule for the least-norm solution of consistent linear equations. It is a pioneering work. In this chapter, we survey the recent results in the field. Its application to parallel computation of the generalized inverses is presented in Chap. 7. Werner [14] gave the unique solution of a more general consistent restricted linear system. He also presented an alternative condensed Cramer rule. The basic idea is to modify the original matrix so that the new matrix is invertible, then solve the solution of the original problem from the corresponding nonsingular system of linear equations. The method is practical since it is easy to construct the nonsingular matrix of a low order. Further discussions on this method can be found in [3, 10, 11, 33].

The singular values and maximum rank minors of the generalized inverses are given in [25, 29, 34–36]. A Cramer rule for finding the unique W-weighted Drazin inverse solution of special restricted linear equations is discussed in [23].

The generalized inverses of bordered matrices are discussed in [15, 37–39].

References

1. S.M. Robinson, A short proof of Cramer's rule. Math. Mag. **43**, 94–95 (1970). Reprinted in Selected Papers on Algebra (S. Montgomery et al. eds.) Math. Assoc. Amer. 313–314 (1977)
2. A. Ben-Israel, A Cramer rule for least-norm solution of consistent linear equations. Linear Algebra Appl. **43**, 223–226 (1982)
3. Y. Chen, A Cramer rule for solution of the general restricted linear equation. Linear Multilinear Algebra **34**, 177–186 (1993)
4. Y. Chen, Expressions and determinantal formulas for solution of a restricted matrix equation. Acta Math. Appl. Sinica **18**(1), 65–73 (1995)
5. J. Ji, An alternative limit expression of Drazin inverse and its applications. Appl. Math. Comput. **61**, 151–156 (1994)
6. W. Sun, Cramer rule for weighted system. J. Nanjing Univ. **3**, 117–121 (1986). in Chinese
7. G.C. Verghes, A "Cramer rule" for least-norm least-square-error solution of inconsistent linear equations. Linear Algebra Appl. **48**, 315–316 (1982)
8. G. Wang, A Cramer rule for minimum-norm(T) least-squares(S) solution of inconsistent equations. Linear Algebra Appl. **74**, 213–218 (1986)
9. G. Wang, A Cramer rule for finding the solution of a class of singular equations. Linear Algebra Appl. **116**, 27–34 (1989)
10. G. Wang, On extensions of cramer rule. J. Shanghai Normal Univ. **21**, 1–7 (1992)

11. G. Wang, Z. Wang, More condensed determinant representation for the best approximate solution of the matrix equation $AXH = K$. J. Shanghai Normal Univ. **19**, 27–31 (1990). in Chinese
12. H.J. Werner, When is $B^- A^-$ a generalized inverse of AB? Linear Algebra Appl. **210**, 255–263 (1994)
13. G. Wang, On the singularity of a certain bordered matrix and its applications to the computation of generalized inverses. J. Shanghai Normal Univ. **18**, 7–14 (1989). in Chinese
14. H.J. Werner, On extensions of Cramer's rule for solutions of restricted linear systems. Linear Multilinear Algebra **15**, 319–330 (1984)
15. Y. Chen, B. Zhou, On g-inverses and the nonsingularity of a bordered matrix $\begin{pmatrix} A & B \\ C & 0 \end{pmatrix}$. Linear Algebra Appl. **133**, 133–151 (1990)
16. Y. Wei, A characterization and representation of the generalized inverse $A_{T,S}^{(2)}$ and its applications. Linear Algebra Appl. **280**, 87–96 (1998)
17. R. Penrose, On best approximate solution of linear matrix equations. Proc. Cambridge Philos. Soc. **52**, 17–19 (1956)
18. J. Ji, The cramer rule for the best approximate solution of the matrix equation $AXB = D$. J. Shanghai Normal Univ. **14**, 19–23 (1985). in Chinese
19. M. Marcus, M. Minc, *A Survey of Matrix Theory and Matrix Inequalities* (Mass, Allyn and Bacon, Boston, 1964)
20. G. Wang, *The Generalized Inverses of Matrices and Operators* (Science Press, Beijing, 1994). in Chinese
21. G. Wang, Weighted Moore-Penrose, Drazin, and group inverses of the Kronecker product $A \otimes B$ and some applications. Linear Algebra Appl. **250**, 39–50 (1997)
22. J. Ji, Explicit expressions of the generalized inverses and condensed cramer rules. Linear Algebra Appl. **404**, 183–192 (2005)
23. Y. Wei, A characterization for the W-weighted Drazin inverse and Cramer rule for W-weighted Drazin inverse solution. Appl. Math. Comput. **125**, 303–310 (2002)
24. R.A. Brualdi, H. Schneider, Determinantal: gauss, schur, cauchy, sylvester, kronecker, jacobi, binet, laplace, muir and cayley. Linear Algebra Appl. **52**, 769–791 (1983)
25. A. Ben-Israel, Minors of the Moore-Penrose inverse. Linear Algebra Appl. **195**, 191–207 (1993)
26. A. Ben-Israel, A volume associated with $m \times n$ matrices. Linear Algebra Appl. **167**, 87–111 (1992)
27. G. Wang, J. Sun, J. Gao. Minors of the weighted Moore-Penrose inverse. Numer. Math., J. Chinese Univ. **19**(4), 343–348 (1997). in Chinese
28. R.A. Horn, C.R. Johnson, *Matrix Analysis*, 2nd edn. (Cambridge University Press, Cambridge, 2012)
29. J. Miao, Reflexive generalized inverses and their minors. Linear Multilinear Algebra **35**, 153–163 (1993)
30. G. Wang, J. Gao, J. Sun, Minors of Drazin inverses. J. Shanghai Normal Univ. **28**, 12–15 (1998). in Chinese
31. G. Wang, J. Gao, Minors of the generalized inverse $A_{T,S}^{(2)}$. Math. Numer. Sinica **23**(4), 439–446 (2001). in Chinese
32. Y. Yu, The generalized inverse $A_{T,S}^{(2)}$ on an associative ring: theory and computation. Ph.D. thesis, Shanghai Normal University, China, 2006. in Chinese
33. Y. Chen, Finite algorithm for the (2)-generalized inverse $A_{T,S}^{(2)}$. Linear Multilinear Algebra **40**, 61–68 (1995)
34. R.B. Bapat, A. Ben-Israel, Singular values and maximum rank minors of generalized inverses. Linear Multilinear Algebra **40**, 153–161 (1995)
35. R.E. Hartwig, Singular value decomposition and the Moore-Penrose inverse of bordered matices. SIAM J. Appl. Math. **31**, 31–41 (1976)
36. R.E. Hartwig, Block generalized inverses. Arch Rational Mech. Anal. **61**, 197–251 (1976)
37. J. Miao, Representations for the weighted Moore-Penrose inverse of a partitioned matrix. J. Comput. Math. **7**, 320–323 (1989)

38. J. Miao, Some results for computing the Drazin inverse of a partitioned matrix. J. Shanghai Normal Univ. **18**, 25–31 (1989). in Chinese
39. M. Wei, W. Guo, On g-inverses of a bordered matrix: revisited. Linear Algebra Appl. **347**, 189–204 (2002)

Chapter 4
Reverse Order and Forward Order Laws for $A_{T,S}^{(2)}$

4.1 Introduction

The reverse order law for the generalized inverses of a matrix product yields a class of interesting fundamental problems in the theory of the generalized inverses of matrices. They have attracted considerable attention since the middle 1960s. Greville [1] first studied the Moore-Penrose inverse of the product of two matrices A and B, and gave a necessary and sufficient condition for reverse order law:

$$(AB)^\dagger = B^\dagger A^\dagger.$$

Since then more equivalent conditions for $(AB)^\dagger = B^\dagger A^\dagger$ have been discovered. Hartwig [2] and Tian [3, 4] studied reverse order law for the Moore-Penrose inverse of a product of three and n matrices respectively.

On the other hand, the reverse order law for the weighted Moore-Penrose inverse of a product of two and three matrices was considered by Sun and Wei [5] and Wang [6], respectively. Greville [1] first studied the reverse order law for the Drazin inverse of a product of two matrices A and B. He proved that

$$(AB)_d = B_d A_d$$

holds under the condition $AB = BA$. The necessary and sufficient conditions for the reverse order law for the Drazin inverse of the products of two and n matrices were considered by Tian [7] and Wang [8] respectively. Djordjević [9] considered the reverse order law of the form

$$(AB)_{KL}^{(2)} = B_{TS}^{(2)} A_{MN}^{(2)}$$

for outer generalized inverse with prescribed range and null space. Reverse order law for the generalized inverse $A_{T,S}^{(2)}$ of multiple matrix products has not been studied yet in literature.

© Springer Nature Singapore Pte Ltd. and Science Press 2018
G. Wang et al., *Generalized Inverses: Theory and Computations*,
Developments in Mathematics 53, https://doi.org/10.1007/978-981-13-0146-9_4

In this chapter, by using the ranks of matrices, the necessary and sufficient condition is given for the reverse order law

$$(A_1 A_2 \cdots A_n)_{T,S}^{(2)} = (A_n)_{T_n,S_n}^{(2)} \cdots (A_2)_{T_2,S_2}^{(2)} (A_1)_{T_1,S_1}^{(2)} \qquad (4.1.1)$$

to hold. From the above equation, we can obtain the necessary and sufficient conditions for the reverse order law of the Moore-Penrose inverse, the weighted Moore-Penrose inverse, the Drazin inverse and the group inverse.

Throughout this chapter, all matrices are over the complex number field $\mathbb{C}$ and the symbol $RS(A)$ denotes the row space of A.

From Theorem 1.3.8, we have the following lemma.

Lemma 4.1.1 ([10]) *Let $A \in \mathbb{C}_r^{m \times n}$, and the columns of $U \in \mathbb{C}_t^{n \times t}$ form a basis for T and the columns of $V^* \in \mathbb{C}_t^{m \times t}$ form a basis for $S^\perp$ ($t \leq r$), that is,*

$$\mathcal{R}(U) = T \quad and \quad \mathcal{N}(V) = S,$$

then $\mathrm{rank}(VAU) = \mathrm{rank}(V) = \mathrm{rank}(U) = t$, *that is, VAU is nonsingular, and*

$$X = U(VAU)^{-1}V \qquad (4.1.2)$$

is a $\{2\}$ inverse of A having range T and null space S, that is,

$$X = A_{T,S}^{(2)}. \qquad (4.1.3)$$

The next lemma shows that the common four kinds of generalized inverses $A^\dagger$, $A_{MN}^\dagger$, A_d and A_g are all the generalized inverse $A_{T,S}^{(2)}$.

Lemma 4.1.2 ([10, 11])

(1) *Let $A \in \mathbb{C}^{m \times n}$, then*

    (a) $A^\dagger = A_{\mathcal{R}(A^*),\mathcal{N}(A^*)}^{(2)} = A^*(A^*AA^*)^\dagger A^*$;

    (b) $A_{MN}^\dagger = A_{\mathcal{R}(A^\#),\mathcal{N}(A^\#)}^{(2)} = A^\#(A^\# AA^\#)^\dagger A^\#$,
    where M and N are Hermitian positive definite matrices of order m and n, respectively. In addition, $A^\# = N^{-1}A^*M$.

(2) *Let $A \in \mathbb{C}^{n \times n}$, then*

    (a) *If* $\mathrm{Ind}(A) = k$, *then* $A_d = A_{\mathcal{R}(A^k),\mathcal{N}(A^k)}^{(2)} = A^k(A^{2k+1})^\dagger A^k$;

    (b) *If* $\mathrm{Ind}(A) = 1$, *then* $A_g = A_{\mathcal{R}(A),\mathcal{N}(A)}^{(2)} = A(A^3)^\dagger A$.

Lemma 4.1.3 ([12]) *Suppose that A, B, C and D satisfy the conditions:*

$$\mathcal{R}(B) \subset \mathcal{R}(A) \quad and \quad RS(C) \subset RS(A)$$

or

$$\mathcal{R}(C) \subset \mathcal{R}(D) \quad and \quad RS(B) \subset RS(D),$$

then

$$\mathrm{rank} \begin{bmatrix} A & B \\ C & D \end{bmatrix} = \mathrm{rank}(A) + \mathrm{rank}(D - CA^\dagger B) \qquad (4.1.4)$$

or

$$\mathrm{rank} \begin{bmatrix} A & B \\ C & D \end{bmatrix} = \mathrm{rank}(D) + \mathrm{rank}(A - BD^\dagger C).$$

Lemma 4.1.4 ([13]) *Suppose B, C and D satisfy*

$$\mathcal{R}(D) \subset \mathcal{R}(C) \quad and \quad RS(D) \subset RS(B),$$

then

$$M^\dagger = \begin{bmatrix} O & B \\ C & D \end{bmatrix}^\dagger = \begin{bmatrix} -C^\dagger D B^\dagger & C^\dagger \\ B^\dagger & O \end{bmatrix}.$$

Lemma 4.1.5 *Suppose* $A_i \in \mathbb{C}^{m_i \times n_i}$, $i = 0, 1, 2, \ldots, n+1$, $B_i \in \mathbb{C}^{m_i \times n_{i-1}}$, $i = 1, 2, \ldots, n+1$, *satisfy*

$$\mathcal{R}(B_i) \subset \mathcal{R}(A_i), \; RS(B_i) \subset RS(A_{i-1}), i = 1, 2, \ldots, n+1, \qquad (4.1.5)$$

then the Moore-Penrose inverse of the $(n+2) \times (n+2)$ *block matrix*

$$J_{n+2} = \begin{bmatrix} & & & & A_0 \\ & & & A_1 & B_1 \\ & & \cdot^{\textstyle\cdot^{\textstyle\cdot}} & \cdot^{\textstyle\cdot^{\textstyle\cdot}} & \\ & A_n & B_n & & \\ A_{n+1} & B_{n+1} & & & \end{bmatrix} \qquad (4.1.6)$$

can be expressed as

$$J_{n+2}^\dagger = \begin{bmatrix} F(n+1, 0) & F(n+1, 1) & \cdots & F(n+1, n) & F(n+1, n+1) \\ F(n, 0) & F(n, 1) & \cdots & F(n, n) & \\ \vdots & & \cdot^{\textstyle\cdot^{\textstyle\cdot}} & & \\ \vdots & \cdot^{\textstyle\cdot^{\textstyle\cdot}} & & & \\ F(0, 0) & & & & \end{bmatrix}, \qquad (4.1.7)$$

where

$$F(i, i) = A_i^\dagger,$$
$$F(i, j) = (-1)^{i-j} A_i^\dagger B_i A_{i-1}^\dagger B_{i-1} \cdots A_{j+1}^\dagger B_{j+1} A_j^\dagger, \ 0 \le j \le i \le n+1. \tag{4.1.8}$$

Proof We use induction on n. For $n = 0$, from the conditions (4.1.5) and Lemma 4.1.4, the Moore-Penrose inverse of J_2 in (4.1.7) is

$$J_2^\dagger = \begin{bmatrix} O & A_0 \\ A_1 & B_1 \end{bmatrix}^\dagger = \begin{bmatrix} -A_1^\dagger B_1 A_0^\dagger & A_1^\dagger \\ A_0^\dagger & O \end{bmatrix} = \begin{bmatrix} F(1,0) & F(1,1) \\ F(0,0) & O \end{bmatrix},$$

which shows that the lemma holds when $n = 0$. Now we assume that it is also true for $n + 1$. In other words, under the conditions (4.1.5), the Moore-Penrose inverse of J_{n+1} in (4.1.6) is given by

$$J_{n+1}^\dagger = \begin{bmatrix} F(n,0) & F(n,1) & \cdots & F(n, n-1) & F(n, n) \\ F(n-1,0) & F(n-1,1) & \cdots & F(n-1, n-1) & \\ \vdots & & \ddots & & \\ \vdots & & \ddots & & \\ F(0,0) & & & & \end{bmatrix}. \tag{4.1.9}$$

Next we consider the Moore-Penrose inverse of J_{n+2} in (4.1.6). First, partition J_{n+2} in (4.1.6) into the form:

$$J_{n+2} = \begin{bmatrix} O & J_{n+1} \\ A_{n+1} & H \end{bmatrix},$$

where $H = [B_{n+1} \ O]$. Then it is easy to see from the conditions (4.1.5) that the three submatrices in the above J_{n+2} satisfy the inclusions

$$\mathcal{R}(H) \subset \mathcal{R}(A_{n+1}) \quad \text{and} \quad RS(H) \subset RS(J_{n+1}).$$

Hence by Lemma 4.1.4,

$$J_{n+2}^\dagger = \begin{bmatrix} -A_{n+1}^\dagger H J_{n+1}^\dagger & A_{n+1}^\dagger \\ J_{n+1}^\dagger & O \end{bmatrix}. \tag{4.1.10}$$

Using the induction hypothesis (4.1.9), the structure of H, $F(i, i)$ and $F(i, j)$ in $J_{n+1}^\dagger$, and (4.1.8), we obtain

$$-A_{n+1}^{\dagger} H J_{n+1}^{\dagger}$$
$$= [-A_{n+1}^{\dagger} B_{n+1} F(n,0) \quad -A_{n+1}^{\dagger} B_{n+1} F(n,1) \cdots -A_{n+1}^{\dagger} B_{n+1} F(n,n)]$$
$$= [F(n+1,0) \ F(n+1,1) \ \cdots \ F(n+1,n)] \tag{4.1.11}$$

and

$$A_{n+1}^{\dagger} = F(n+1, n+1). \tag{4.1.12}$$

Finally, substituting (4.1.9), (4.1.11), and (4.1.12) into (4.1.10) directly produces (4.1.7). $\qquad\square$

Next, we consider the product $A = A_0 A_1 A_2 \cdots A_n A_{n+1}$. Let

$$n_i = m_{i+1}, \ i = 0, 1, \ldots, n, \ \text{and} \ m_0 = m_1, \ n_{n+1} = m_{n+1},$$

then

$$A_0 \in \mathbb{C}^{m_1 \times m_1}, \ A_{n+1} \in \mathbb{C}^{m_{n+1} \times m_{n+1}}, \ A_i \in \mathbb{C}^{m_i \times m_{i+1}}, \ i = 1, 2, \ldots, n,$$

and

$$B_i \in \mathbb{C}^{m_i \times m_i}, \ i = 1, 2, \ldots, n+1.$$

Let I_{m_1} and $I_{m_{n+1}}$ be identity matrices of orders m_1 and m_{n+1} respectively and

$$P = [I_{m_{n+1}} \ O], \quad Q = [I_{m_1} \ O]^*, \tag{4.1.13}$$

then the $(1, 1)$-block in (4.1.7) has the form

$$F(n+1, 0) = P J_{n+2}^{\dagger} Q = (-1)^{n+1} A_{n+1}^{\dagger} B_{n+1} A_n^{\dagger} B_n \cdots A_1^{\dagger} B_1 A_0^{\dagger}. \tag{4.1.14}$$

If we replace A_0 and A_{n+1} in (4.1.6) with I_{m_1} and $I_{m_{n+1}}$ respectively, and A_i with the nonsingular matrices $V_i A_i U_i$, where

$$V_i^* \in \mathbb{C}_{t_i}^{m_i \times t_i}, \quad U_i \in \mathbb{C}_{t_i}^{m_{i+1} \times t_i}$$

and

$$\mathcal{R}(U_i) = T_i \quad \text{and} \quad \mathcal{N}(V_i) = S_i$$

satisfy the conditions in Lemma 4.1.1 and

$$\text{rank}(V_i A_i U_i) = \text{rank}(V_i) = \text{rank}(U_i) = t_i, \quad i = 1, 2, \ldots, n,$$

then, when

$$B_1 = V_1, \quad B_{n+1} = U_n, \quad \text{and} \quad B_i = V_i U_{i-1}, \quad i = 2, 3, \ldots, n,$$

we have

$$\mathcal{R}(B_1) = \mathcal{R}(V_1) = \mathcal{R}(V_1 A_1 U_1),$$

$$\mathcal{R}(B_i) = \mathcal{R}(V_i U_{i-1}) \subset \mathcal{R}(V_i) = \mathcal{R}(V_i A_i U_i), \quad i = 2, 3, \ldots, n,$$

$$\mathcal{R}(B_{n+1}) = \mathcal{R}(U_n) \subset \mathcal{R}(I_{m_{n+1}}),$$

$$RS(B_1) = RS(V_1) \subset RS(I_{m_1}), \tag{4.1.15}$$

$$RS(B_i) = RS(V_i U_{i-1}) \subset RS(U_{i-1}) = RS(V_{i-1} A_{i-1} U_{i-1}),$$

$$i = 2, 3, \ldots, n,$$

$$RS(B_{n+1}) = RS(U_n) = RS(V_n A_n U_n).$$

Thus it is obvious from the above equations that the conditions (4.1.5) in Lemma 4.1.5 are satisfied. Hence the Moore-Penrose inverse of the $(n + 2) \times (n + 2)$ block matrix

$$M = \begin{bmatrix} & & & & I_{m_1} \\ & & & V_1 A_1 U_1 & V_1 \\ & & V_2 A_2 U_2 & V_2 U_1 \\ & \cdot\cdot & \cdot\cdot & \\ & V_n A_n U_n & V_n U_{n-1} & \\ I_{m_{n+1}} & U_n & & \end{bmatrix}$$

$$= \begin{bmatrix} O & E_1 \\ E_2 & N \end{bmatrix}, \tag{4.1.16}$$

can be expressed as the form (4.1.7), where

$$E_1 = [O \ I_{m_1}] \quad \text{and} \quad E_2 = [O \ I_{m_{n+1}}]^T,$$

According to (4.1.7), (4.1.8), and (4.1.14), the $(1, 1)$-block $F(n + 1, 0)$ in (4.1.7) becomes

$$F(n + 1, 0)$$

$$= PM^\dagger Q$$

$$= (-1)^{n+1} I_{m_{n+1}}^\dagger U_n (V_n A_n U_n)^\dagger V_n U_{n-1} (V_{n-1} A_{n-1} U_{n-1})^\dagger V_{n-1} U_{n-2} \cdots$$

$$V_2 U_1 (V_1 A_1 U_1)^\dagger V_1 I_{m_1}^\dagger$$

$$= (-1)^{n+1} U_n (V_n A_n U_n)^{-1} V_n U_{n-1} (V_{n-1} A_{n-1} U_{n-1})^{-1} V_{n-1} U_{n-2} \cdots$$

$$V_2 U_1 (V_1 A_1 U_1)^{-1} V_1$$

$$= (-1)^{n+1} (A_n)_{T_n, S_n}^{(2)} (A_{n-1})_{T_{n-1}, S_{n-1}}^{(2)} \cdots (A_1)_{T_1, S_1}^{(2)}. \tag{4.1.17}$$

From the structure of M in (4.1.16), we see that it has the following properties to be used in the next section.

Lemma 4.1.6 *Let* $M, U_i, V_i, P,$ *and* Q *be given in* (4.1.16) *and* (4.1.13), *and let* $A_i \in \mathbb{C}^{m_i \times m_{i+1}}, i = 1, 2, \ldots, n, A_0 = I_{m_1}$ *and* $A_{n+1} = I_{m_{n+1}}$. *Then*

$$\text{rank}(M) = m_1 + m_{n+1} + \sum_{i=1}^{n} \text{rank}(V_i A_i U_i)$$

$$= m_1 + m_{n+1} + \sum_{i=1}^{n} \text{rank}(V_i) \tag{4.1.18}$$

$$= m_1 + m_{n+1} + \sum_{i=1}^{n} \text{rank}(U_i),$$

and

$$\mathcal{R}(Q) \subset \mathcal{R}(M), \quad RS(P) \subset RS(M). \tag{4.1.19}$$

4.2 Reverse Order Law

In this section, we first give a sufficient and necessary condition for the reverse order law (4.1.1). Then we discuss special case for the Moore-Penrose inverse, the Drazin inverse, and the group inverse.

Theorem 4.2.1 ([14]) *Suppose that* $A_i \in \mathbb{C}^{m_i \times m_{i+1}},$ *for* $i = 1, 2, \ldots, n,$ $A = A_1 A_2 \cdots A_n \in \mathbb{C}^{m_1 \times m_{n+1}}, X = (A_n)_{T_n, S_n}^{(2)} (A_{n-1})_{T_{n-1}, S_{n-1}}^{(2)} \cdots (A_1)_{T_1, S_1}^{(2)}$ *and*

$$\mathcal{R}(U) = T, \ \mathcal{N}(V) = S,$$
$$\mathcal{R}(U_i) = T_i, \ \mathcal{N}(V_i) = S_i, i = 1, 2, \ldots, n, \tag{4.2.1}$$

and

$$\text{rank}(VAU) = \text{rank}(U) = \text{rank}(V) = t,$$
$$\text{rank}(V_i A_i U_i) = \text{rank}(U_i) = \text{rank}(V_i) = t_i, i = 1, 2, \ldots, n.$$

Then $X = A_{T,S}^{(2)}$, *that is, the reverse order law* (4.1.1) *holds, if and only if* A, U, V *and* $A_i, U_i, V_i, i = 1, 2, \ldots, n,$ *satisfy the following rank conditions:*

$$\text{rank} \begin{bmatrix} (-1)^n VAU & VE_1 \\ E_2 U & N \end{bmatrix}$$

$$= \text{rank}(U) + \sum_{i=1}^{n} \text{rank}(U_i)$$

$$= \text{rank}(V) + \sum_{i=1}^{n} \text{rank}(V_i) \tag{4.2.2}$$

$$= \text{rank}(VAU) + \sum_{i=1}^{n} \text{rank}(V_i A_i U_i),$$

where E_1, E_2 and N are defined in (4.1.16).

Proof From (4.1.17), $X = (-1)^{n+1} P M^\dagger Q$, where

$$M = \begin{bmatrix} O & E_1 \\ E_2 & N \end{bmatrix}.$$

It is obvious that $X = A_{T,S}^{(2)}$ holds if and only if

$$0 = \text{rank}(A_{T,S}^{(2)} - X) = \text{rank}((-1)^n A_{T,S}^{(2)} + P M^\dagger Q). \tag{4.2.3}$$

Now using the matrices in (4.2.1) and (4.1.13), we construct the following 3×3 block matrix:

$$G = \begin{bmatrix} M & O & Q \\ O & (-1)^n VAU & V \\ P & U & O \end{bmatrix}.$$

It follows from (4.1.19) that

$$\mathcal{R}\left(\begin{bmatrix} Q \\ V \end{bmatrix} \right) \subset \mathcal{R}\left(\begin{bmatrix} M & O \\ O & (-1)^n VAU \end{bmatrix} \right),$$

$$RS([P \ U]) \subset RS\left(\begin{bmatrix} M & O \\ O & (-1)^n VAU \end{bmatrix} \right).$$

Hence, by the rank formula (4.1.4), we have

$$\text{rank}(G) = \text{rank} \begin{bmatrix} M & O \\ O & (-1)^n VAU \end{bmatrix}$$

$$+ \text{rank} \left([P \ U] \begin{bmatrix} M & O \\ O & (-1)^n VAU \end{bmatrix}^\dagger \begin{bmatrix} Q \\ V \end{bmatrix} \right)$$

$$= \text{rank}(M) + \text{rank}(VAU) + \text{rank}(PM^\dagger Q + (-1)^n U(VAU)^{-1}V)$$

$$= \text{rank}(M) + \text{rank}(VAU) + \text{rank}(PM^\dagger Q + (-1)^n A_{T,S}^{(2)}). \qquad (4.2.4)$$

Substituting the complete expression (4.1.16) of M and then calculating the rank of G, we get

$$\text{rank}(G) = \text{rank} \left(\begin{bmatrix} O & E_1 & O & I_{m_1} \\ E_2 & N & O & O \\ O & O & (-1)^n VAU & V \\ I_{m_{n+1}} & O & U & O \end{bmatrix} \right)$$

$$= \text{rank} \left(\begin{bmatrix} O & O & O & I_{m_1} \\ O & N & -E_2 U & O \\ O & -VE_1 & (-1)^n VAU & O \\ I_{m_{n+1}} & O & O & O \end{bmatrix} \right)$$

$$= m_1 + m_{n+1} + \text{rank} \left(\begin{bmatrix} (-1)^n VAU & VE_1 \\ E_2 U & N \end{bmatrix} \right).$$

Combining (4.1.18), (4.2.3), (4.2.4), and the above equation yields (4.2.2). □

Corollary 4.2.1 ([4])   Suppose $A_i \in \mathbb{C}^{m_i \times m_{i+1}}$,   for $i = 1, 2, \ldots, n$, $A = A_1 A_2 \cdots A_n \in \mathbb{C}^{m_1 \times m_{n+1}}$, $X = A_n^\dagger \cdots A_2^\dagger A_1^\dagger$, then $X = A^\dagger$, that is, the reverse order law

$$(A_1 A_2 \cdots A_n)^\dagger = A_n^\dagger \cdots A_2^\dagger A_1^\dagger$$

holds, if and only if

$$\text{rank} \begin{bmatrix} (-1)^n A^* AA^* & A^* E_1 \\ E_2 A^* & N \end{bmatrix}$$

$$= \text{rank}(A^*) + \sum_{i=1}^{n} \text{rank}(A_i^*)$$

$$= \text{rank}(A) + \sum_{i=1}^{n} \text{rank}(A_i),$$

where E_1 and E_2 are defined in (4.1.16) and

$$N = \begin{bmatrix} & & & & A_1^* A_1 A_1^* \ A_1^* \\ & & & \iddots & A_2^* A_1^* \\ & & A_{n-1}^* A_{n-1} A_{n-1}^* & \iddots & \\ A_n^* A_n A_n^* & A_n^* A_{n-1}^* & & & \\ A_n^* & & & & \end{bmatrix}.$$

Corollary 4.2.2 *Suppose that $A_i \in \mathbb{C}^{m_i \times m_{i+1}}$, for $i = 1, 2, \ldots, n$, M_i are $m_i \times m_i$ Hermitian positive definite matrices, $i = 1, 2, \ldots, n+1$, $A = A_1 A_2 \cdots A_n \in \mathbb{C}^{m_1 \times m_{n+1}}$, $X = (A_n)_{M_n, M_{n+1}}^\dagger \cdots (A_2)_{M_2, M_3}^\dagger (A_1)_{M_1, M_2}^\dagger$, then $X = A_{M_1, M_{n+1}}^\dagger$, that is, the reverse order law*

$$(A_1 A_2 \cdots A_n)_{M_1, M_{n+1}}^\dagger = (A_n)_{M_n, M_{n+1}}^\dagger \cdots (A_2)_{M_2, M_3}^\dagger (A_1)_{M_1, M_2}^\dagger$$

holds, if and only if

$$\mathrm{rank} \begin{bmatrix} (-1)^n A^\# A A^\# & A^\# E_1 \\ E_2 A^\# & N \end{bmatrix}$$

$$= \mathrm{rank}(A^\#) + \sum_{i=1}^n \mathrm{rank}(A_i^\#)$$

$$= \mathrm{rank}(A) + \sum_{i=1}^n \mathrm{rank}(A_i),$$

where E_1 and E_2 are defined in (4.1.16),

$$A^\# = M_{n+1}^{-1} A^* M_1, \quad A_i^\# = M_{i+1}^{-1} A_i^* M_i, \ i = 1, 2, \ldots, n,$$

and

$$N = \begin{bmatrix} & & & & A_1^\# A_1 A_1^\# \ A_1^\# \\ & & & \iddots & A_2^\# A_1^\# \\ & & A_{n-1}^\# A_{n-1} A_{n-1}^\# & \iddots & \\ A_n^\# A_n A_n^\# & A_n^\# A_{n-1}^\# & & & \\ A_n^\# & & & & \end{bmatrix}.$$

Corollary 4.2.3 ([8]) *Let $A_i \in \mathbb{C}^{m \times m}$, for $i = 1, 2, \ldots, n$, $A = A_1 A_2 \cdots A_n \in \mathbb{C}^{m \times m}$, $X = (A_n)_d \cdots (A_2)_d (A_1)_d$, and $k = \max_i \{\mathrm{Ind}(A), \ \mathrm{Ind}(A_i)\}$, then $X = A_d$, that is, the reverse order law*

$$(A_1 A_2 \cdots A_n)_d = (A_n)_d \cdots (A_2)_d (A_1)_d$$

holds, if and only if

$$\text{rank}\begin{bmatrix} (-1)^n A^k A A^k & A^k E_1 \\ E_2 A^k & N \end{bmatrix} = \text{rank}(A^k) + \sum_{i=1}^{n} \text{rank}(A_i^k),$$

where E_1 and E_2 are defined in (4.1.16) and

$$N = \begin{bmatrix} & & & A_1^{2k+1} & A_1^k \\ & & & \cdot^{\cdot^{\cdot}} & A_2^k A_1^k \\ & & A_{n-1}^{2k+1} & \cdot^{\cdot^{\cdot}} & \\ A_n^{2k+1} & A_n^k A_{n-1}^k & & & \\ A_n^k & & & & \end{bmatrix}.$$

Corollary 4.2.4 *Suppose $A_i \in \mathbb{C}^{m\times m}$, $i = 1, 2, \ldots, n$, $A = A_1 A_2 \cdots A_n \in \mathbb{C}^{m\times m}$, $\text{Ind}(A_i) = \text{Ind}(A) = 1$, $X = (A_n)_g \cdots (A_2)_g (A_1)_g$, then $X = A_g$, that is, the reverse order law*

$$(A_1 A_2 \cdots A_n)_g = (A_n)_g \cdots (A_2)_g (A_1)_g$$

holds, if and only if

$$\text{rank}\begin{bmatrix} (-1)^n A^3 & A E_1 \\ E_2 A & N \end{bmatrix} = \text{rank}(A) + \sum_{i=1}^{n} \text{rank}(A_i),$$

where E_1 and E_2 are defined in (4.1.16) and

$$N = \begin{bmatrix} & & & A_1^3 & A_1 \\ & & & \cdot^{\cdot^{\cdot}} & A_2 A_1 \\ & & A_{n-1}^3 & \cdot^{\cdot^{\cdot}} & \\ A_n^3 & A_n A_{n-1} & & & \\ A_n & & & & \end{bmatrix}.$$

From the above results, the conclusion of the reverse order law given in [3, 6, 7] can be obtained easily. In [15], Sun and Wei give a triple reverse order law for weighted generalized inverses.

4.3 Forward Order Law

In this section, by using the concept of ranks, the necessary and sufficient condition is given for the forward order law:

$$(A_1 A_2 \cdots A_n)_{T,S}^{(2)} = (A_1)_{T_1,S_1}^{(2)} (A_2)_{T_2,S_2}^{(2)} \cdots (A_n)_{T_n,S_n}^{(2)}, \tag{4.3.1}$$

from which we can obtain the necessary and sufficient conditions for the forward order laws of the Moore-Penrose inverse, the weighted Moore-Penrose inverse, the Drazin inverse, and the group inverse.

Apparently, the matrices $A_1, A_2, \ldots, A_n$ must be square and of the same order, say m.

Lemma 4.3.1 *Suppose $A_i \in \mathbb{C}^{m \times m}, i = 0, 1, 2, \ldots, n + 1, B_i \in \mathbb{C}^{m \times m}, i = 0, 1, 2, \ldots, n$, satisfy*

$$\mathcal{R}(B_i) \subset \mathcal{R}(A_i), \quad RS(B_i) \subset RS(A_{i+1}), \ i = 0, 1, 2, \ldots, n, \tag{4.3.2}$$

then the Moore-Penrose inverse of the $(n + 2) \times (n + 2)$ block matrix

$$J_{n+2} = \begin{bmatrix} & & & & A_{n+1} \\ & & & A_n & B_n \\ & & \cdot{\cdot}^{\cdot} & \cdot{\cdot}^{\cdot} & \\ & A_1 & B_1 & & \\ A_0 & B_0 & & & \end{bmatrix} \tag{4.3.3}$$

can be expressed as

$$J_{n+2}^{\dagger} = \begin{bmatrix} F(0, n+1) & F(0, n) & \cdots & F(0, 1) & F(0, 0) \\ F(1, n+1) & F(1, n) & \cdots & F(1, 1) & \\ \vdots & & \cdot{\cdot}^{\cdot} & & \\ F(n, n+1) & F(n, n) & & & \\ F(n+1, n+1) & & & & \end{bmatrix}, \tag{4.3.4}$$

where

$$F(i, i) = A_i^{\dagger}, \quad i = 0, 1, \ldots, n+1,$$
$$F(i, j) = (-1)^{j-i} A_i^{\dagger} B_i A_{i+1}^{\dagger} B_{i+1} \cdots A_{j-1}^{\dagger} B_{j-1} A_j^{\dagger},$$
$$0 \leq i \leq j \leq n+1. \tag{4.3.5}$$

Proof We use induction on n. For $n = 0$, according (4.3.2) and Lemma 4.1.4, the Moore-Penrose inverse of J_2 in (4.3.3) is

$$J_2^\dagger = \begin{bmatrix} O & A_1 \\ A_0 & B_0 \end{bmatrix}^\dagger = \begin{bmatrix} -A_0^\dagger B_0 A_1^\dagger & A_0^\dagger \\ A_1^\dagger & O \end{bmatrix} = \begin{bmatrix} F(0,1) & F(0,0) \\ F(1,1) & O \end{bmatrix},$$

which shows that the lemma is true for $n = 0$. Now suppose that it is true for $n + 1$, that is, under the condition (4.3.2), the Moore-Penrose inverse of J_{n+1} in (4.3.3) is

$$J_{n+1}^\dagger = \begin{bmatrix} F(0,n) & F(0, n-1) & \cdots & F(0,1) & F(0,0) \\ F(1,n) & F(1, n-1) & \cdots & F(1,1) & \\ \vdots & & \cdots & & \\ \vdots & & \cdots & & \\ F(n,n) & & & & \end{bmatrix}. \tag{4.3.6}$$

Next, we consider the Moore-Penrose inverse of J_{n+2} in (4.3.3). First, we partition J_{n+2} in (4.3.3) into the form

$$J_{n+2} = \begin{bmatrix} O & A_{n+1} \\ J_{n+1} & H \end{bmatrix},$$

where $H = [B_n^* \ O]^*$. Then it is easy to see from the conditions (4.3.2) that the three submatrices in the above J_{n+2} satisfy

$$\mathcal{R}(H) \subset \mathcal{R}(J_{n+1}) \quad \text{and} \quad RS(H) \subset RS(A_{n+1}).$$

Hence by Lemma 4.1.4,

$$J_{n+2}^\dagger = \begin{bmatrix} -J_{n+1}^\dagger H A_{n+1}^\dagger & J_{n+1}^\dagger \\ A_{n+1}^\dagger & O \end{bmatrix}. \tag{4.3.7}$$

According to the induction hypothesis (4.3.6) for $J_{n+1}^\dagger$ and the structure of H, $F(i, i)$ and $F(i, j)$ in $J_{n+1}^\dagger$, and (4.3.5), we obtain

$$-J_{n+1}^\dagger H A_{n+1}^\dagger = \begin{bmatrix} -F(0,n) B_n A_{n+1}^\dagger \\ -F(1,n) B_n A_{n+1}^\dagger \\ \vdots \\ -F(n,n) B_n A_{n+1}^\dagger \end{bmatrix} = \begin{bmatrix} F(0, n+1) \\ F(1, n+1) \\ \vdots \\ F(n, n+1) \end{bmatrix} \tag{4.3.8}$$

and

$$A_{n+1}^\dagger = F(n+1, n+1).$$

Finally, substituting (4.3.6), (4.3.8) and the above equation into (4.3.7) directly produces (4.3.4).   □

Let

$$P = [I_m \ O] \quad \text{and} \quad Q = [I_m \ O]^T, \tag{4.3.9}$$

then the $(1, 1)$-block in (4.3.4) has the form

$$F(0, n+1) = P J_{n+2}^\dagger Q = (-1)^{n+1} A_0^\dagger B_0 A_1^\dagger B_1 \cdots A_n^\dagger B_n A_{n+1}^\dagger. \tag{4.3.10}$$

If we set $A_0 = I_m$, $A_{n+1} = I_m$, and A_i to the nonsingular matrix $V_i A_i U_i$, where

$$V_i^* \in \mathbb{C}_{t_i}^{m \times t_i} \quad \text{and} \quad U_i \in \mathbb{C}_{t_i}^{m \times t_i}$$

satisfying the conditions

$$\mathcal{R}(U_i) = T_i \quad \text{and} \quad \mathcal{N}(V_i) = S_i$$

in Lemma 4.1.1 and

$$\text{rank}(V_i A_i U_i) = \text{rank}(V_i) = \text{rank}(U_i) = t_i, \quad i = 1, 2, \ldots, n,$$

then, when

$$B_0 = U_1, \quad B_i = V_i U_{i+1}, \ i = 2, \ldots, n-1, \quad \text{and} \quad B_n = V_n,$$

we have

$$\mathcal{R}(B_0) = \mathcal{R}(U_1) \subset \mathcal{R}(I_m) = \mathcal{R}(A_0),$$
$$\mathcal{R}(B_i) = \mathcal{R}(V_i U_{i+1}) \subset \mathcal{R}(V_i) = \mathcal{R}(V_i A_i U_i),$$
$$i = 1, 2, \ldots, n-1,$$
$$\mathcal{R}(B_n) = \mathcal{R}(V_n) \subset \mathcal{R}(V_n A_n U_n),$$
$$RS(B_0) = RS(U_1) = RS(V_1 A_1 U_1),$$
$$RS(B_i) = RS(V_i U_{i+1}) \subset RS(U_{i+1}) = RS(V_{i+1} A_{i+1} U_{i+1}),$$
$$i = 2, 3, \ldots, n-1,$$
$$RS(B_n) = RS(V_n) \subset RS(I_m) = RS(A_{n+1}).$$

It then follows from the above equations that the conditions (4.3.2) in Lemma 4.3.1 are satisfied. Hence the Moore-Penrose inverse of the $(n+2) \times (n+2)$ block matrix

$$
M = \begin{bmatrix}
& & & & & & I_m \\
& & & & & V_n A_n U_n & V_n \\
& & & & V_{n-1} A_{n-1} U_{n-1} & V_{n-1} U_n & \\
& & & \iddots & \iddots & & \\
& & V_2 A_2 U_2 & V_2 U_3 & & & \\
& V_1 A_1 U_1 & V_1 U_2 & & & & \\
I_m & U_1 & & & & &
\end{bmatrix}
$$

$$
= \begin{bmatrix} O & E_1 \\ E_2 & N \end{bmatrix}
\tag{4.3.11}
$$

can be expressed in the form (4.3.4), where

$$
E_1 = [O \ \ I_m] \quad \text{and} \quad E_2 = [O \ \ I_m]^T.
$$

According to (4.1.2), (4.1.3), and (4.3.10), the $(1,1)$-block $F(0, n+1)$ in (4.3.4) becomes

$$
F(0, n+1)
$$
$$
= P M^\dagger Q
$$
$$
= (-1)^{n+1} I_m^\dagger U_1 (V_1 A_1 U_1)^\dagger V_1 U_2 (V_2 A_2 U_2)^\dagger V_2 U_3 \cdots
$$
$$
V_{n-2} U_{n-1} (V_{n-1} A_{n-1} U_{n-1})^\dagger V_{n-1} U_n (V_n A_n U_n)^\dagger V_n I_m^\dagger
$$
$$
= (-1)^{n+1} I_m U_1 (V_1 A_1 U_1)^{-1} V_1 U_2 (V_2 A_2 U_2)^{-1} V_2 U_3 \cdots \tag{4.3.12}
$$
$$
V_{n-2} U_{n-1} (V_{n-1} A_{n-1} U_{n-1})^{-1} V_{n-1} U_n (V_n A_n U_n)^{-1} V_n I_m
$$
$$
= (-1)^{n+1} (A_1)^{(2)}_{T_1, S_1} (A_2)^{(2)}_{T_2, S_2} \cdots (A_{n-1})^{(2)}_{T_{n-1}, S_{n-1}} (A_n)^{(2)}_{T_n, S_n}.
$$

From the structure of M in (4.3.11), we see that it has the following simple properties to be used in the proof of the following Theorem 4.3.1.

Lemma 4.3.2 *Let M, P, and Q be given in (4.3.11) and (4.3.9), and $A_0 = I_m$, $A_{n+1} = I_m$, $A_i \in \mathbb{C}^{m \times m}$, $i = 1, 2, \ldots, n$, and $A = A_0 A_1 \cdots A_n \in \mathbb{C}^{m \times m}$, then*

$$
\text{rank}(M) = 2m + \sum_{i=1}^{n} \text{rank}(V_i A_i U_i)
$$

$$
= 2m + \sum_{i=1}^{n} \text{rank}(V_i)
$$

$$
= 2m + \sum_{i=1}^{n} \text{rank}(U_i) \tag{4.3.13}
$$

and

$$\mathcal{R}(Q) \subset \mathcal{R}(M) \quad and \quad RS(P) \subset RS(M). \tag{4.3.14}$$

Theorem 4.3.1 *Suppose* $A_i \in \mathbb{C}^{m \times m}$, $i = 1, 2, \ldots, n$, $A = A_1 A_2 \cdots A_n \in \mathbb{C}^{m \times m}$, $X = (A_1)_{T_1, S_1}^{(2)} (A_2)_{T_2, S_2}^{(2)} \cdots (A_n)_{T_n, S_n}^{(2)}$,

$$\begin{aligned}
\mathcal{R}(U) = T, \ \mathcal{N}(V) = S, \\
\text{rank}(VAU) = \text{rank}(V) = \text{rank}(U) = t,
\end{aligned} \tag{4.3.15}$$

and

$$\begin{aligned}
\mathcal{R}(U_i) = T_i, \ \mathcal{N}(V_i) = S_i, \\
\text{rank}(V_i A_i U_i) = \text{rank}(V_i) = \text{rank}(U_i) = t_i,
\end{aligned} \quad i = 1, 2, \ldots, n,$$

then $X = A_{T,S}^{(2)}$, *that is, the forward order law* (4.3.1) *holds, if and only if* A, U, V *and* A_i, U_i, V_i *satisfy the following rank conditions:*

$$\text{rank} \begin{bmatrix} (-1)^n VAU & VE_1 \\ E_2 U & N \end{bmatrix}$$

$$= \text{rank}(U) + \sum_{i=1}^{n} \text{rank}(U_i)$$

$$= \text{rank}(V) + \sum_{i=1}^{n} \text{rank}(V_i) \tag{4.3.16}$$

$$= \text{rank}(VAU) + \sum_{i=1}^{n} \text{rank}(V_i A_i U_i),$$

where E_1, E_2, *and* N *are defined in* (4.3.11).

Proof From (4.3.12), $X = (-1)^{n+1} P M^\dagger Q$, where

$$M = \begin{bmatrix} O & E_1 \\ E_2 & N \end{bmatrix}.$$

It is obvious that $X = A_{T,S}^{(2)}$ if and only if

$$0 = \text{rank}(A_{T,S}^{(2)} - X) = \text{rank}((-1)^n A_{T,S}^{(2)} + P M^\dagger Q). \tag{4.3.17}$$

Now using the matrices in (4.3.9) and (4.3.15), we construct the following 3×3 block matrix

$$G = \begin{bmatrix} M & O & Q \\ O & (-1)^n VAU & V \\ P & U & O \end{bmatrix}.$$

It follows from (4.3.16), we have

$$\mathcal{R}\left(\begin{bmatrix} Q \\ V \end{bmatrix}\right) \subset \mathcal{R}\left(\begin{bmatrix} M & O \\ O & (-1)^n VAU \end{bmatrix}\right)$$

and

$$RS([P \ U]) \subset RS\left(\begin{bmatrix} M & O \\ O & (-1)^n VAU \end{bmatrix}\right).$$

Hence, by the rank formula (4.1.4), we have

$$
\begin{aligned}
&\text{rank}(G) \\
&= \text{rank}\left(\begin{bmatrix} M & O \\ O & (-1)^n VAU \end{bmatrix}\right) \\
&\quad + \text{rank}\left([P \ U]\begin{bmatrix} M & O \\ O & (-1)^n VAU \end{bmatrix}^\dagger \begin{bmatrix} Q \\ V \end{bmatrix}\right) \\
&= \text{rank}(M) + \text{rank}(VAU) + \text{rank}(PM^\dagger Q + (-1)^n U(VAU)^{-1}V) \\
&= \text{rank}(M) + \text{rank}(VAU) + \text{rank}(PM^\dagger Q + (-1)^n A_{T,S}^{(2)}). \quad (4.3.18)
\end{aligned}
$$

Substituting the complete expression of M in (4.3.11) and then calculating the rank of G will produce

$$
\begin{aligned}
\text{rank}(G) &= \text{rank}\left(\begin{bmatrix} O & E_1 & O & I_m \\ E_2 & N & O & O \\ O & O & (-1)^n VAU & V \\ I_m & O & U & O \end{bmatrix}\right) \\
&= \text{rank}\left(\begin{bmatrix} O & O & O & I_m \\ O & N & -E_2 U & O \\ O & -VE_1 & (-1)^n VAU & O \\ I_m & O & O & O \end{bmatrix}\right) \\
&= 2m + \text{rank}\left(\begin{bmatrix} (-1)^n VAU & VE_1 \\ E_2 U & N \end{bmatrix}\right).
\end{aligned}
$$

Finally, combining (4.3.13), (4.3.17), (4.3.18), and the above equation yields (4.3.16). $\square$

Applying the above theorem, from Lemma 4.1.2, we have the following forward order laws for various generalized inverses.

Corollary 4.3.1 *Suppose* $A_i \in \mathbb{C}^{m \times m}$, $i = 1, 2, \ldots, n$, $A = A_1 A_2 \cdots A_n \in \mathbb{C}^{m \times m}$, $X = A_1^\dagger A_2^\dagger \cdots A_n^\dagger$, *then* $X = A^\dagger$, *that is, the forward order law*

$$(A_1 A_2 \cdots A_n)^\dagger = A_1^\dagger A_2^\dagger \cdots A_n^\dagger$$

holds, if and only if

$$\mathrm{rank}\left(\begin{bmatrix} (-1)^n A^* A A^* & A^* E_1 \\ E_2 A^* & N \end{bmatrix}\right)$$

$$= \mathrm{rank}(A^*) + \sum_{i=1}^{n} \mathrm{rank}(A_i^*)$$

$$= \mathrm{rank}(A) + \sum_{i=1}^{n} \mathrm{rank}(A_i),$$

where E_1 *and* E_2 *are defined in (4.3.11) and*

$$N = \begin{bmatrix} & & & & A_n^* A_n A_n^* & A_n^* \\ & & & A_{n-1}^* A_{n-1} A_{n-1}^* & A_{n-1}^* A_n^* \\ & & \iddots & \iddots & \\ & A_2^* A_2 A_2^* & A_2^* A_3^* & & \\ A_1^* A_1 A_1^* & A_1^* A_2^* & & & \\ A_1^* & & & & \end{bmatrix}.$$

Corollary 4.3.2 *Suppose* $A_i \in \mathbb{C}^{m \times m}$, $i = 1, 2, \ldots, n$, $A = A_1 A_2 \cdots A_n \in \mathbb{C}^{m \times m}$, M_i *are* $m \times m$ *Hermitian positive definite matrices*, $i = 1, \ldots, n+1$, *and*

$$X = (A_1)_{M_1, M_2}^\dagger (A_2)_{M_2, M_3}^\dagger \cdots (A_n)_{M_n, M_{n+1}}^\dagger,$$

then $X = A_{M_1, M_{n+1}}^\dagger$, *that is, the forward order law*

$$(A_1 A_2 \cdots A_n)_{M_1, M_{n+1}}^\dagger = (A_1)_{M_1, M_2}^\dagger (A_2)_{M_2, M_3}^\dagger \cdots (A_n)_{M_n, M_{n+1}}^\dagger$$

holds, if and only if

$$\text{rank} \begin{bmatrix} (-1)^n A^\# A A^\# & A^\# E_1 \\ E_2 A^\# & N \end{bmatrix}$$

$$= \text{rank}(A^\#) + \sum_{i=1}^{n} \text{rank}(A_i^\#)$$

$$= \text{rank}(A) + \sum_{i=1}^{n} \text{rank}(A_i),$$

where

$$A^\# = M_{n+1}^{-1} A^* M_1, \quad A_i^\# = M_{i+1}^{-1} A_i^* M_i, \ i = 1, 2, \ldots, n,$$

E_1 *and* E_2 *are defined in (4.3.11), and*

$$N = \begin{bmatrix} & & & & A_n^\# A_n A_n^\# & A_n^\# \\ & & & A_{n-}^\# A_{n-1} A_{n-1}^\# & A_{n-1}^\# A_n^\# \\ & & \iddots & \iddots & \\ & A_2^\# A_2 A_2^\# & A_2^\# A_3^\# & & \\ A_1^\# A_1 A_1^\# & A_1^\# A_2^\# & & & \\ A_1^\# & & & & \end{bmatrix}.$$

Corollary 4.3.3 *Suppose* $A_i \in \mathbb{C}^{m \times m}$, $i = 1, 2, \ldots, n$, $A = A_1 A_2 \cdots A_n \in \mathbb{C}^{m \times m}$, $X = (A_1)_d (A_2)_d \cdots (A_n)_d$, *and* $k = \max_i \{\text{Ind}(A), \text{Ind}(A_i)\}$, *then* $X = A_d$, *that is, the forward order law*

$$(A_1 A_2 \cdots A_n)_d = (A_1)_d (A_2)_d \cdots (A_n)_d$$

holds, if and only if

$$\text{rank} \begin{bmatrix} (-1)^n A^{2k+1} & A^k E_1 \\ E_2 A^k & N \end{bmatrix} = \text{rank}(A^k) + \sum_{i=1}^{n} \text{rank}(A_i^k),$$

where E_1 *and* E_2 *are defined in (4.3.11) and*

$$N = \begin{bmatrix} & & & & A_n^{2k+1} & A_n^k \\ & & & \iddots & A_{n-1}^k A_n^k \\ & & A_2^{2k+1} & \iddots & \\ A_1^{2k+1} & A_1^k A_2^k & & \\ A_1^k & & & \end{bmatrix}.$$

Corollary 4.3.4 *Suppose* $A_i \in \mathbb{C}^{m \times m}$, $i = 1, 2, \ldots, n$, $A = A_1 A_2 \cdots A_n \in \mathbb{C}^{m \times m}$, $\mathrm{Ind}(A_i) = \mathrm{Ind}(A) = 1$, *and* $X = (A_1)_g (A_2)_g \cdots (A_n)_g$, *then* $X = A_g$, *that is, the forward order law*

$$(A_1 A_2 \cdots A_n)_g = (A_1)_g (A_2)_g \cdots (A_n)_g$$

holds, if and only if

$$r \begin{bmatrix} (-1)^n A^3 & A E_1 \\ E_2 A & N \end{bmatrix} = \mathrm{rank}(A) + \sum_{i=1}^{n} \mathrm{rank}(A_i),$$

where E_1 *and* E_2 *are defined in* (4.3.11) *and*

$$N = \begin{bmatrix} & & & A_n^3 & A_n \\ & & \cdot^{\cdot^{\cdot}} & A_{n-1} A_n & \\ & A_2^3 & \cdot^{\cdot^{\cdot}} & & \\ A_1^3 & A_1 A_2 & & & \\ A_1 & & & & \end{bmatrix}.$$

Corollary 4.3.5 *Let* $A_1, A_2 \in \mathbb{C}_m^{m \times m}$, $A = A_1 A_2 \in \mathbb{C}_m^{m \times m}$, *and* $X = A_1^{-1} A_2^{-1}$, *then* $X = A^{-1}$, *that is, the forward order law*

$$(A_1 A_2)^{-1} = A_1^{-1} A_2^{-1}$$

holds, if and only if

$$A_1 A_2 = A_2 A_1.$$

Proof Since

$$A^{-1} = A_{\mathcal{R}(I_m), \mathcal{N}(I_m)}^{(2)} \quad \text{and} \quad A_i^{-1} = (A_i)_{\mathcal{R}(I_m), (I_m)}^{(2)}, \quad i = 1, 2,$$

that is, $U = V = U_i = V_i = I_m$. It follows from Theorem 4.3.1 that

$$(A_1 A_2)^{-1} = A_1^{-1} A_2^{-1}$$

holds if and only if

$$\mathrm{rank} \left(\begin{bmatrix} (-1)^2 A_1 A_2 & E_1 \\ E_2 & N \end{bmatrix} \right)$$

$$= \mathrm{rank}(A_1 A_2) + \mathrm{rank}(A_1) + \mathrm{rank}(A_2)$$

$$= 3m, \tag{4.3.19}$$

where

$$E_1 = [O \ O \ I_m], \ E_2 = [O \ O \ I_m]^T, \text{ and } N = \begin{bmatrix} O & A_2 & I_m \\ A_1 & I_m & O \\ I_m & O & O \end{bmatrix}.$$

From (4.3.19), we have

$$3m = \text{rank} \left(\begin{bmatrix} A_1 A_2 & O & O & I_m \\ O & O & A_2 & I_m \\ O & A_1 & I_m & O \\ I_m & I_m & O & O \end{bmatrix} \right)$$

$$= \text{rank} \left(\begin{bmatrix} O & O & O & I_m \\ O & A_1 A_2 & A_2 & O \\ O & A_1 & I_m & O \\ I_m & O & O & O \end{bmatrix} \right)$$

$$= 2m + \text{rank} \left(\begin{bmatrix} A_1 A_2 & A_2 \\ A_1 & I_m \end{bmatrix} \right)$$

$$= 2m + \text{rank} \left(\begin{bmatrix} A_1 A_2 - A_2 A_1 & O \\ O & I_m \end{bmatrix} \right)$$

$$= 3m + \text{rank}(A_1 A_2 - A_2 A_1),$$

which implies that $\text{rank}(A_1 A_2 - A_2 A_1) = 0$, that is, $A_1 A_2 = A_2 A_1$. □

Remarks
The reverse order law for {1}-inverse or {1, 2}-inverse of products of two or more matrices can be found in [16–19]. A generalized triple reverse order law is presented in [20]. More results on the reverse order law are given in [9, 21–23]. The reverse order law for the Drazin inverse and the weighted M-P inverse is discussed in [5, 15, 24] and [25] respectively.

References

1. T.N.E. Greville, Note on the generalized inverse of a matrix product. SIAM Rev. **8**, 518–521 (1966)
2. R.E. Hartwig, The reverse order law revisited. Linear Algebra Appl. **76**, 241–246 (1986)
3. Y. Tian, The Moore-Penrose inverse of a triple matrix product. Math. Pract. Theory **1**, 64–70 (1992). in Chinese
4. Y. Tian, Reverse order laws for the generalized inverse of multiple matrix products. Linear Algebra Appl. **211**, 85–100 (1994)

5. W. Sun, Y. Wei, Inverse order rule for weighted generalized inverse. SIAM J. Matrix Anal. Appl. **19**, 772–775 (1998)
6. G. Wang, J. Gao, Reverse order laws for weighted Moore-Penrose inverse of a triple matrix product. J. Shanghai Normal Univ. **29**, 1–8 (2000). in Chinese
7. H. Tian, On the reverse order laws $(AB)^D = B^D A^D$. J. Math. Res. Expo. **19**, 355–358 (1999)
8. G. Wang, The reverse order law for Drazin inverses of multiple matrix products. Linear Algebra Appl. **348**, 265–272 (2002)
9. D.S. Djordjević, Unified approach to the reverse order rule for generalized inverses. Acta Sci. Math. (Szeged) **67**, 761–776 (2001)
10. A. Ben-Israel, T.N.E. Greville, *Generalized Inverses: Theory and Applications*, 2nd edn. (Springer, New York, 2003)
11. S.L. Campbell, C.D. Meyer Jr., *Generalized Inverses of Linear Transformations* (Pitman, London, 1979)
12. G. Marsaglia, G.P.H. Styan, Equalities and inequalities for ranks of matrices. Linear Multilinear Algebra **2**, 269–292 (1974)
13. R.E. Hartwig, Block generalized inverses. Arch. Ration. Mech. Anal. **61**, 197–251 (1976)
14. G. Wang, B. Zheng, The reverse order law for the generalized inverse $A_{T,S}^{(2)}$. Appl. Math. Comput. **157**, 295–305 (2004)
15. W. Sun, Y. Wei, Triple reverse-order law for weighted generalized inverses. Appl. Math. Comput. **25**, 221–229 (2002)
16. M. Wei, Equivalent conditions for generalized inverses of products. Linear Algebra Appl. **266**, 347–363 (1997)
17. A.R. Depierro, M. Wei, Reverse order laws for reflexive generalized inverses of products of matrices. Linear Algebra Appl. **277**, 299–311 (1996)
18. M. Wei, Reverse order laws for generalized inverses of multiple matrix products. Linear Algebra Appl. **293**, 273–288 (1999)
19. H.J. Werner, When is $B^- A^-$ a generalized inverse of AB? Linear Algebra Appl. **210**, 255–263 (1994)
20. Q. Xu, C. Song, G. Wang, Multiplicative perturbations of matrices and the generalized triple reverse order law for the Moore-Penrose inverse. Linear Algebra Appl. **530**, 366–383 (2017)
21. A.R. De Pierro, M. Wei, Reverse order law for reflexive generalized inverses of products of matrices. Linear Algebra Appl. **277**(1–3), 299–311 (1998)
22. Y. Takane, Y. Tian, H. Yanai, On reverse-order laws for least-squares g-inverses and minimum norm g-inverses of a matrix product. Aequ. Math. **73**(1–2), 56–70 (2007)
23. M. Wei, W. Guo, Reverse order laws for least squares g-inverses and minimum norm g-inverses of products of two matrices. Linear Algebra Appl. **342**, 117–132 (2002)
24. Y. Tian, Reverse order laws for the Drazin inverse of a triple matrix product. Publ. Math. Debr. **63**(3), 261–277 (2003)
25. Y. Tian, Reverse order laws for the weighted Moore-Penrose inverse of a triple matrix product with applications. Int. Math. J. **3**(1), 107–117 (2003)

Chapter 5
Computational Aspects

It follows from Chap. 1 that the six important kinds of generalized inverse: the M-P inverse $A^\dagger$, the weighted M-P inverse $A^\dagger_{MN}$, the group inverse A_g, the Drazin inverse A_d, the Bott-Duffin inverse $A^{(-1)}_{(L)}$ and the generalized Bott-Duffin inverse $A^{(\dagger)}_{(L)}$ are all the generalized inverse $A^{(2)}_{T,S}$, which is the {2}-inverse of A with the prescribed range T and null space S. Specifically, let $A \in \mathbb{C}^{m \times n}$, then

$$A^\dagger = A^{(2)}_{\mathcal{R}(A^*),\mathcal{N}(A^*)},$$

$$A^\dagger_{MN} = A^{(2)}_{\mathcal{R}(A^\#),\mathcal{N}(A^\#)},$$

where $A^\# = N^{-1}A^*M$, M and N are Hermitian positive definite matrices of orders m and n respectively.

Let $A \in \mathbb{C}^{n \times n}$, then

$$A_g = A^{(2)}_{\mathcal{R}(A),\mathcal{N}(A)},$$

$$A_d = A^{(2)}_{\mathcal{R}(A^k),\mathcal{N}(A^k)},$$

where $k = \mathrm{Ind}(A)$,

$$A^{(-1)}_{(L)} = A^{(2)}_{L,L^\perp},$$

where L is a subspace of $\mathbb{C}^n$ and satisfies $AL \oplus L^\perp = \mathbb{C}^n$,

$$A^{(\dagger)}_{(L)} = A^{(2)}_{S,S^\perp},$$

where L is a subspace of $\mathbb{C}^n$, P_L is the orthogonal projector on L, $S = \mathcal{R}(P_L A)$, and A is L-p.s.d. matrix.

In this chapter, the direct methods for computing the generalized inverse $A^{(2)}_{T,S}$ are discussed. A direct method means that the solution for a problem is computed in

© Springer Nature Singapore Pte Ltd. and Science Press 2018
G. Wang et al., *Generalized Inverses: Theory and Computations*,
Developments in Mathematics 53, https://doi.org/10.1007/978-981-13-0146-9_5

finite steps. In practice, we can only compute in finite precision. So a direct method can only compute an approximation of the exact solution.

If the matrices involved are small, then it is possible to carry out the calculations exactly; if, however, the matrices are large, it is impractical to perform exact computation, then the conditioning of matrices and accumulation of rounding errors must be considered. The purpose of this chapter is to offer some useful suggestions for those who wish to compute the generalized inverses numerically.

The direct methods discussed in this chapter are based on the full rank factorization, singular value decomposition and (M, N)-singular value decomposition, a variety of partitioned methods, the embedding methods, the finite methods, and splitting methods.

5.1   Methods Based on the Full Rank Factorization

Let $A \in \mathbb{C}_r^{m \times n}$, two subspaces $T \subset \mathbb{C}^n$ and $S \subset \mathbb{C}^m$, $\dim(T) = \dim(S^\perp) = t \le r$, and $G \in \mathbb{C}^{n \times m}$ such that

$$\mathcal{R}(G) = T \quad \text{and} \quad \mathcal{N}(G) = S.$$

Suppose that $G = UV$ is a full rank factorization of G, then

$$T = \mathcal{R}(G) = \mathcal{R}(U) \quad \text{and} \quad S = \mathcal{N}(G) = \mathcal{N}(V).$$

It follows from (1.3.15) that

$$A_{T,S}^{(2)} = U(VAU)^{-1}V. \tag{5.1.1}$$

Let $A \in \mathbb{C}^{m \times n}$ and $A^* = C^*B^*$ be a full rank factorization of A^*. Using (5.1.1), we have

$$A^\dagger = C^*(B^*AC^*)^{-1}B^* = C^*(CC^*)^{-1}(B^*B)^{-1}B^*. \tag{5.1.2}$$

If $A^\# = N^{-1}A^*M = (N^{-1}C^*)(B^*M)$ is a full rank factorization of $A^\#$, then

$$\begin{aligned} A_{MN}^\dagger &= N^{-1}C^*(B^*MAN^{-1}C^*)^{-1}B^*M \\ &= N^{-1}C^*(CN^{-1}C^*)^{-1}(B^*MB)^{-1}B^*M. \end{aligned} \tag{5.1.3}$$

Now, suppose that $A \in \mathbb{C}^{n \times n}$ and $\text{Ind}(A) = k$, we have a sequence of full rank factorizations:

$$A_1 = A = B_1 C_1,$$
$$A_2 = C_1 B_1 = B_2 C_2,$$
$$\vdots$$
$$A_k = C_{k-1} B_{k-1} = B_k C_k,$$

then

$$A^k = (B_1 C_1)^k$$
$$= B_1 (C_1 B_1)^{k-1} C_1$$
$$= B_1 (B_2 C_2)^{k-1} C_1$$
$$= B_1 B_2 (C_2 B_2)^{k-2} C_2 C_1$$
$$\cdots$$
$$= (B_1 B_2 \cdots B_{k-1} B_k)(C_k C_{k-1} \cdots C_2 C_1)$$

is a full rank factorization of A^k. It then follows from (5.1.1) that

$$A_d = B_1 B_2 \cdots B_k (C_k C_{k-1} \cdots C_2 C_1 A B_1 B_2 \cdots B_k)^{-1} C_k C_{k-1} \cdots C_2 C_1$$
$$= B_1 B_2 \cdots B_k (C_k C_{k-1} \cdots C_2 C_1 B_1 C_1 B_1 B_2 \cdots B_k)^{-1} C_k C_{k-1} \cdots C_2 C_1$$
$$= B_1 B_2 \cdots B_k (C_k B_k)^{-(k+1)} C_k C_{k-1} \cdots C_2 C_1. \tag{5.1.4}$$

Especially, when $A \in \mathbb{C}^{n \times n}$, $\mathrm{Ind}(A) = 1$ and $A = B_1 C_1$ is a full rank factorization of A, then

$$A_g = B_1 (C_1 B_1)^{-2} C_1. \tag{5.1.5}$$

Let $A \in \mathbb{C}^{n \times n}$, $U \in \mathbb{C}^{n \times t}$ and $V \in \mathbb{C}^{n \times t}$ be matrices whose columns form the bases for $\mathcal{R}(U) = L$ and $\mathcal{R}(V) = \mathcal{R}(P_L A) = S$, respectively, then

$$A_{(L)}^{(-1)} = U(U^* A U)^{-1} U^* \tag{5.1.6}$$

and

$$A_{(L)}^{(\dagger)} = V(V^* A V)^{-1} V^*. \tag{5.1.7}$$

Thus, by applying the algorithm for the full rank factorization, we can compute the generalized inverses $A^\dagger, A_{MN}^\dagger, A_g, A_{(L)}^{(-1)}$, and $A_{(L)}^{(\dagger)}$ using the formulas (5.1.2)–(5.1.7).

There are several methods for performing full rank factorization. Since the matrix A can always be row reduced to the row echelon form by elementary row operations, one method is based on the row echelon form. The other two methods are based on Gaussian elimination with complete pivoting and Householder transformation. The method based on the row-echelon form is suitable for the case when the order of the matrix is low and the calculations are done manually. The others are suitable for the case when the calculations are carried out by a computer.

5.1.1 Row Echelon Forms

In this subsection, we give a row echelon form based method for the full rank factorization.

Definition 5.1.1 Let $E \in \mathbb{C}_r^{m \times n}$. If E is of the form

$$E = \begin{bmatrix} C \\ O \end{bmatrix}, \tag{5.1.8}$$

where O is an $(m - r) \times n$ zero matrix and $C = [c_{ij}] \in \mathbb{C}^{r \times n}$ satisfies the following three conditions:

1. $c_{ij} = 0$, when $i > j$;
2. The first non-zero entry in each row of C is 1;
3. If $c_{ij} = 1$ is the first non-zero entry of the ith row, then the jth column of C is the unit vector $\mathbf{e}_i$ whose only non-zero entry is in the ith position;

then E is said to have the row echelon form.

For example, the matrix

$$\begin{bmatrix} 1 & 2 & 0 & 3 & 3 \\ 0 & 0 & 1 & 1 & -2 \\ 0 & 0 & 0 & 0 & 0 \\ 0 & 0 & 0 & 0 & 0 \end{bmatrix} \tag{5.1.9}$$

is of the row echelon form. Below we state some properties of the row echelon form. Their proofs can be found in [1].

Let $A \in \mathbb{C}_r^{m \times n}$, then

(1) A can always be row reduced to the row echelon form by elementary row operations, that is, there always exists a non-singular matrix $P \in \mathbb{C}^{m \times m}$ such that $PA = E_A$ is in the row echelon form;
(2) for a given A, the row echelon form E_A obtained by reducing the rows of A is unique;
(3) if E_A is the row echelon form of A and the unit vectors in E_A appear in columns $i_1, i_2, \cdots, i_r$, then the corresponding columns $\mathbf{a}_{i_1}, \mathbf{a}_{i_2}, \cdots, \mathbf{a}_{i_r}$ of A form a basis for $\mathcal{R}(A)$. This particular basis is called the set of distinguished columns of A. The remaining columns are called undistinguished columns of A. For example, if A is a matrix whose row echelon form is given by (5.1.9), then the first and third columns of A are distinguished columns;
(4) if E_A is the row echelon form (5.1.8) for A, then $\mathcal{N}(A) = \mathcal{N}(E_A) = \mathcal{N}(C)$.
(5) if (5.1.8) is the row echelon form of A and $B \in \mathbb{C}^{m \times r}$ is the matrix consisting of the distinguished columns of A, i.e., $B = [\mathbf{a}_{i_1}, \mathbf{a}_{i_2}, \cdots, \mathbf{a}_{i_r}]$, then $A = BC$, where C is obtained from the row echelon form, is a full rank factorization.

Now we show, through an example, how to find $A^\dagger$ using the row echelon form and (5.1.2). Suppose

$$A = \begin{bmatrix} 1 & 2 & 1 & 4 & 1 \\ 2 & 4 & 0 & 6 & 6 \\ 1 & 2 & 0 & 3 & 3 \\ 2 & 4 & 0 & 6 & 6 \end{bmatrix},$$

we reduce A to the row echelon form:

$$E_A = \begin{bmatrix} 1 & 2 & 0 & 3 & 3 \\ 0 & 0 & 1 & 1 & -2 \\ 0 & 0 & 0 & 0 & 0 \\ 0 & 0 & 0 & 0 & 0 \end{bmatrix}.$$

Next, we select the distinguished columns of A and place them as the columns of a matrix B in the same order as they appear in A. In particular, the first and third columns are distinguished. Thus

$$B = \begin{bmatrix} 1 & 1 \\ 2 & 0 \\ 1 & 0 \\ 2 & 0 \end{bmatrix}.$$

Selecting the non-zero rows of E_A and placing them as the rows of a matrix C in the same order they appear in E_A, we have

$$C = \begin{bmatrix} 1 & 2 & 0 & 3 & 3 \\ 0 & 0 & 1 & 1 & -2 \end{bmatrix}.$$

Finally, computing

$$(B^*B)^{-1} = \frac{1}{9}\begin{bmatrix} 1 & -1 \\ -1 & 10 \end{bmatrix} \quad \text{and} \quad (CC^*)^{-1} = \frac{1}{129}\begin{bmatrix} 6 & 3 \\ 3 & 23 \end{bmatrix}$$

and applying the formula (5.1.2) for $A^\dagger$, we get

$$A^\dagger = C^*(CC^*)^{-1}(B^*B)^{-1}B^*$$

$$= \frac{1}{1161}\begin{bmatrix} 27 & 6 & 3 & 6 \\ 54 & 12 & 6 & 12 \\ 207 & -40 & -20 & -40 \\ 288 & -22 & -11 & -22 \\ -333 & 98 & 49 & 98 \end{bmatrix}.$$

5.1.2   Gaussian Elimination with Complete Pivoting

We describe a full rank factorization method based on the Gaussian elimination with complete pivoting. Let $A \in \mathbb{C}_r^{m \times n}$. The basic idea of the method is as follows. The matrix $A = A_0$ is transformed in succession to matrices $A_1, A_2, \cdots, A_r$, where A_k, $1 \leq k \leq r$, is of the form

$$A_k = \begin{bmatrix} U_k & V_k \\ O & W_k \end{bmatrix} \in \mathbb{C}^{m \times n}, \qquad (5.1.10)$$

where U_k is a nonsingular upper triangular matrix of order k. Denoting the (i, j)-element in W_k by w_{ij}, we determine the element in W_k with the largest modulus. Suppose that w_{pq} is such element. Interchange the pth and $(k + 1)$th rows, and qth and $(k + 1)$th columns, so w_{pq} is now in the $(k + 1, k + 1)$ position. We denote the elementary permutations $I_{k+1,p}$ and $I_{k+1,q}$ by P_{k+1} and Q_{k+1}, respectively. Forming an elementary lower triangular matrix

$$M_{k+1} = \begin{bmatrix} 1 & & & & & \\ & \ddots & & & 0 & \\ & & 1 & & & \\ & & -l_{k+2,k+1} & 1 & & \\ & & \vdots & & \ddots & \\ & & -l_{m,k+1} & & & 1 \end{bmatrix}, \quad l_{i,k+1} = \frac{w_{i,k+1}}{w_{k+1,k+1}},$$

we have

$$A_{k+1} = M_{k+1} P_{k+1} A_k Q_{k+1} = \begin{bmatrix} U_{k+1} & V_{k+1} \\ O & W_{k+1} \end{bmatrix},$$

where U_{k+1} is a nonsingular upper triangular matrix of order $k + 1$. With exact computation, the process will terminate with A_r:

$$A_r = \begin{bmatrix} U_r & V_r \\ O & W_r \end{bmatrix} = \begin{bmatrix} U_r & V_r \\ O & O \end{bmatrix} = \begin{bmatrix} U \\ O \end{bmatrix},$$

where U is $r \times n$ upper trapezoidal with diagonal elements $u_{i,i} \neq 0, i = 1, 2, \cdots, r$, that is,

$$U = \begin{bmatrix} u_{1,1} & \cdots & u_{1,r} & u_{1,r+1} & \cdots & u_{1,n} \\ & \ddots & \vdots & \vdots & & \vdots \\ & & u_{r,r} & u_{r,r+1} & \cdots & u_{r,n} \end{bmatrix}.$$

It follows from [2, 3] that

$$A_r = M_r P_r M_{r-1} P_{r-1} \cdots M_1 P_1 A Q_1 Q_2 \cdots Q_r$$
$$= M_r M_{r-1} \cdots M_1 P_r P_{r-1} \cdots P_1 A Q_1 Q_2 \cdots Q_r.$$

Let

$$P_r P_{r-1} \cdots P_1 = P, \quad Q_1 Q_2 \cdots Q_r = Q,$$

and

$$(M_r M_{r-1} \cdots M_1)^{-1} = M_1^{-1} M_2^{-1} \cdots M_r^{-1} = [L \, \tilde{L}],$$

where L is $m \times r$ lower trapezoidal, with zeros above the principal diagonal, 1s on the principal diagonal, and elements $l_{i,j}$ below the diagonal in the jth column, that is,

$$L = \begin{bmatrix} 1 & & & \\ l_{2,1} & 1 & & \\ \vdots & & \ddots & \\ l_{r,1} & \cdots & l_{r,r-1} & 1 \\ l_{r+1,1} & \cdots & \cdots & l_{r+1,r} \\ \vdots & & & \vdots \\ l_{m,1} & \cdots & \cdots & l_{m,r} \end{bmatrix} \in \mathbb{C}_r^{m \times r}.$$

Thus

$$PAQ = LU$$

and

$$A = (P^* L)(U Q^*)$$

is a full rank factorization of A. Consequently, the corresponding Moore-Penrose inverse is

$$A^\dagger = Q U^\dagger L^\dagger P,$$

where

$$U^\dagger = U^* (U U^*)^{-1} \quad \text{and} \quad L^\dagger = (L^* L)^{-1} L^*. \tag{5.1.11}$$

In the above argument it was assumed that rank(A) was exactly r, and that the exact arithmetic was performed. In practice, however, rank(A) may not be known and rounding errors occur. We have to decide, at the kth stage of the reduction, when A is reduced to A_k of the form (5.1.10), whether the elements of W_k should be considered as zero. In many practical situations it is sufficient to regard numbers less than a predetermined small number ϵ as zero. We then say that A has a "computational ϵ-rank" equal to r if the magnitude of all the elements of the computed W_r is less than or equal to ϵ and at least one element of W_{r-1} whose magnitude is greater than ϵ.

If P and Q are identity matrices, then a stable method for computing (5.1.11) is to solve the following two sets of Hermitian positive definite equations in turn

$$
\begin{aligned}
(L^*L)X &= L^*, \\
(UU^*)Y &= X
\end{aligned}
\tag{5.1.12}
$$

and then form

$$
A^\dagger = U^*Y.
$$

Equations (5.1.12) involve the solution of two sets of equations with m different right-hand sides. Another method for computing (5.1.11) solves only one set of equations. We set $U = DU_1$, where D is diagonal and U_1 has unit diagonal. It follows from the second equation in (5.1.12) that

$$
DU_1U_1^*D^*Y = X.
$$

Setting $D^*Y = \widetilde{Y}$ and substituting it into the first equation in (5.1.12), we get one set of equations:

$$
(L^*LDU_1U_1^*)\widetilde{Y} = L^*.
\tag{5.1.13}
$$

The coefficient matrix is no longer Hermitian. The solution $\widetilde{Y}$ of (5.1.13) can be obtained by using Gaussian elimination with partial pivoting. We then form

$$
A^\dagger = U_1^*\widetilde{Y}.
$$

5.1.3 Householder Transformation

Now, we give a full rank factorization method using the Householder transformations.

Definition 5.1.2 Let $\mathbf{u} \in \mathbb{C}^n$, $\mathbf{u}^*\mathbf{u} = 1$, then the $n \times n$ matrix

$$
H = I - 2\mathbf{u}\mathbf{u}^*
$$

is called a Householder transformation.

It is easy to show that $H^* = H = H^{-1}$. Hence H is a unitary matrix.

Theorem 5.1.1 Let $\mathbf{v} \in \mathbb{C}^n$, $\mathbf{v} \neq \mathbf{0}$, $\mathbf{u} = \mathbf{v} + \sigma\|\mathbf{v}\|_2\mathbf{e}_1$, and

$$
\sigma = \begin{cases} +1, & v_1 \geq 0; \\ -1, & v_1 < 0, \end{cases}
$$

where v_1 denotes the first element of $\mathbf{v}$, then $H = I - (2/\|\mathbf{u}\|_2^2)\mathbf{u}\mathbf{u}^*$ is a Householder transformation and $H\mathbf{v} = -\sigma\|\mathbf{v}\|_2\mathbf{e}_1$.

Proof See for example [2]. □

The above theorem shows that given a nonzero n-vector $\mathbf{v}$ we can find an $n \times n$ Householder transformation H such that $H\mathbf{v} = -\sigma \|\mathbf{v}\|_2 \mathbf{e}_1$, a scalar multiple of the first unit vector $\mathbf{e}_1$.

Applying the above theorem, we can premultiply an arbitrary $m \times n$ matrix A with an $m \times m$ Householder transformation H to produce a matrix HA, in which all the elements in first column are zero except the first.

Theorem 5.1.2 *Let $A \in \mathbb{C}_r^{m \times n}$. There is an $m \times m$ unitary matrix H and an $n \times n$ permutation matrix P such that*

$$HAP = \begin{bmatrix} U_r & V_r \\ O & O \end{bmatrix},$$

where U_r is $r \times r$ upper triangular with nonzero diagonal elements. The appropriate null matrices are absent if $r = m$ or n.

Proof Choose P_1 so that the first r columns of A are linearly independent. Apply an $m \times m$ Householder transformation H_1 to AP_1 such that the first column of the resulting matrix is a scalar multiple of the first unit vector:

$$H_1 A P_1 = \begin{bmatrix} u_{11} & \mathbf{v}_1^T \\ O & W_1 \end{bmatrix}.$$

Let H_2 be an $(m-1) \times (m-1)$ Householder transformation that transforms the first column of W_1 to a multiple of an $(m-1)$ unit column vector $\mathbf{e}_1$. Thus

$$\begin{bmatrix} 1 & \mathbf{0}^T \\ \mathbf{0} & H_2 \end{bmatrix} H_1 A P_1 = \begin{bmatrix} U_2 & V_2 \\ O & W_2 \end{bmatrix},$$

where U_2 is a 2×2 upper triangular matrix. Proceeding in this way, we find an $m \times m$ matrix

$$H = \begin{bmatrix} I_{r-1} & O \\ O & H_r \end{bmatrix} \cdots \begin{bmatrix} 1 & \mathbf{0}^T \\ \mathbf{0} & H_2 \end{bmatrix} H_1$$

such that

$$HAP = \begin{bmatrix} U_r & V_r \\ O & W_r \end{bmatrix}, \tag{5.1.14}$$

where U_r is an $r \times r$ upper triangular matrix and $P = P_1 P_2 \cdots P_r$. Since the first r column of AP are linearly independent, the columns of U_r must be linearly independent, i.e., the diagonal elements of U_r must be nonzero. The matrix W_r must be zero since otherwise the rank of HAP would be greater than r, which contradicts the assumption that the rank of A is r. □

Let

$$H^{-1} = H_1 \begin{bmatrix} 1 & 0^T \\ 0 & H_2 \end{bmatrix} \cdots \begin{bmatrix} I_{r-1} & O \\ O & H_r \end{bmatrix} = [Q \ \tilde{Q}], \qquad (5.1.15)$$

where $Q \in \mathbb{C}^{m \times r}$ and $Q^*Q = I_r$, and

$$\begin{bmatrix} U_r & V_r \\ O & O \end{bmatrix} = \begin{bmatrix} R \\ O \end{bmatrix}. \qquad (5.1.16)$$

From (5.1.14)–(5.1.16), we have

$$AP = QR.$$

Thus

$$A = Q(RP^*)$$

is a full rank factorization of A and

$$A^\dagger = PR^*(RR^*)^{-1}Q^*. \qquad (5.1.17)$$

A stable method for computing (5.1.17) is to solve the set of Hermitian positive definite equations:

$$(RR^*)X = Q^*$$

and then form

$$A^\dagger = PR^*X.$$

Set $R = DR_1$, where D is diagonal matrix of order r and R_1 has unit on the diagonal and $|r_{ij}| \le 1$. Thus

$$A^\dagger = PR_1^*(R_1R_1^*)^{-1}D^{-1}Q^*.$$

Another method for computing (5.1.17) is to solve the set of Hermitian positive definite equations:

$$(R_1R_1^*)X = D^{-1}Q^*$$

and then form

$$A^\dagger = PR_1^*X.$$

The full rank factorization based methods for computing other generalized inverses $A_{T,S}^{(2)}$ are omitted here and left as exercises.

5.2 Singular Value Decompositions and (M, N) Singular Value Decompositions

This section discusses the singular value decomposition (SVD) based methods for computing the generalized inverses.

5.2.1 Singular Value Decomposition

Definition 5.2.1 Let $A \in \mathbb{C}^{m \times n}$, $\mathbf{u} \in \mathbb{C}^m$, $\mathbf{v} \in \mathbb{C}^n$, and $\sigma \geq 0$ such that

$$A\mathbf{v} = \sigma\mathbf{u} \quad \text{and} \quad A^*\mathbf{u} = \sigma\mathbf{v}, \tag{5.2.1}$$

then σ is called the singular value of A, $\mathbf{u}$ and $\mathbf{v}$ are called the left and right singular vectors of A respectively.

By (5.2.1),
$$A^*A\mathbf{v} = \sigma^2\mathbf{v} \quad \text{and} \quad AA^*\mathbf{u} = \sigma^2\mathbf{u},$$

thus σ^2 is an eigenvalue of A^*A or AA^*.

Theorem 5.2.1 *Let $A \in \mathbb{C}_r^{m \times n}$, then there are unitary matrices $U \in \mathbb{C}^{m \times m}$ and $V \in \mathbb{C}^{n \times n}$ such that*

$$A = U \begin{bmatrix} \Sigma & O \\ O & O \end{bmatrix} V^*, \tag{5.2.2}$$

where $\Sigma = \mathrm{diag}(\sigma_1, \sigma_2, \cdots, \sigma_r)$, $\sigma_i = \sqrt{\lambda_i}$ and $\lambda_1 \geq \lambda_2 \geq \cdots \geq \lambda_r > 0$ are the nonzero eigenvalues of A^*A, then $\sigma_1 \geq \sigma_2 \geq \cdots \geq \sigma_r > 0$ are the nonzero singular value of A and (5.2.2) is called the singular value decomposition (SVD) of A, and

$$\|A\|_2 = \sigma_1 \quad \text{and} \quad \|A^\dagger\|_2 = 1/\sigma_r.$$

Proof Since $A^*A \in \mathbb{C}_r^{n \times n}$ is Hermitian positive semidefinite, its eigenvalues are nonnegative. Suppose they are $\sigma_1^2, \sigma_2^2, \cdots, \sigma_n^2$, where $\sigma_1 \geq \sigma_2 \geq \cdots \geq \sigma_r > \sigma_{r+1} = \cdots = \sigma_n = 0$. Let $\mathbf{v}_1, \mathbf{v}_2, \cdots, \mathbf{v}_n$ be the orthonormal eigenvectors corresponding to $\sigma_1^2, \sigma_2^2, \cdots, \sigma_n^2$ respectively and

$$V_1 = [\mathbf{v}_1 \ \mathbf{v}_2 \ \cdots \ \mathbf{v}_r], \quad V_2 = [\mathbf{v}_{r+1} \ \mathbf{v}_{r+2} \ \cdots \ \mathbf{v}_n],$$

and

$$\Sigma = \mathrm{diag}(\sigma_1, \sigma_2, \cdots, \sigma_r),$$

then

$$V_1^*(A^*A)V_1 = \Sigma^2 \quad \text{and} \quad V_2^*(A^*A)V_2 = O,$$

and consequently

$$\Sigma^{-1} V_1^* A^* A V_1 \Sigma^{-1} = I \quad \text{and} \quad A V_2 = O.$$

Now, let

$$U_1 = A V_1 \Sigma^{-1},$$

then $U_1^* U_1 = I$, that is, the columns of U_1 are orthonormal. Let U_2 be chosen so that $U = [U_1 \ U_2]$ is unitary, then

$$U^* A V = \begin{bmatrix} U_1^* A V_1 & U_1^* A V_2 \\ U_2^* A V_1 & U_2^* A V_2 \end{bmatrix} = \begin{bmatrix} (A V_1 \Sigma^{-1})^* A V_1 & O \\ U_2^*(U_1 \Sigma) & O \end{bmatrix} = \begin{bmatrix} \Sigma & O \\ O & O \end{bmatrix}.$$

Thus (5.2.2) holds. From (5.2.2), we have

$$A^* A = V \begin{bmatrix} \Sigma^2 & O \\ O & O \end{bmatrix} V^*.$$

Thus the eigenvalues of $A^* A$ are $\sigma_i^2 = \lambda_i(A^* A)$, $i = 1, 2, \cdots, n$, and

$$\|A\|_2^2 = \|A^* A\|_2 = |\lambda_1(A^* A)| = \sigma_1^2.$$

So $\|A\|_2 = \sigma_1$. It is easy to verify that

$$A^\dagger = V \begin{bmatrix} \Sigma^{-1} & O \\ O & O \end{bmatrix} U^*. \tag{5.2.3}$$

Hence the non-zero singular values of $A^\dagger$ are

$$\frac{1}{\sigma_r} \geq \frac{1}{\sigma_{r-1}} \geq \cdots \geq \frac{1}{\sigma_1} > 0.$$

Thus $\|A^\dagger\|_2 = 1/\sigma_r$.   □

The following perturbation theorem for the singular values is useful in Chap. 8.

Lemma 5.2.1 *Let A and $B = A + E \in \mathbb{C}^{m \times n}$ have singular values $\sigma_1(A) \geq \sigma_2(A) \geq \cdots \geq \sigma_n(A)$ and $\sigma_1(B) \geq \sigma_2(B) \geq \cdots \geq \sigma_n(B)$ respectively, then*

$$\sigma_i(A) - \|E\|_2 \leq \sigma_i(B) \leq \sigma_i(A) + \|E\|_2, \quad i = 1, 2, \cdots, n.$$

Proof See for example [2].   □

5.2.2 (M, N) *Singular Value Decomposition*

Definition 5.2.2 Let $A \in \mathbb{C}_r^{m \times n}$, M and N be Hermitian positive definite matrices of orders m and n respectively, then the (M, N) singular values of A are the elements of the set $\mu_{MN}(A)$ defined by

$$\mu_{MN}(A) = \left\{ \mu : \mu \geq 0, \ \mu \text{ is a stationary value of } \frac{\|A\mathbf{x}\|_M}{\|\mathbf{x}\|_N} \right\}.$$

Next is a theorem of the (M, N) singular value decomposition.

Theorem 5.2.2 ([4, 5]) *Let $A \in \mathbb{C}_r^{m \times n}$, M and N be Hermitian positive definite matrices of orders m and n respectively, then there are matrices $U \in \mathbb{C}^{m \times m}$ and $V \in \mathbb{C}^{n \times n}$ satisfying*

$$U^* M U = I_m \quad and \quad V^* N^{-1} V = I_n,$$

such that

$$A = U \begin{bmatrix} D & O \\ O & O \end{bmatrix} V^*, \tag{5.2.4}$$

where $D = \text{diag}(\mu_1, \mu_2, \cdots, \mu_r)$, $\mu_i = \sqrt{\lambda_i}$, $\lambda_1 \geq \lambda_2 \geq \cdots \geq \lambda_r > 0$ are the nonzero eigenvalues of $A^\# A = (N^{-1} A^* M)A$. Then $\mu_1 \geq \mu_2 \geq \cdots \geq \mu_r > 0$ are the nonzero (M, N) singular values of A and (5.2.4) is called the (M, N) singular value decomposition of A, and

$$\|A\|_{MN} = \mu_1 \quad and \quad \|A_{MN}^\dagger\|_{NM} = \mu_r^{-1}. \tag{5.2.5}$$

Proof Let $M = LL^*$ and $N = KK^*$ be the Cholesky factorizations of M and N respectively. Set $C = L^* A K^{-*} \in \mathbb{C}_r^{m \times n}$ (where $K^{-*} = (K^{-1})^* = (K^*)^{-1}$) and let

$$C = Q \begin{bmatrix} D & O \\ O & O \end{bmatrix} Z^*$$

represent the singular value decomposition of C, where

$$Q^* Q = I_m, \quad Z^* Z = I_n, \quad and \quad D = \text{diag}(\mu_1, \mu_2, \cdots, \mu_r), \ \mu_i > 0.$$

By defining $U = L^{-*} Q$ and $V = KZ$, we have

$$U^* M U = I_m, \quad V^* N^{-1} V = I_n,$$

and

$$A = U \begin{bmatrix} D & O \\ O & O \end{bmatrix} V^*.$$

Thus (5.2.4) holds. It follows from (5.2.4) that

$$AA^\# U = AN^{-1}A^*MU = U \begin{bmatrix} D^2 & O \\ O & O \end{bmatrix}.$$

So the squares of the (M, N) singular values μ_i of A are equal to the eigenvalues of $AA^\#$. Moreover,

$$\begin{aligned}
\|A\|_{MN}^2 &= \|M^{\frac{1}{2}}AN^{-\frac{1}{2}}\|_2^2 \\
&= \|M^{\frac{1}{2}}AN^{-\frac{1}{2}}N^{-\frac{1}{2}}A^*M^{\frac{1}{2}}\|_2 \\
&= \|M^{\frac{1}{2}}(AN^{-1}A^*M)M^{-\frac{1}{2}}\|_2 \\
&= \|AA^\#\|_2 \\
&= |\lambda_1(AA^\#)| \\
&= \mu_1^2.
\end{aligned}$$

Thus $\|A\|_{MN} = \mu_1$. By using (5.2.4), it is easy to verify that

$$A_{MN}^\dagger = N^{-1}V \begin{bmatrix} D^{-1} & O \\ O & O \end{bmatrix} U^*M. \tag{5.2.6}$$

Hence the non-zero (M, N) singular values of $A_{MN}^\dagger$ are $\mu_r^{-1} \geq \mu_{r-1}^{-1} \geq \cdots \geq \mu_1^{-1} > 0$. Thus $\|A_{MN}^\dagger\|_{NM} = \mu_r^{-1}$. $\qquad\square$

The following perturbation theorem for (M, N) singular values is useful in Chap. 8.

Lemma 5.2.2 *Let $A, E \in \mathbb{C}^{m\times n}$, $\mu_i(A + E)$ and $\mu_i(A)$ be the (M, N) singular values of $A + E$ and A respectively, then*

$$\mu_i(A) - \|E\|_{MN} \leq \mu_i(A + E) \leq \mu_i(A) + \|E\|_{MN}.$$

Proof Let $\widetilde{E} = M^{\frac{1}{2}}EN^{-\frac{1}{2}}$ and $\widetilde{A} = M^{\frac{1}{2}}AN^{-\frac{1}{2}}$, then $\widetilde{A + E} = M^{\frac{1}{2}}(A + E)N^{-\frac{1}{2}} = \widetilde{A} + \widetilde{E}$. It follows from Lemma 5.2.1 that the singular values $\mu_i(\widetilde{A})$ and $\mu_i(\widetilde{A} + \widetilde{E}) = \mu_i(\widetilde{A + E})$ of $\widetilde{A}$ and $\widetilde{A} + \widetilde{E}$ satisfy

$$\mu_i(\widetilde{A}) - \|\widetilde{E}\|_2 \leq \mu_i(\widetilde{A + E}) \leq \mu_i(\widetilde{A}) + \|\widetilde{E}\|_2.$$

This completes the proof, since $\|\widetilde{E}\|_2 = \|E\|_{MN}$, $\mu_i(\widetilde{A}) = \mu_i(A)$, and $\mu_i(\widetilde{A + E}) = \mu_i(A + E)$. $\qquad\square$

5.2.3 Methods Based on SVD and (M, N) SVD

Let $A \in \mathbb{C}^{m \times n}$. If the singular value decomposition of A is (5.2.2), then (5.2.3) gives:

$$A^\dagger = V \begin{bmatrix} \Sigma^{-1} & O \\ O & O \end{bmatrix} U^*.$$

If the (M, N) singular value decomposition of A is (5.2.4), then we have (5.2.6):

$$A_{MN}^\dagger = N^{-1} V \begin{bmatrix} D^{-1} & O \\ O & O \end{bmatrix} U^* M.$$

The algorithms and programs for computing (5.2.2) and (5.2.3) can be found in [2]. From these results and Theorem 5.2.2, the algorithms and programs for computing (5.2.4) and (5.2.6) can be obtained easily.

When $A \in \mathbb{C}^{n \times n}$, the methods for finding A_d and A_g are given as follows.

Lemma 5.2.3 *Let $A \in \mathbb{C}^{n \times n}$, $\mathrm{Ind}(A) = k$, $W \in \mathbb{C}_n^{n \times n}$, and $WAW^{-1} = R$, then*

$$
\begin{aligned}
A_d &= W^{-1} R_d W, \quad \mathrm{Ind}(A) = \mathrm{Ind}(R), \\
&\mathrm{Core\text{-}rank}(A) = \mathrm{Core\text{-}rank}(R),
\end{aligned}
\tag{5.2.7}
$$

where $\mathrm{Core\text{-}rank}(A) = \mathrm{rank}(A^k)$ *is called the core-rank of A.*

Proof See [6]. □

Lemma 5.2.4 ([7]) *Let*

$$M = \begin{bmatrix} A & O \\ B & C \end{bmatrix}$$

be a block lower triangular matrix. If A is nonsingular, then

$$\mathrm{Ind}(M) = \mathrm{Ind}(C).$$

Proof Let $\mathrm{Ind}(M) = k$, then $\mathrm{rank}(M^k) = \mathrm{rank}(M^{k+1})$, i.e.,

$$\mathrm{rank}(A^k) + \mathrm{rank}(C^k) = \mathrm{rank}(A^{k+1}) + \mathrm{rank}(C^{k+1}).$$

By the assumption, we have $\mathrm{rank}(A^k) = \mathrm{rank}(A^{k+1})$. Thus $\mathrm{rank}(C^k) = \mathrm{rank}(C^{k+1})$ and so $\mathrm{Ind}(C) = k$. □

The algorithm for the Drazin inverse of block lower triangular matrices are given as follows.

Theorem 5.2.3 *Let a block triangular matrix*

$$M = \begin{bmatrix} B_1 & O \\ B_2 & N \end{bmatrix} \in \mathbb{C}^{n \times n},$$

where $B_1 \in \mathbb{C}_s^{s \times s}$ and $N = [n_{ij}] \in \mathbb{C}^{t \times t}$ is a strictly lower triangular matrix. Then

$$M_d = \begin{bmatrix} B_1^{-1} & O \\ XB_1^{-1} & O \end{bmatrix},$$

where X is the solution of the Sylvester equation

$$XB_1 - NX = B_2. \tag{5.2.8}$$

Let $\mathbf{r}_i$ be the ith row of B_2, then the rows $\mathbf{x}_i$ of X can be recursively solved by:

$$\mathbf{x}_1 = \mathbf{r}_1 B_1^{-1} \quad and \quad \mathbf{x}_i = \left(\mathbf{r}_i + \sum_{j=1}^{i-1} n_{ij} \mathbf{x}_j \right) B_1^{-1}.$$

Proof Let

$$Y = \begin{bmatrix} B_1^{-1} & O \\ XB_1^{-1} & O \end{bmatrix},$$

then we have

$$MY = \begin{bmatrix} B_1 & O \\ B_2 & N \end{bmatrix} \begin{bmatrix} B_1^{-1} & O \\ XB_1^{-1} & O \end{bmatrix} = \begin{bmatrix} I & O \\ X & O \end{bmatrix} = YM$$

and

$$YMY = \begin{bmatrix} B_1^{-1} & O \\ XB_1^{-1} & O \end{bmatrix} = Y.$$

Note that for any positive integer p,

$$\begin{bmatrix} B_1 & O \\ B_2 & N \end{bmatrix}^p = \begin{bmatrix} B_1^p & O \\ S(p) & N^p \end{bmatrix}, \quad where \quad S(p) = \sum_{i=0}^{p-1} N^{p-1-i} B_2 B_1^i.$$

If $p \geq t$, then $N^p = O$, because N is a strictly lower triangular matrix of order t. Thus

$$M^{t+1}Y = M^t(MY)$$

$$= \begin{bmatrix} B_1^t & O \\ S(t) & O \end{bmatrix} \begin{bmatrix} I & O \\ X & O \end{bmatrix}$$

$$= \begin{bmatrix} B_1^t & O \\ S(t) & O \end{bmatrix}$$

$$= M^t.$$

Therefore $Y = M_d$ and (5.2.8) follows from $MM_d = M_dM$. $\qquad\qquad \square$

By using Theorem 5.2.3 and the singular value decomposition of $A \in \mathbb{C}^{n \times n}$, an orthogonal deflation method for calculating A_d is given as follows.

Algorithm 5.2.1 ([6]) Given a matrix $A \in \mathbb{C}^{n \times n}$, this algorithm computes its Drazin inverse A_d.

1. Compute the singular value decomposition of A:

$$A = U \begin{bmatrix} \Sigma & O \\ O & O \end{bmatrix} V^*.$$

If $\Sigma \in \mathbb{R}_n^{n \times n}$, then $A_d = A^{-1} = V\Sigma^{-1}U^*$. If $\text{Ind}(A) > 0$, write

$$V^*AV = V^*U \begin{bmatrix} \Sigma & O \\ O & O \end{bmatrix} = \begin{bmatrix} A_{11}^{(1)} & O \\ A_{21}^{(1)} & O \end{bmatrix}.$$

2. Now compute the singular value decomposition of $A_{11}^{(1)}$:

$$A_{11}^{(1)} = U_1 \begin{bmatrix} \Sigma_1 & O \\ O & O \end{bmatrix} V_1^*.$$

If $A_{11}^{(1)}$ is nonsingular, go to step 4, otherwise

$$V_1^*A_{11}^{(1)}V_1 = V_1^*U_1 \begin{bmatrix} \Sigma_1 & O \\ O & O \end{bmatrix} = \begin{bmatrix} A_{11}^{(2)} & O \\ A_{21}^{(2)} & O \end{bmatrix}.$$

Thus

$$\begin{bmatrix} V_1^* & O \\ O & I \end{bmatrix} V^*AV \begin{bmatrix} V_1 & O \\ O & I \end{bmatrix}$$

$$= \begin{bmatrix} V_1^*A_{11}^{(1)}V_1 & O \\ A_{21}^{(1)}V_1 & O \end{bmatrix}$$

$$= \begin{bmatrix} A_{11}^{(2)} & O & O \\ A_{21}^{(2)} & O & O \\ A_{31}^{(2)} & A_{32}^{(2)} & O \end{bmatrix}, \qquad (5.2.9)$$

where $A_{21}^{(1)} V_1 = [A_{31}^{(2)} \ A_{32}^{(2)}]$ is partitioned accordingly.

3. Continue the procedure in step 2, at each step, compute the singular value decomposition of $A_{11}^{(m)}$ and perform the appropriate multiplication as in (5.2.9) to get $A_{11}^{(m+1)}$. If some $A_{11}^{(k)} = O$, then $A_d = O$. If some $A_{11}^{(k)}$ is nonsingular, go to step 4.

4. By Lemma 5.2.4, we now have $k = \text{Ind}(A)$ and

$$
WAW^* = \begin{bmatrix}
A_{11}^{(k)} & O & \cdots & \cdots & O \\
A_{21}^{(k)} & O & \cdots & \cdots & O \\
A_{31}^{(k)} & A_{32}^{(k)} & \ddots & & \vdots \\
\vdots & \vdots & & \ddots & \vdots \\
A_{k+1,1}^{(k)} & A_{k+1,2}^{(k)} & \cdots & A_{k+1,k}^{(k)} & O
\end{bmatrix}
= \begin{bmatrix} B_1 & O \\ B_2 & N \end{bmatrix},
$$

where $B_1 = A_{11}^{(k)}$ is nonsingular,

$$
B_2 = \begin{bmatrix}
A_{21}^{(k)} \\
A_{31}^{(k)} \\
\vdots \\
A_{k+1,1}^{(k)}
\end{bmatrix}
\quad \text{and} \quad
N = \begin{bmatrix}
O & \cdots & \cdots & O \\
A_{32}^{(k)} & \ddots & & \vdots \\
\vdots & & \ddots & \vdots \\
A_{k+1,2}^{(k)} & \cdots & A_{k+1,k}^{(k)} & O
\end{bmatrix}
$$

is a strictly lower triangular matrix, thus $N^k = O$, and

$$
W = \begin{bmatrix} V_k^* & O \\ O & I \end{bmatrix} \begin{bmatrix} V_{k-1}^* & O \\ O & I \end{bmatrix} \cdots \begin{bmatrix} V_1^* & O \\ O & I \end{bmatrix} V^*
$$

is unitary, that is, $W^* = W^{-1}$. It then follows from Lemma 5.2.3 and Theorem 5.2.3 that

$$
A_d = W^{-1} \begin{bmatrix} B_1^{-1} & O \\ XB_1^{-1} & O \end{bmatrix} W,
$$

where X is the solution of $XB_1 - NX = B_2$. Let $N = [n_{ij}]$ and $\mathbf{r}_i$ be the ith row of B_2, then the rows $\mathbf{x}_i$ of X can be recursively solved by

$$
\mathbf{x}_1 = \mathbf{r}_1 B_1^{-1} \quad \text{and} \quad \mathbf{x}_i = \left(\mathbf{r}_i + \sum_{j=1}^{i-1} n_{ij} \mathbf{x}_j \right) B_1^{-1}.
$$

5.3 Generalized Inverses of Sums and Partitioned Matrices

Let

$$M = \begin{bmatrix} A & C \\ B & D \end{bmatrix},$$

where A, B, C, D are four matrices of appropriate dimensions, then A, B, C, D are called conformable. Matrix M can be viewed in two ways. One is to view M as a bordered matrix built from A. In this case, the blocks are considered fixed. If one is trying to build M with certain properties, then one might choose a particular kind of blocks. For example, in Chap. 3, if $A \in \mathbb{C}_r^{m \times n}$, and $C \in \mathbb{C}_{(m-r)}^{m \times (m-r)}$ whose columns form a basis for $\mathcal{N}(A^*)$, $B \in \mathbb{C}_{(n-r)}^{(n-r) \times n}$ whose columns form a basis for $\mathcal{N}(A)$, and D is an $(n-r) \times (m-r)$ zero matrix, then M is nonsingular. Another is to view M as a partitioned matrix. In this case, M is considered fixed. Different partitions may be considered to compute M from its blocks.

Consider the partitioned matrix M above. Suppose that the matrix M and the block A are nonsingular and

$$M^{-1} = \begin{bmatrix} E & G \\ F & H \end{bmatrix},$$

then

$$H = (D - BA^{-1}C)^{-1},$$
$$G = -A^{-1}CH,$$
$$F = -HBA^{-1},$$
$$E = A^{-1}(I - CF).$$

As a special case, let the bordered matrix

$$A_n = \begin{bmatrix} a_{11} & \cdots & a_{1,n-1} & a_{1n} \\ \vdots & & \vdots & \vdots \\ a_{n-1,1} & \cdots & a_{n-1,n-1} & a_{n-1,n} \\ a_{n1} & \cdots & a_{n,n-1} & a_{n,n} \end{bmatrix} = \begin{bmatrix} A_{n-1} & \mathbf{u}_n \\ \mathbf{v}_n^* & a_{nn} \end{bmatrix}$$

and A_{n-1} be nonsingular, then

$$A_n^{-1} = \begin{bmatrix} A_{n-1}^{-1} + \alpha_n^{-1} A_{n-1}^{-1} \mathbf{u}_n \mathbf{v}_n^* A_{n-1}^{-1} & -\alpha_n^{-1} A_{n-1}^{-1} \mathbf{u}_n \\ -\alpha_n^{-1} \mathbf{v}_n^* A_{n-1}^{-1} & \alpha_n^{-1} \end{bmatrix}, \tag{5.3.1}$$

where

$$\alpha_n = a_{nn} - \mathbf{v}_n^* A_{n-1}^{-1} \mathbf{u}_n. \tag{5.3.2}$$

Thus, the inverse of A_n can be computed from the inverse of its submatrix A_{n-1} and its last column and last row.

Can we partition a matrix and compute its generalized inverses from the blocks? This section is concerned with the computation of the generalized inverses of a matrix from its blocks using various partitions. A related problem of computing the generalized inverses of a sum of matrices is also discussed.

5.3.1 Moore-Penrose Inverse of Rank-One Modified Matrix

Let $A_k \in \mathbb{C}^{m \times k}$ be partitioned and decomposed as the following:

$$A_k = [A_{k-1} \quad \mathbf{o}_1] + \mathbf{a}_k[\mathbf{o}_2^T \quad 1],$$

where $\mathbf{o}_1 \in \mathbb{C}^m$ and $\mathbf{o}_2 \in \mathbb{C}^{k-1}$ are zero vectors and $\mathbf{a}_k$ is the last column of A_k. Let $A = [A_{k-1} \quad \mathbf{o}_1]$, $\mathbf{a}_k = \mathbf{c}$, $\mathbf{d}^* = [\mathbf{o}_2^T \quad 1]$, then $A_k = A + \mathbf{cd}^*$ is a rank-one modified matrix of A.

In order to obtain an expression for $A_k^\dagger = [A_{k-1} \quad \mathbf{a}_k]^\dagger$, we will now develop a formulation for the Moore-Penrose inverse of a rank-one modified matrix $A + \mathbf{cd}^*$.

Theorem 5.3.1 ([6]) *Let $A \in \mathbb{C}^{m \times n}$, $\mathbf{c} \in \mathbb{C}^m$, $\mathbf{d} \in \mathbb{C}^n$,*

$$\begin{aligned}
\mathbf{k} &= A^\dagger \mathbf{c}, \\
\mathbf{h}^* &= \mathbf{d}^* A^\dagger, \\
\mathbf{u} &= (I - AA^\dagger)\mathbf{c}, \\
\mathbf{v}^* &= \mathbf{d}^*(I - A^\dagger A), \\
\beta &= 1 + \mathbf{d}^* A^\dagger \mathbf{c}.
\end{aligned}$$

(Notice that $\mathbf{c} \in \mathcal{R}(A)$ if and only if $\mathbf{u} = 0$ and $\mathbf{d} \in \mathcal{R}(A^*)$ if and only if $\mathbf{v} = 0$.) Then the Moore-Penrose inverse of $A + \mathbf{cd}^*$ is as follows.

(1) *If $\mathbf{u} \neq 0$ and $\mathbf{v} \neq 0$, then*

$$(A + \mathbf{cd}^*)^\dagger = A^\dagger - \mathbf{ku}^\dagger - \mathbf{v}^{*\dagger}\mathbf{h} + \beta \mathbf{v}^{*\dagger}\mathbf{u}^\dagger.$$

(2) *If $\mathbf{u} = 0$, $\mathbf{v} \neq 0$ and $\beta = 0$, then*

$$(A + \mathbf{cd}^*)^\dagger = A^\dagger - \mathbf{kk}^\dagger A^\dagger - \mathbf{v}^{*\dagger}\mathbf{h}^*.$$

(3) *If $\mathbf{u} = 0$ and $\beta \neq 0$, then*

$$(A + \mathbf{cd}^*)^\dagger = A^\dagger + \frac{1}{\beta}\mathbf{vk}^* A^\dagger - \frac{\bar{\beta}}{\sigma_1}\mathbf{p}_1\mathbf{q}_1^*,$$

where

$$\mathbf{p}_1 = -\left(\frac{\|\mathbf{k}\|^2}{\bar{\beta}}\mathbf{v} + \mathbf{k}\right), \quad \mathbf{q}_1^* = -\left(\frac{\|\mathbf{v}\|^2}{\bar{\beta}}\mathbf{k}^*A^\dagger + \mathbf{h}^*\right),$$

and $\sigma_1 = \|\mathbf{k}\|^2\|\mathbf{v}\|^2 + |\beta|^2$.

(4) *If $\mathbf{u} \neq \mathbf{0}$, $\mathbf{v} = \mathbf{0}$ and $\beta = 0$, then*

$$(A + \mathbf{cd}^*)^\dagger = A^\dagger - A^\dagger \mathbf{h}^{*\dagger}\mathbf{h}^* - \mathbf{k}\mathbf{u}^\dagger.$$

(5) *If $\mathbf{v} = \mathbf{0}$ and $\beta \neq 0$, then*

$$(A + \mathbf{cd}^*)^\dagger = A^\dagger + \frac{1}{\beta}A^\dagger\mathbf{h}\mathbf{u}^* - \frac{\bar{\beta}}{\sigma_2}\mathbf{p}_2\mathbf{q}_2^*,$$

where

$$\mathbf{p}_2 = -\left(\frac{\|\mathbf{u}\|^2}{\bar{\beta}}A^\dagger\mathbf{h} + \mathbf{k}\right), \quad \mathbf{q}_2^* = -\left(\frac{\|\mathbf{h}\|^2}{\bar{\beta}}\mathbf{u}^* + \mathbf{h}^*\right),$$

and $\sigma_2 = \|\mathbf{h}\|^2\|\mathbf{u}\|^2 + |\beta|^2$.

(6) *If $\mathbf{u} = \mathbf{0}$, $\mathbf{v} = \mathbf{0}$ and $\beta = 0$, then*

$$(A + \mathbf{cd}^*)^\dagger = A^\dagger - \mathbf{k}\mathbf{k}^\dagger A^\dagger - A^\dagger \mathbf{h}^{*\dagger}\mathbf{h}^* + (\mathbf{k}^\dagger A^\dagger \mathbf{h}^{*\dagger})\mathbf{k}\mathbf{h}^*.$$

Before proving Theorem 5.3.1, we state the following lemma.

Lemma 5.3.1 *Let A, $\mathbf{c}$, $\mathbf{d}$, $\mathbf{u}$, $\mathbf{v}$, and β be the same as in Theorem 5.3.1, then*

$$\mathrm{rank}(A + \mathbf{cd}^*) = \mathrm{rank}\begin{bmatrix} A & \mathbf{u} \\ \mathbf{v}^* & -\beta \end{bmatrix} - 1.$$

Proof This follows immediately from the factorization

$$\begin{bmatrix} A + \mathbf{cd}^* & \mathbf{c} \\ \mathbf{0}^T & -1 \end{bmatrix} = \begin{bmatrix} I & 0 \\ \mathbf{h}^* & 1 \end{bmatrix}\begin{bmatrix} A & \mathbf{u} \\ \mathbf{v}^* & -\beta \end{bmatrix}\begin{bmatrix} I & \mathbf{k} \\ \mathbf{0}^T & 1 \end{bmatrix}\begin{bmatrix} I & 0 \\ \mathbf{d}^* & 1 \end{bmatrix},$$

which can be verified by the definitions of $\mathbf{k}$, $\mathbf{h}$, $\mathbf{u}$, $\mathbf{v}$, and β. □

We now proceed to the proof of Theorem 5.3.1. Throughout, we assume $\mathbf{c} \neq \mathbf{0}$ and $\mathbf{d} \neq \mathbf{0}$.

Proof (1) Let X_1 denote the right-hand side of the equation in (1) and

$$M = A + \mathbf{cd}^*.$$

The proof consists of showing that X_1 satisfies the four Penrose conditions. From the definitions of $\mathbf{k}$, $\mathbf{v}$, and β, we have $A\mathbf{v}^{*\dagger} = \mathbf{0}$, $\mathbf{d}^*\mathbf{v}^{*\dagger} = 1$, $\mathbf{d}^*\mathbf{k} = \beta - 1$, and

$\mathbf{c} - A\mathbf{k} = \mathbf{u}$, which implies that

$$MX_1 = AA^\dagger + \mathbf{u}\mathbf{u}^\dagger.$$

So the third Penrose condition holds. Using $\mathbf{u}^\dagger A = \mathbf{0}^T$, $\mathbf{u}^\dagger \mathbf{c} = 1$, $\mathbf{h}^* \mathbf{c} = \beta - 1$, and $\mathbf{d}^* - \mathbf{h}^* A = \mathbf{v}^*$, one obtains

$$X_1 M = A^\dagger A = \mathbf{v}^{*\dagger} \mathbf{v}^*$$

and hence the fourth condition holds. The first and second conditions follow easily.
(2) Let X_2 denote the right-hand side of the equation in (2). By using $A\mathbf{k} = \mathbf{c}$, $A\mathbf{v}^{*\dagger} = \mathbf{0}$, $\mathbf{d}^* \mathbf{v}^{*\dagger} = 1$, and $\mathbf{d}^* \mathbf{k} = \beta - 1$, we can see that

$$(A + \mathbf{c}\mathbf{d}^*)X_2 = AA^\dagger,$$

which is Hermitian. From the equalities $\mathbf{k}^\dagger A^\dagger A = \mathbf{k}^\dagger$, $\mathbf{h}^* \mathbf{c} = -1$, and $\mathbf{d}^* - \mathbf{h}^* A = \mathbf{v}^*$, it follows that

$$X_2(A + \mathbf{c}\mathbf{d}^*) = A^\dagger A - \mathbf{k}\mathbf{k}^\dagger + \mathbf{v}^{*\dagger} \mathbf{v}^*,$$

which is also Hermitian. The first and second Penrose conditions can now be easily verified.
(3) This is the most difficult case. Since $\mathbf{u} = \mathbf{0}$, $\mathbf{c} \in \mathcal{R}(A)$ implying that $\mathcal{R}(A + \mathbf{c}\mathbf{d}^*) \subset \mathcal{R}(A)$. Since $\beta \neq 0$, from Lemma 5.3.1, we have $\text{rank}(A + \mathbf{c}\mathbf{d}^*) = \text{rank}(A)$. Therefore, $\mathcal{R}(A + \mathbf{c}\mathbf{d}^*) = \mathcal{R}(A)$ and

$$(A + \mathbf{c}\mathbf{d}^*)(A + \mathbf{c}\mathbf{d}^*)^\dagger = AA^\dagger, \tag{5.3.3}$$

because $AA^\dagger$ is the unique orthogonal projector onto $\mathcal{R}(A)$. Let X_3 denote the right-hand side of the equation in (3) and $M = A + \mathbf{c}\mathbf{d}^*$. Because $\mathbf{q}_1^* AA^\dagger = \mathbf{q}_1^*$, it follows from (5.3.3) that

$$X_3 MM^\dagger = X_3 AA^\dagger = X_3.$$

We claim that

$$M^\dagger M = X_3 M. \tag{5.3.4}$$

Thus, we have

$$M^\dagger = (M^\dagger M)M^\dagger = X_3 MM^\dagger = X_3.$$

Now we prove (5.3.4) by showing

$$M^\dagger M = A^\dagger A - \mathbf{k}\mathbf{k}^\dagger + \mathbf{p}_1 \mathbf{p}_1^\dagger \tag{5.3.5}$$

and

$$X_3 M = A^\dagger A - \mathbf{k}\mathbf{k}^\dagger + \mathbf{p}_1 \mathbf{p}_1^\dagger. \tag{5.3.6}$$

The matrix $A^{\dagger}A - \mathbf{k}\mathbf{k}^{\dagger} + \mathbf{p}_1\mathbf{p}_1^{\dagger}$ is Hermitian and idempotent. It is obvious that it is Hermitian. The fact that it is idempotent can be verified by using the identities $A^{\dagger}A\mathbf{k} = A^{\dagger}\mathbf{c} = \mathbf{k}$, $A^{\dagger}A\mathbf{p}_1 = -\mathbf{k}$, and $\mathbf{k}\mathbf{k}^{\dagger}\mathbf{p}_1 = -\mathbf{k}$. Since the rank of an idempotent matrix is equal to its trace,

$$\begin{aligned} \operatorname{rank}(A^{\dagger}A - \mathbf{k}\mathbf{k}^{\dagger} + \mathbf{p}_1\mathbf{p}_1^{\dagger}) &= \operatorname{tr}(A^{\dagger}A - \mathbf{k}\mathbf{k}^{\dagger} + \mathbf{p}_1\mathbf{p}_1^{\dagger}) \\ &= \operatorname{tr}(A^{\dagger}A) - \operatorname{tr}(\mathbf{k}\mathbf{k}^{\dagger}) + \operatorname{tr}(\mathbf{p}_1\mathbf{p}_1^{\dagger}). \end{aligned}$$

Now, $\mathbf{k}\mathbf{k}^{\dagger}$ and $\mathbf{p}_1\mathbf{p}_1^{\dagger}$ are idempotent matrices of rank one and $A^{\dagger}A$ is an idempotent matrix whose rank is $\operatorname{rank}(A)$, so

$$\operatorname{rank}(A^{\dagger}A - \mathbf{k}\mathbf{k}^{\dagger} + \mathbf{p}_1\mathbf{p}_1^{\dagger}) = \operatorname{rank}(A + \mathbf{c}\mathbf{d}^*). \tag{5.3.7}$$

Using the equalities $A\mathbf{k} = \mathbf{c}$, $A\mathbf{p}_1 = -\mathbf{c}$, $\mathbf{d}^*\mathbf{k} = \beta - 1$, $\mathbf{d}^*\mathbf{p}_1 = 1 - \sigma_1\bar{\beta}^{-1}$, and $\mathbf{d}^*A^{\dagger}A = \mathbf{d}^* - \mathbf{v}^*$, we obtain

$$(A + \mathbf{c}\mathbf{d}^*)(A^{\dagger}A - \mathbf{k}\mathbf{k}^{\dagger} + \mathbf{p}_1\mathbf{p}_1^{\dagger}) = A + \mathbf{c}\mathbf{d}^* - \mathbf{c}(\mathbf{v}^* + \beta\mathbf{k}^{\dagger} + \sigma_1\bar{\beta}^{-1}\mathbf{p}_1^{\dagger}).$$

Now, $\|\mathbf{p}_1\|^2 = \|\mathbf{k}\|^2\sigma_1|\beta|^{-2}$, so $\sigma_1\bar{\beta}^{-1}\|\mathbf{p}_1\|^{-2} = \beta\|\mathbf{k}\|^{-2}$. Hence,

$$\sigma_1\bar{\beta}^{-1}\mathbf{p}_1^{\dagger} = \beta\|\mathbf{k}\|^{-2}\mathbf{p}_1^* = -\mathbf{v}^* - \beta\mathbf{k}^{\dagger}.$$

Thus,

$$(A + \mathbf{c}\mathbf{d}^*)(A^{\dagger}A - \mathbf{k}\mathbf{k}^{\dagger} + \mathbf{p}_1\mathbf{p}_1^{\dagger}) = A + \mathbf{c}\mathbf{d}^*.$$

Because $A^{\dagger}A - \mathbf{k}\mathbf{k}^{\dagger} + \mathbf{p}_1\mathbf{p}_1^{\dagger}$ is an orthogonal projector,

$$\mathcal{R}(A^* + \mathbf{d}\mathbf{c}^*) \subset \mathcal{R}(A^{\dagger}A - \mathbf{k}\mathbf{k}^{\dagger} + \mathbf{p}_1\mathbf{p}_1^{\dagger}).$$

From (5.3.7), we conclude that

$$\mathcal{R}(A^* + \mathbf{d}\mathbf{c}^*) = \mathcal{R}(A^{\dagger}A - \mathbf{k}\mathbf{k}^{\dagger} + \mathbf{p}_1\mathbf{p}_1^{\dagger}).$$

Consequently,

$$(A + \mathbf{c}\mathbf{d}^*)^{\dagger}(A + \mathbf{c}\mathbf{d}^*) = A^{\dagger}A - \mathbf{k}\mathbf{k}^{\dagger} + \mathbf{p}_1\mathbf{p}_1^{\dagger},$$

which is equivalent to (5.3.7).

To show the Eq. (5.3.6), using the identities $\mathbf{k}^*A^{\dagger}A = \mathbf{k}^*$, $\mathbf{q}_1^*\mathbf{c} = 1 - \sigma_1\bar{\beta}^{-1}$, and $\mathbf{q}_1^*A + \mathbf{d}^* = -\|\mathbf{v}\|^2\bar{\beta}^{-1}\mathbf{k}^* + \mathbf{v}^*$, we get

$X_3 M$

$= X_3(A + \mathbf{cd}^*)$

$= A^\dagger A + \bar{\beta}^{-1}\mathbf{vk}^* - \bar{\beta}\sigma_1^{-1}\mathbf{p}_1\mathbf{q}_1^* A + (\mathbf{k} + \bar{\beta}^{-1}\|\mathbf{k}\|^2\mathbf{v})\mathbf{d}^* - \bar{\beta}\sigma_1^{-1}\mathbf{p}_1\mathbf{q}_1^*\mathbf{cd}^*$

$= A^\dagger A + \bar{\beta}^{-1}\mathbf{vk}^* - \bar{\beta}\sigma_1^{-1}\mathbf{p}_1\mathbf{q}_1^* A - \mathbf{p}_1\mathbf{d}^* - \bar{\beta}\sigma_1^{-1}\mathbf{p}_1\mathbf{d}^* + \mathbf{p}_1\mathbf{d}^*$

$= A^\dagger A + \bar{\beta}^{-1}\mathbf{vk}^* - \bar{\beta}\sigma_1^{-1}\mathbf{p}_1(\mathbf{q}_1^* A + \mathbf{d}^*)$

$= A^\dagger A + \bar{\beta}^{-1}\mathbf{vk}^* - \bar{\beta}\sigma_1^{-1}\mathbf{p}_1(\mathbf{v}^* - \bar{\beta}^{-1}\|\mathbf{v}\|^2\mathbf{k}^*).$

Writing $\mathbf{v}^*$ as $\mathbf{v}^* = -\beta\|\mathbf{k}\|^{-2}(\mathbf{p}_1^* + \mathbf{k}^*)$, substituting this into the above expression in parentheses, and using the fact that $\|\mathbf{p}_1\|^{-2} = |\beta|^2\sigma_1^{-1}\|\mathbf{k}\|^{-2}$, we obtain

$$X_3 M = A^\dagger A + \bar{\beta}^{-1}\mathbf{vk}^* + \mathbf{p}_1\mathbf{p}_1^\dagger + \|\mathbf{k}\|^{-2}\mathbf{p}_1\mathbf{k}^*.$$

Since

$$\bar{\beta}^{-1}\mathbf{v} + \|\mathbf{k}\|^{-2}\mathbf{p}_1 = \bar{\beta}^{-1}\mathbf{v} - \bar{\beta}^{-1}\mathbf{v} - \|\mathbf{k}\|^{-2}\mathbf{k} = -\|\mathbf{k}\|^{-2}\mathbf{k},$$

we have

$$X_3 M = A^\dagger A + \mathbf{p}_1\mathbf{p}_1^\dagger - \mathbf{kk}^\dagger,$$

that is, (5.3.6) holds.

(4) and (5) It is easy to see that (4) follows from (2) and (5) follows from (3) by taking conjugate transposes and using the fact that for any matrix M, $(M^\dagger)^* = (M^*)^\dagger$.

(6) Both matrices $AA^\dagger - \mathbf{h}^{*\dagger}\mathbf{h}^*$ and $A^\dagger A - \mathbf{kk}^\dagger$ are orthogonal projectors. The fact that they are idempotent follows from $AA^\dagger\mathbf{h}^{*\dagger} = \mathbf{h}^{*\dagger}$, $\mathbf{h}^* AA^\dagger = \mathbf{h}^*$, $A^\dagger A\mathbf{k} = \mathbf{k}$, and $\mathbf{k}^\dagger A^\dagger A = \mathbf{k}^\dagger$. It is clear that they are Hermitian.

Moreover

$$\begin{aligned}
\mathrm{rank}(AA^\dagger - \mathbf{h}^{*\dagger}\mathbf{h}^*) &= \mathrm{tr}(AA^\dagger - \mathbf{h}^{*\dagger}\mathbf{h}^*) \\
&= \mathrm{tr}(AA^\dagger) - \mathrm{tr}(\mathbf{h}^{*\dagger}\mathbf{h}^*) \\
&= \mathrm{rank}(AA^\dagger) - \mathrm{rank}(\mathbf{h}^{*\dagger}\mathbf{h}^*) \\
&= \mathrm{rank}(A) - 1.
\end{aligned}$$

Similarly

$$\mathrm{rank}(A^\dagger A - \mathbf{kk}^\dagger) = \mathrm{rank}(A) - 1.$$

Since $\mathbf{u} = \mathbf{0}$, $\mathbf{v} = \mathbf{0}$, and $\beta = 0$, from Lemma 5.3.1,

$$\mathrm{rank}(A + \mathbf{cd}^*) = \mathrm{rank}(A) - 1.$$

Hence

$$\mathrm{rank}(A + \mathbf{cd}^*) = \mathrm{rank}(AA^\dagger - \mathbf{h}^{*\dagger}\mathbf{h}^*) = \mathrm{rank}(A^\dagger A - \mathbf{kk}^\dagger). \tag{5.3.8}$$

From $AA^\dagger c = c$, $h^*c = -1$, and $h^*A = d^*$, it is easy to see that

$$(AA^\dagger - h^{*\dagger}h^*)(A + cd^*) = A + cd^*.$$

Hence $\mathcal{R}(A + cd^*) \subset \mathcal{R}(AA^\dagger - h^{*\dagger}h^*)$. Likewise, using $d^*A^\dagger A = d^*$, $d^*k = -1$, and $Ak = c$, we have

$$(A + cd^*)(A^\dagger A - kk^\dagger) = A + cd^*.$$

Hence $\mathcal{R}(A^* + dc^*) \subset \mathcal{R}(A^\dagger A - kk^\dagger)$. It now follows from (5.3.8) that

$$(A + cd^*)(A + cd^*)^\dagger = AA^\dagger - h^{*\dagger}h^* \tag{5.3.9}$$

and

$$(A^* + dc^*)(A^* + dc^*)^\dagger = A^\dagger A - kk^\dagger.$$

Taking conjugate transposes on the both sides of the above equation, we get

$$(A + cd^*)^\dagger(A + cd^*) = A^\dagger A - kk^\dagger. \tag{5.3.10}$$

Let X_4 denote the right-hand side of (6) and $M = A + cd^*$. Using (5.3.9) and $h^*AA^\dagger = h^*$, we obtain

$$X_4MM^\dagger = X_4(AA^\dagger - h^{*\dagger}h^*) = X_4.$$

On the other hand, Using $k^\dagger A^\dagger A = k^\dagger$, $h^*A = d^*$, $h^*c = -1$, and (5.3.10), we obtain

$$X_4M = X_4(A + cd^*) = A^\dagger A - kk^\dagger = M^\dagger M.$$

Hence

$$M^\dagger = (M^\dagger M)M^\dagger = X_4MM^\dagger = X_4,$$

which completes the proof. □

Corollary 5.3.1 *Let $c \in \mathcal{R}(A)$, $d \in \mathcal{R}(A^*)$, and $\beta \neq 0$, then*

$$(A + cd^*)^\dagger = A^\dagger - \beta^{-1}A^\dagger cd^*A^\dagger = A^\dagger - \beta^{-1}kh^*.$$

Proof Set $v = 0$ in (3) or $u = 0$ in (5) of Theorem 5.3.1. □

The following is a special case when A and $A + cd^*$ are nonsingular.

Corollary 5.3.2 *Let A and $A + cd^*$ be nonsingular, then*

$$(A + cd^*)^{-1} = A^{-1} - \beta^{-1}A^{-1}cd^*A^{-1}.$$

The Moore-Penrose inverse of a rank-r modified matrix $M = A - CB$ is discussed in [8], where $\text{rank}(S) = r$ and $S = BC$ is a full rank factorization. The Drazin inverse of a rank-one modified matrix $M = A + \mathbf{cd}^*$ is discussed in [9].

By using the generalized singular value decomposition, a more general weighted Moore-Penrose inverse of a rank-r modified matrix $A - CB$ is discussed in [10], where CB need not be a full rank factorization. The Drazin inverse of a rank-r modified matrix $M = A - CB$ is discussed in [11]. Cases similar to those in Theorem 5.3.1 are considered in the derivation of the expressions of the weighted Moore-Penrose inverse of a rank-r modified matrix or the Drazin inverse of a rank-r modified matrix. See [10, 11] for details.

5.3.2 Greville's Method

Let $A_k \in \mathbb{C}^{m \times k}$ be partitioned as $A_k = [A_{k-1} \ \mathbf{a}_k]$, where $A_{k-1} \in \mathbb{C}^{m \times (k-1)}$ and $\mathbf{a}_k \in \mathbb{C}^m$ is the last column. It is a simple partitioned matrix called the Greville's partitioned matrix.

The Greville's method [12] for computing the Moore-Penrose inverse $A^\dagger$ is a recursive algorithm. At the kth iteration ($k = 1, 2, \cdots, n$) it computes $A_k^\dagger$, where A_k is the submatrix of A consisting of its first k columns. For $k = 2, 3, \cdots, n$, the matrix A_k is partitioned as

$$A_k = [A_{k-1} \ \mathbf{a}_k],$$

where $\mathbf{a}_k$ is the kth column of A. For $k = 2, 3, \cdots, n$, let the vector $\mathbf{d}_k$ and $\mathbf{c}_k$ be defined by

$$\mathbf{d}_k = A_{k-1}^\dagger \mathbf{a}_k,$$

$$\mathbf{c}_k = \mathbf{a}_k - A_{k-1} \mathbf{d}_k = (I - A_{k-1} A_{k-1}^\dagger) \mathbf{a}_k,$$

then the following theorem gives an expression of $A_k^\dagger$ in terms of $A_{k-1}^\dagger$, $\mathbf{d}_k$, and $\mathbf{c}_k$.

Theorem 5.3.2 *Let $A \in \mathbb{C}^{m \times n}$, using the above notations, then the Moore-Penrose inverse of A_k is given by*

$$A_k^\dagger = [A_{k-1} \ \mathbf{a}_k]^\dagger = \begin{bmatrix} A_{k-1}^\dagger - \mathbf{d}_k \mathbf{b}_k^* \\ \mathbf{b}_k^* \end{bmatrix}, \quad k = 2, 3, \cdots, n,$$

and

$$A_n^\dagger = A^\dagger,$$

where

$$\mathbf{b}_k^* = \begin{cases} \mathbf{c}_k^\dagger & \text{if } \mathbf{c}_k \neq 0, \\ (1 + \mathbf{d}_k^* \mathbf{d}_k)^{-1} \mathbf{d}_k^* A_{k-1}^\dagger & \text{if } \mathbf{c}_k = 0. \end{cases}$$

Proof Since A_k can be written as

$$A_k = [A_{k-1} \ \mathbf{a}_k] = [A_{k-1} \ \mathbf{0}_1] + \mathbf{a}_k[\mathbf{0}_2^T \ 1],$$

where $\mathbf{0}_1 \in \mathbb{C}^m$ and $\mathbf{0}_2 \in \mathbb{C}^{k-1}$ are zero vectors, A_k is in the rank-one modified matrix form in Theorem 5.3.1. Let $\widetilde{A} = [A_{k-1} \ \mathbf{0}_1]$, $\widetilde{\mathbf{c}} = \mathbf{a}_k$, and $\widetilde{\mathbf{d}}^* = [\mathbf{0}_2^T, 1]$, then

$$A_k^\dagger = (\widetilde{A} + \widetilde{\mathbf{c}}\widetilde{\mathbf{d}}^*)^\dagger.$$

Using the notations in Theorem 5.3.1 and the equation

$$\widetilde{A}^\dagger = \begin{bmatrix} A_{k-1}^\dagger \\ \mathbf{0}^T \end{bmatrix},$$

we have $\widetilde{\mathbf{h}}^* = \widetilde{\mathbf{d}}^*\widetilde{A}^\dagger = \mathbf{0}^T$, so $\widetilde{\beta} = 1 + \widetilde{\mathbf{d}}^*\widetilde{A}^\dagger\widetilde{\mathbf{c}} = 1$ and $\widetilde{\mathbf{v}}^* = \widetilde{\mathbf{d}}^*(I - \widetilde{A}^\dagger\widetilde{A}) = \widetilde{\mathbf{d}}^* \neq \mathbf{0}^T$, $\widetilde{\mathbf{u}} = (I - \widetilde{A}\widetilde{A}^\dagger)\widetilde{\mathbf{c}} = (I - A_{k-1}A_{k-1}^\dagger)\mathbf{a}_k = \mathbf{c}_k$. Thus, there are only two cases to consider: $\mathbf{c}_k \neq \mathbf{0}$ and $\mathbf{c}_k = \mathbf{0}$.

Case (1): If $\widetilde{\mathbf{u}} = \mathbf{c}_k \neq \mathbf{0}$ and $\widetilde{\mathbf{v}} \neq \mathbf{0}$, then the case (1) in Theorem 5.3.1 is applied to obtain $A_k^\dagger$. It is clear that

$$\widetilde{\mathbf{k}} = \widetilde{A}^\dagger\widetilde{\mathbf{c}} = \begin{bmatrix} A_{k-1}^\dagger\mathbf{a}_k \\ 0 \end{bmatrix} \quad \text{and} \quad (\widetilde{\mathbf{v}}^*)^\dagger = (\widetilde{\mathbf{d}}^*)^\dagger = \begin{bmatrix} 0 \\ 1 \end{bmatrix}.$$

Thus

$$\begin{aligned}
A_k^\dagger &= \widetilde{A}^\dagger - \widetilde{\mathbf{k}}\widetilde{\mathbf{u}}^\dagger - (\widetilde{\mathbf{v}}^*)^\dagger\widetilde{\mathbf{h}}^* + \widetilde{\beta}(\widetilde{\mathbf{v}}^*)^\dagger\widetilde{\mathbf{u}}^\dagger \\
&= \begin{bmatrix} A_{k-1}^\dagger \\ \mathbf{0}^T \end{bmatrix} - \begin{bmatrix} A_{k-1}^\dagger\mathbf{a}_k\mathbf{c}_k^\dagger \\ \mathbf{0}^T \end{bmatrix} - O + \begin{bmatrix} O \\ \mathbf{c}_k^\dagger \end{bmatrix} \\
&= \begin{bmatrix} A_{k-1}^\dagger - \mathbf{d}_k\mathbf{c}_k^\dagger \\ \mathbf{c}_k^\dagger \end{bmatrix}.
\end{aligned}$$

Case (2): If $\widetilde{\mathbf{u}} = \mathbf{c}_k = \mathbf{0}$ and $\widetilde{\beta} \neq 0$, then the case (3) in Theorem 5.3.1 is applied to obtain $A_k^\dagger$. It is clear that

$$\widetilde{\mathbf{k}} = \widetilde{A}^\dagger\widetilde{\mathbf{c}} = \begin{bmatrix} A_{k-1}^\dagger\mathbf{a}_k \\ 0 \end{bmatrix} = \begin{bmatrix} \mathbf{d}_k \\ 0 \end{bmatrix}, \quad \widetilde{\sigma}_1 = 1 + \mathbf{d}_k^*\mathbf{d}_k,$$

$$\widetilde{\mathbf{p}}_1 = -\begin{bmatrix} \mathbf{d}_k \\ \mathbf{d}_k^*\mathbf{d}_k \end{bmatrix}, \quad \text{and} \quad \widetilde{\mathbf{q}}_1^* = -\mathbf{d}_k^*A_{k-1}^\dagger.$$

Thus

$$A_k^\dagger = \tilde{A}^\dagger + (\bar{\bar{\beta}})^{-1}\widetilde{\mathbf{vk}}^*\tilde{A}^\dagger - \bar{\bar{\beta}}\,\tilde{\sigma}_1^{-1}\widetilde{\mathbf{p}}_1\widetilde{\mathbf{q}}_1^*$$

$$= \begin{bmatrix} A_{k-1}^\dagger - (1 + \mathbf{d}_k^*\mathbf{d}_k)^{-1}\mathbf{d}_k\mathbf{d}_k^*A_{k-1}^\dagger \\ (1 + \mathbf{d}_k^*\mathbf{d}_k)^{-1}\mathbf{d}_k^*A_{k-1}^\dagger \end{bmatrix}.$$

The proof is completed by noting that $\mathbf{b}_k^* = (1 + \mathbf{d}_k^*\mathbf{d}_k)^{-1}\mathbf{d}_k^*A_{k-1}^\dagger$. $\square$

There are other proofs of the above theorem, see [13–16].

A method for computing the weighted Moore-Penrose inverse of Greville's partitioned matrix is given as follows.

Theorem 5.3.3 ([17]) *Let* $A \in \mathbb{C}^{m\times n}$, *M and N be Hermitian positive definite matrices of orders m and n respectively. Denote* A_k *as the submatrix consisting of the first k columns of A,* $\mathbf{a}_k$ *as the kth column of A, and* $A_k = [A_{k-1}\ \mathbf{a}_k]$. *Suppose that*

$$N_k = \begin{bmatrix} N_{k-1} & \mathbf{l}_k \\ \mathbf{l}_k^* & n_{kk} \end{bmatrix} \in \mathbb{C}^{k\times k}$$

is the kth leading principal submatrix of N and

$$X_{k-1} = (A_{k-1})_{M,N_{k-1}}^\dagger \quad and \quad X_k = (A_k)_{M,N_k}^\dagger$$

are the weighted Moore-Penrose inverses of A_{k-1} *and* A_k *respectively. For* $k = 2, 3, \cdots, n$, *define*

$$\mathbf{d}_k = X_{k-1}\mathbf{a}_k,$$
$$\mathbf{c}_k = \mathbf{a}_k - A_{k-1}\mathbf{d}_k = (I - A_{k-1}X_{k-1})\mathbf{a}_k,$$

then

$$X_k = \begin{bmatrix} X_{k-1} - (\mathbf{d}_k + (I - X_{k-1}A_{k-1})N_{k-1}^{-1}\mathbf{l}_k)\mathbf{b}_k^* \\ \mathbf{b}_k^* \end{bmatrix}$$

and

$$X_n = A_{MN}^\dagger,$$

where

$$\mathbf{b}_k^* = \begin{cases} (\mathbf{c}_k^*M\mathbf{c}_k)^{-1}\mathbf{c}_k^*M, & if\ \mathbf{c}_k \neq 0, \\ \delta_k^{-1}(\mathbf{d}_k^*N_{k-1} - \mathbf{l}_k^*)X_{k-1}, & if\ \mathbf{c}_k = 0, \end{cases}$$

and

$$\delta_k = n_{kk} + \mathbf{d}_k^*N_k\mathbf{d}_k - (\mathbf{d}_k^*\mathbf{l}_k + \mathbf{l}_k^*\mathbf{d}_k) - \mathbf{l}_k^*(I - X_{k-1}A_{k-1})N_{k-1}^{-1}\mathbf{l}_k.$$

Note that the initial $X_1 = (\mathbf{a}_1^*M\mathbf{a}_1)^{-1}\mathbf{a}_1^*M$ and it is easy to compute N_{k-1}^{-1} by (5.3.1) and (5.3.2).

5.3.3 Cline's Method

An extension of the Greville's partitioned matrix is $A = [U \ \ V]$, where U and V are general conformable matrices and V no longer is just a column. The matrix $[U \ \ V]$ is called the Cline's partitioned matrix.

The Cline's method for computing the Moore-Penrose inverse of partitioned matrix $A = [U \ \ V]$ is given by the following theorem.

Theorem 5.3.4 ([18]) *Let* $A = [U \ \ V] \in \mathbb{C}^{m \times n}$, *where* $U \in \mathbb{C}^{m \times p}$ $(p < n)$ *and* $V \in \mathbb{C}^{m \times (n-p)}$, *then*

$$
A^\dagger = [U \ \ V]^\dagger \tag{5.3.11}
$$
$$
= \begin{bmatrix} U^\dagger - U^\dagger V C^\dagger - U^\dagger V (I - C^\dagger C) K_1^{-1} V^* U^{\dagger*} U^\dagger (I - V C^\dagger) \\ C^\dagger + (I - C^\dagger C) K_1^{-1} V^* U^{\dagger*} U^\dagger (I - V C^\dagger) \end{bmatrix},
$$

where

$$
\begin{aligned}
C &= (I - UU^\dagger)V, \\
K_1 &= I + (I - C^\dagger C) V^* U^{\dagger*} U^\dagger V (I - C^\dagger C).
\end{aligned} \tag{5.3.12}
$$

If

$$
(C^\dagger C)(V^* U^{\dagger*} U^\dagger V) = (V^* U^{\dagger*} U^\dagger V)(C^\dagger C),
$$

then from (5.3.11) and (5.3.12) we have

$$
A^\dagger = [U \ \ V]^\dagger
$$
$$
= \begin{bmatrix} U^\dagger - U^\dagger V C^\dagger - U^\dagger V (I - C^\dagger C) K_2^{-1} V^* U^{\dagger*} U^\dagger \\ C^\dagger + (I - C^\dagger C) K_2^{-1} V^* U^{\dagger*} U^\dagger \end{bmatrix}, \tag{5.3.13}
$$

where

$$
\begin{aligned}
C &= (I - UU^\dagger)V, \\
K_2 &= I + V^* U^{\dagger*} U^\dagger V.
\end{aligned}
$$

In 1971, an alternative representation of the Moore-Penrose inverse of the partitioned matrix $A = [U \ \ V]$ was given by Mihalyffy in [19]:

$$
A^\dagger = [U \ \ V]^\dagger = \begin{bmatrix} KU^\dagger(I - VC^\dagger) \\ T^* KU^\dagger(I - VC^\dagger) + C^\dagger \end{bmatrix},
$$

where

$$
\begin{aligned}
C &= (I - UU^\dagger)V, \\
T &= U^\dagger V (I - C^\dagger C), \\
K &= (I + TT^*)^{-1}.
\end{aligned}
$$

The reader should be aware that there are other ways of representing the Moore-Penrose inverse for the partitioned matrix $A = [U \quad V]$. The interested reader is referred to [20, 21].

The method for computing the weighted Moore-Penrose inverse of the partitioned matrix $A = [U \quad V]$ is given by the following theorem.

Theorem 5.3.5 ([22]) *Let $A = [U \quad V] \in \mathbb{C}^{m \times n}$, where $U \in \mathbb{C}^{m \times p}$ ($p < n$) and $V \in \mathbb{C}^{m \times (n-p)}$, and M and N be Hermitian positive definite matrices of orders m and n respectively. Partition N as*

$$N = \begin{bmatrix} N_1 & L \\ L^* & N_2 \end{bmatrix},$$

where $N_1 \in \mathbb{C}^{p \times p}$ and let

$$D = U^{\dagger}_{MN_1} V,$$
$$C = (I - UU^{\dagger}_{MN_1})V,$$
$$K = N_2 + D^* N_1 D - (D^* L + L^* D) - L^*(I - U^{\dagger}_{MN_1} U)N_1^{-1}L,$$

then K is a Hermitian positive definite matrix and

$$A^{\dagger}_{MN} = \begin{bmatrix} U^{\dagger}_{MN_1} - DH - (I - U^{\dagger}_{MN_1} U)N_1^{-1}LH \\ H \end{bmatrix}, \qquad (5.3.14)$$

where

$$H = C^{\dagger}_{MK} + (I - C^{\dagger}_{MK}C)K^{-1}(D^* N_1 - L^*)U^{\dagger}_{MN_1}.$$

Proof Let X denote the right-hand side of (5.3.14). It is easy to verify that the following four conditions are satisfied:

$$AXA = A, \qquad XAX = X,$$
$$(MAX)^* = MAX, \quad (NXA)^* = NXA,$$

which completes the proof.   □

The following corollary follows from Theorem 5.3.5.

Corollary 5.3.3 *Suppose that A, U and V are the same as in Theorem 5.3.5. Then*

$$A^{\dagger} = [U \quad V]^{\dagger} \qquad (5.3.15)$$
$$= \begin{bmatrix} U^{\dagger} - U^{\dagger}VC^{\dagger}_{IK_1} - U^{\dagger}V(I - C^{\dagger}_{IK_1}C)K_1^{-1}V^*U^{\dagger *}U^{\dagger} \\ C^{\dagger}_{IK_1} + (I - C^{\dagger}_{IK_1}C)K_1^{-1}V^*U^{\dagger *}U^{\dagger} \end{bmatrix},$$

where

$$C = (I - UU^{\dagger})V,$$
$$K_1 = I + V^*U^{\dagger *}U^{\dagger}V.$$

Comparing (5.3.13) with (5.3.15), $C^\dagger$ in (5.3.13) is replaced by $C^\dagger_{IK_1}$ in (5.3.15). No additional conditions are necessary here.

5.3.4 Noble's Method

An extension of the Cline's partitioned matrix is the Noble's partitioned matrix of the form:

$$M = \begin{bmatrix} A & C \\ B & D \end{bmatrix}, \tag{5.3.16}$$

where A, B, C, and D are general conformable matrices.

The answer to "What is the Moore-Penrose inverse of M?" is difficult. However, if we place some restrictions on the blocks in M, we can obtain some useful results.

Lemma 5.3.2 *If A, B, C, and D in (5.3.16) are conformable matrices such that A is square and nonsingular, and* $\mathrm{rank}(M) = \mathrm{rank}(A)$, *then $D = BA^{-1}C$. Furthermore, if $P = BA^{-1}$ and $Q = A^{-1}C$, then*

$$M = \begin{bmatrix} A & C \\ B & D \end{bmatrix} = \begin{bmatrix} I \\ P \end{bmatrix} A [I \ \ Q]. \tag{5.3.17}$$

Proof The factorization

$$\begin{bmatrix} I & O \\ -BA^{-1} & I \end{bmatrix} \begin{bmatrix} A & C \\ B & D \end{bmatrix} \begin{bmatrix} I & -A^{-1}C \\ O & I \end{bmatrix} = \begin{bmatrix} A & O \\ O & D - BA^{-1}C \end{bmatrix} \tag{5.3.18}$$

yields

$$\mathrm{rank}(M) = \mathrm{rank}(A) + \mathrm{rank}(D - BA^{-1}C).$$

Therefore, it can be concluded that $\mathrm{rank}(D - BA^{-1}C) = 0$, equivalently, $D = BA^{-1}C$. Then (5.3.17) follows from (5.3.18). □

Theorem 5.3.6 *Let A, B, C, and D in (5.3.16) be conformable matrices such that A is square and nonsingular, and* $\mathrm{rank}(M) = \mathrm{rank}(A)$, $P = BA^{-1}$, *and* $Q = A^{-1}C$, *then*

$$M^\dagger = \begin{bmatrix} A & C \\ B & D \end{bmatrix}^\dagger = \begin{bmatrix} I \\ Q^* \end{bmatrix} ((I + P^*P)A(I + QQ^*))^{-1} [I \ \ P^*].$$

Proof Let $A \in \mathbb{C}^{r \times r}_r$,

$$F = \begin{bmatrix} I_r \\ P \end{bmatrix} A \quad \text{and} \quad G = [I_r \ \ Q].$$

then

$$\mathrm{rank}(F) = \mathrm{rank}(G) = \mathrm{rank}(A) = \mathrm{rank}(M) = r.$$

It follows from (5.3.17) that $M = FG$. By Theorem 1.1.5 and the nonsingularity of A, we have

$$
\begin{aligned}
M^\dagger &= G^\dagger F^\dagger \\
&= G^*(GG^*)^{-1}(F^*F)^{-1}F^* \\
&= \begin{bmatrix} I \\ Q^* \end{bmatrix} (I + QQ^*)^{-1}(A^*(I + P^*P)A)^{-1}A^*[I \ \ P^*] \\
&= \begin{bmatrix} I \\ Q^* \end{bmatrix} (I + QQ^*)^{-1}((I + P^*P)A)^{-1}[I \ \ P^*] \\
&= \begin{bmatrix} I \\ Q^* \end{bmatrix} ((I + P^*P)A(I + QQ^*))^{-1}[I \ \ P^*],
\end{aligned}
$$

which completes the proof. $\square$

Let A, B, C, and D in (5.3.16) be conformable matrices such that M and A are square. The Drazin inverse of the Noble's partitioned matrix M is given by the following theorem. First, we give two lemmas.

Lemma 5.3.3 Let $T \in \mathbb{C}^{n \times n}$, $R \in \mathbb{C}_n^{r \times n}$, and $S \in \mathbb{C}_n^{n \times s}$, then

$$
\mathrm{rank}(RTS) = \mathrm{rank}(T).
$$

Proof Since $R^\dagger R = I_n$ and $SS^\dagger = I_n$, we have

$$
\mathrm{rank}(RTS) \le \mathrm{rank}(T) = \mathrm{rank}(R^\dagger RTSS^\dagger) \le \mathrm{rank}(RTS),
$$

implying that $\mathrm{rank}(RTS) = \mathrm{rank}(T)$. $\square$

Lemma 5.3.4 Let $A \in \mathbb{C}_r^{r \times r}$ and

$$
M = \begin{bmatrix} A & C \\ B & D \end{bmatrix}.
$$

If $\mathrm{rank}(M) = \mathrm{rank}(A) = r$, then

$$
\mathrm{Ind}(M) = \mathrm{Ind}(A(I + QP)) + 1 = \mathrm{Ind}((I + QP)A) + 1,
$$

where $P = BA^{-1}$ and $Q = A^{-1}C$.

Proof From Lemma 5.3.2,

$$
D = BA^{-1}C \quad \text{and} \quad M = \begin{bmatrix} I \\ P \end{bmatrix} A[I \ \ Q].
$$

Thus for any integer $i > 0$,

$$M^i = \begin{bmatrix} I \\ P \end{bmatrix} (A(I + QP))^{i-1} A[I \ Q] \qquad (5.3.19)$$

$$= \begin{bmatrix} I \\ P \end{bmatrix} A((I + QP)A)^{i-1} [I \ Q].$$

Since the matrix

$$\begin{bmatrix} I \\ P \end{bmatrix}$$

is of full column rank and $A[I \ Q]$ is of full row rank, from Lemma 5.3.3, we have $\mathrm{rank}(M^{m+1}) = \mathrm{rank}((A(I + QP))^m)$. Thus

$$\mathrm{rank}((A(I + QP))^m) = \mathrm{rank}((A(I + QP))^{m-1})$$
$$\Leftrightarrow \mathrm{rank}(M^{m+1}) = \mathrm{rank}(M^m).$$

It then follows that $\mathrm{Ind}(A(I + QP)) + 1 = \mathrm{Ind}(M)$. The proof of $\mathrm{Ind}((I + QP)A) + 1 = \mathrm{Ind}(M)$ is similar. □

Corollary 5.3.4 *Under the same assumptions as Lemma 5.3.4,*

$$\mathrm{Ind}(M) = 1 \quad \Leftrightarrow \quad I + QP \text{ is nonsingular.}$$

Proof From Lemma 5.3.4,

$$\mathrm{Ind}(M) = 1 \Leftrightarrow \mathrm{Ind}(A(I + QP)) = 0$$
$$\Leftrightarrow A(I + QP) \text{ is nonsingular}$$
$$\Leftrightarrow I + QP \text{ is nonsingular.}$$

The penultimate equivalence follows from the nonsingularity of A. □

Now we have a theorem on the Drazin inverse.

Theorem 5.3.7 *Let $A \in \mathbb{C}_r^{r \times r}$,*

$$M = \begin{bmatrix} A & C \\ B & D \end{bmatrix},$$

where $\mathrm{Ind}(M) = m$, $\mathrm{rank}(M) = \mathrm{rank}(A) = r$, and $P = BA^{-1}$, $Q = A^{-1}C$, then

$$M_d = \begin{bmatrix} I \\ P \end{bmatrix} ((AS)^2)_d A[I \ Q]$$

$$= \begin{bmatrix} I \\ P \end{bmatrix} A((SA)^2)_d [I \ Q], \qquad (5.3.20)$$

where $S = I + A^{-1}CBA^{-1} = I + QP$.

Proof Setting

$$X = \begin{bmatrix} I \\ P \end{bmatrix} ((AS)^2)_d A[I \ \ Q]$$

and using (5.3.19), we have

$$M^{m+1}X = \begin{bmatrix} I \\ P \end{bmatrix} (AS)^{m+1}((AS)^2)_d A[I \ \ Q].$$

Since $\mathrm{Ind}(M) = m$, we have

$$\mathrm{Ind}(AS) = \mathrm{Ind}(A(I + QP)) = m - 1$$

by Lemma 5.3.4. Thus

$$(AS)^{m+1}((AS)^2)_d = (AS)^{m-1}.$$

Furthermore

$$M^{m+1}X = \begin{bmatrix} I \\ P \end{bmatrix} (AS)^{m-1} A[I \ \ Q] = M^m.$$

It is easy to verify that $MX = XM$ and $XMX = X$, which is left to the reader as an exercise. Therefore $M_d = X$.

The proof of the second equality of (5.3.20) is similar, it is omitted here. □

In Theorem 5.3.7, we have obtained an expression of M_d when A is nonsingular. If A is a square and singular matrix, then an expression of M_d is given as follows.

Theorem 5.3.8 ([23]) *Let*

$$M = \begin{bmatrix} A & C \\ B & D \end{bmatrix},$$

where A is a square and singular matrix, and

$$H = BA_d, \qquad K = A_dC,$$
$$P = (I - AA_d)C, \ Q = B(I - A_dA),$$
$$Z = D - BA_dC.$$

If $P = O$, $Q = O$, and $Z = O$, then

$$M_d = \begin{bmatrix} I \\ H \end{bmatrix} ((AS)_d)^2 A[I \ \ K]$$

$$= \begin{bmatrix} I \\ H \end{bmatrix} A((SA)_d)^2[I \ \ K],$$

where $S = I + A_dCBA_d = I + KH$.

Proof From the assumptions, we have

$$M = \begin{bmatrix} I \\ H \end{bmatrix} A[I \quad K].$$

By using (5.3.19), we have

$$M^i = \begin{bmatrix} I \\ H \end{bmatrix} (AS)^{i-1} A[I \quad K].$$

Setting

$$X = \begin{bmatrix} I \\ H \end{bmatrix} ((AS)_d)^2 A[I \quad K]$$

and noting that $AS \in \mathbb{C}^{m \times m}$ and $\text{Ind}(AS) \leq m$, we have

$$
\begin{aligned}
M^{m+2}X &= \begin{bmatrix} I \\ H \end{bmatrix} (AS)^{m+2} ((AS)_d)^2 A[I \quad K] \\
&= \begin{bmatrix} I \\ H \end{bmatrix} (AS)^m A[I \quad K] \\
&= M^{m+1}, \quad\quad\quad\quad\quad\quad\quad\quad\quad (5.3.21) \\
MX &= \begin{bmatrix} I \\ H \end{bmatrix} AS((AS)_d)^2 A[I \quad K] \\
&= \begin{bmatrix} I \\ H \end{bmatrix} ((AS)_d)^2 ASA[I \quad K] \\
&= \begin{bmatrix} I \\ H \end{bmatrix} ((AS)_d)^2 A[I \quad K] \begin{bmatrix} I \\ H \end{bmatrix} A[I \quad K] \\
&= XM, \quad\quad\quad\quad\quad\quad\quad\quad\quad\quad (5.3.22)
\end{aligned}
$$

and

$$
\begin{aligned}
&XMX \\
&= \begin{bmatrix} I \\ H \end{bmatrix} ((AS)_d)^2 A[I \quad K] \begin{bmatrix} I \\ H \end{bmatrix} A[I \quad K] \begin{bmatrix} I \\ H \end{bmatrix} ((AS)_d)^2 A[I \quad K] \\
&= \begin{bmatrix} I \\ H \end{bmatrix} ((AS)_d)^2 A[I \quad K] \\
&= X. \quad\quad\quad\quad\quad\quad\quad\quad\quad\quad\quad\quad (5.3.23)
\end{aligned}
$$

It then follows from (5.3.21)–(5.3.23) that $M_d = X$. $\quad\square$

An expression of the group inverse for the Noble's partitioned matrix is given as follows.

Theorem 5.3.9 *Let $A \in \mathbb{C}_r^{r \times r}$,*

$$M = \begin{bmatrix} A & C \\ B & D \end{bmatrix},$$

where $\mathrm{Ind}(M) = 1$ *and* $\mathrm{rank}(M) = \mathrm{rank}(A) = r$. *If* $P = BA^{-1}$ *and* $Q = A^{-1}C$, *then*

$$M_g = \begin{bmatrix} I \\ P \end{bmatrix} (SAS)^{-1} [I \quad Q]$$

$$= \begin{bmatrix} I \\ P \end{bmatrix} S^{-1} A^{-1} S^{-1} [I \quad Q]$$

$$= \begin{bmatrix} A \\ B \end{bmatrix} (A^2 + CB)^{-1} A (A^2 + CB)^{-1} [A \quad C],$$

where $S = I + A^{-1}CBA^{-1} = I + QP$.

Proof It follows from Corollary 5.3.4 that

$$\mathrm{Ind}(M) = 1 \quad \Leftrightarrow \quad S = I + QP \text{ is nonsingular.}$$

By using Theorem 5.3.7 and the nonsingularity of AS, we have

$$M_g = \begin{bmatrix} I \\ P \end{bmatrix} ((AS)^2)^{-1} A [I \quad Q]$$

$$= \begin{bmatrix} I \\ P \end{bmatrix} (AS)^{-1} (AS)^{-1} A [I \quad Q]$$

$$= \begin{bmatrix} I \\ P \end{bmatrix} (SAS)^{-1} [I \quad Q]$$

$$= \begin{bmatrix} I \\ BA^{-1} \end{bmatrix} S^{-1} A^{-1} S^{-1} [I \quad A^{-1}C]$$

$$= \begin{bmatrix} A \\ B \end{bmatrix} A^{-1} (A(A^2 + CB)^{-1}A) A^{-1} (A(A^2 + CB)^{-1}A) A^{-1} [A \quad C]$$

$$= \begin{bmatrix} A \\ B \end{bmatrix} (A^2 + CB)^{-1} A (A^2 + CB)^{-1} [A \quad C].$$

This completes the proof.   □

5.4 Embedding Methods

An embedding method for the Moore-Penrose inverse is given in [24] by Kalaba and Rasakhoo. The basic idea of the embedding methods is that the original problem is embedded in a problem with larger range, if we can get the solution of the later problem, then the solution of the original problem can be found.

The embedding methods for the weighted Moore-Penrose inverse, the Moore-Penrose inverse, the Drazin inverse and the group inverse are presented in [25], and these methods have a uniform formula.

5.4.1 Generalized Inverse as a Limit

In this subsection, we will show how the generalized inverses $A_{MN}^\dagger$, $A^\dagger$, A_d and A_g can be characterized in terms of a limiting process.

Theorem 5.4.1 *Let $A \in \mathbb{C}^{m \times n}$, rank$(A) = r$, and M and N be Hermitian positive definite matrices of orders m and n respectively, then*

$$A_{MN}^\dagger = \lim_{z \to 0^-} (N^{-1}A^*MA - zI)^{-1}N^{-1}A^*M, \qquad (5.4.1)$$

where z tends to zero through negative values.

Proof From the (M, N)-singular value decomposition Theorem 5.2.2, there exist an M-unitary matrix $U \in \mathbb{C}^{m \times m}$ and an N^{-1}-unitary matrix $V \in \mathbb{C}^{n \times n}$ such that

$$A = U \begin{bmatrix} D & O \\ O & O \end{bmatrix} V^*,$$

where

$$U^*MU = I_m, \qquad\qquad V^*N^{-1}V = I_n,$$
$$D = \mathrm{diag}(\mu_1, \mu_2, \cdots, \mu_r), \ \mu_i > 0, \ i = 1, 2, \cdots, r,$$

and

$$A_{MN}^\dagger = N^{-1}V \begin{bmatrix} D^{-1} & O \\ O & O \end{bmatrix} U^*M.$$

Let

$$N^{-1/2}V = \widetilde{V} = [\mathbf{v}_1\ \mathbf{v}_2\ \cdots\ \mathbf{v}_n] \ \text{ and}$$
$$M^{1/2}V = \widetilde{U} = [\mathbf{u}_1\ \mathbf{u}_2\ \cdots\ \mathbf{u}_m],$$

then

$$\widetilde{V}^* = \widetilde{V}^{-1}, \quad \widetilde{U}^* = \widetilde{U}^{-1}$$

and

$$A^{\dagger}_{MN} = N^{-1/2} \left(\sum_{i=1}^{r} \mu_i^{-1} \mathbf{v}_i \mathbf{u}_i^* \right) M^{1/2}.$$

Since

$$N^{-1}A^*MA = N^{-1/2} \left(\sum_{i=1}^{r} \mu_i^2 \mathbf{v}_i \mathbf{v}_i^* \right) N^{1/2}$$

and the vectors $\mathbf{v}_1, \mathbf{v}_2, \cdots, \mathbf{v}_n$ form an orthonormal basis for $\mathbb{C}^n$,

$$I = \sum_{i=1}^{n} \mathbf{v}_i \mathbf{v}_i^* = N^{-1/2} \left(\sum_{i=1}^{n} \mathbf{v}_i \mathbf{v}_i^* \right) N^{1/2}.$$

Therefore

$$N^{-1}A^*MA - zI = N^{-1/2} \left(\sum_{i=1}^{r} (\mu_i^2 - z)\mathbf{v}_i \mathbf{v}_i^* - z \sum_{i=r+1}^{n} \mathbf{v}_i \mathbf{v}_i^* \right) N^{1/2}.$$

Let $\widetilde{A} = M^{1/2} A N^{-1/2}$, then

$$\widetilde{A}^* \widetilde{A} = N^{1/2} (N^{-1}A^*MA) N^{-1/2}.$$

Since $\widetilde{A}^* \widetilde{A}$ is Hermitian positive semidefinite and has nonnegative eigenvalues, $N^{-1}A^*MA$ has nonnegative eigenvalues too. The matrix $N^{-1}A^*MA - zI$ is therefore nonsingular for $z < 0$. Its inverse is

$$(N^{-1}A^*MA - zI)^{-1}$$
$$= N^{-1/2} \left(\sum_{i=1}^{r} (\mu_i^2 - z)^{-1} \mathbf{v}_i \mathbf{v}_i^* - \sum_{i=r+1}^{n} z^{-1} \mathbf{v}_i \mathbf{v}_i^* \right) N^{1/2}.$$

Next, form

$$(N^{-1}A^*MA - zI)^{-1}N^{-1}A^*M = N^{-1/2} \left(\sum_{i=1}^{r} \frac{\mu_i}{\mu_i^2 - z} \mathbf{v}_i \mathbf{u}_i^* \right) M^{1/2}.$$

Now, we take the limit and get

$$\lim_{z \to 0} (N^{-1} A^* M A - zI)^{-1} N^{-1} A^* M$$

$$= \lim_{z \to 0} N^{-1/2} \left(\sum_{i=1}^{r} \frac{\mu_i}{\mu_i^2 - z} \mathbf{v}_i \mathbf{u}_i^* \right) M^{1/2}$$

$$= N^{-1/2} \left(\sum_{i=1}^{r} \mu_i^{-1} \mathbf{v}_i \mathbf{u}_i^* \right) M^{1/2}$$

$$= A_{MN}^\dagger,$$

which is (5.4.1).   □

Corollary 5.4.1 *Let $A \in \mathbb{C}^{m \times n}$, then*

$$A^\dagger = \lim_{z \to 0} (A^* A - zI)^{-1} A^*, \tag{5.4.2}$$

where z tends to zero through negative values.

Theorem 5.4.2 *Let $A \in \mathbb{C}^{n \times n}$ with $\mathrm{Ind}(A) = k$, then*

$$A_d = \lim_{z \to 0} (A^{k+1} - zI)^{-1} A^k, \tag{5.4.3}$$

where z tends to zero through negative values.

Proof From Theorem 2.1.2 of the canonical representations of A and A_d, there exists a nonsingular matrix P such that

$$A = P \begin{bmatrix} C & O \\ O & N \end{bmatrix} P^{-1}, \tag{5.4.4}$$

where C is nonsingular and N is nilpotent of index k, i.e., $N^k = O$, and

$$A_d = P \begin{bmatrix} C^{-1} & O \\ O & O \end{bmatrix} P^{-1}.$$

From (5.4.4),

$$A^{k+1} - zI = P \begin{bmatrix} C^{k+1} - zI & O \\ O & -zI \end{bmatrix} P^{-1}.$$

Since C is nonsingular, C^{k+1} is also nonsingular, and z tends to zero from the left, $C^{k+1} - zI$ is also nonsingular. Then

$$\lim_{z \to 0} (A^{k+1} - zI)^{-1} A^k = \lim_{z \to 0} P \begin{bmatrix} (C^{k+1} - zI)^{-1} C^k & O \\ O & O \end{bmatrix} P^{-1}$$

$$= P \begin{bmatrix} C^{-1} & O \\ O & O \end{bmatrix} P^{-1}$$

$$= A_d,$$

which leads to (5.4.3). □

Corollary 5.4.2 *Let $A \in \mathbb{C}^{n \times n}$ with $\mathrm{Ind}(A) = 1$, then*

$$A_g = \lim_{z \to 0^-} (A^2 - zI)^{-1} A. \tag{5.4.5}$$

If A is nonsingular, then

$$A^{-1} = \lim_{z \to 0^-} (A - zI)^{-1}, \tag{5.4.6}$$

where z tends to zero through negative values.

5.4.2 Embedding Methods

In order to find the generalized inverses $A_{MN}^{\dagger}$, $A^{\dagger}$, A_d, A_g, and the regular inverse A^{-1}, from (5.4.1), (5.4.2), (5.4.3), (5.4.5) and (5.4.6), we must find the inverse of the matrix $B_t(z)$, an $n \times n$ matrix of z,

$$B_t(z) = [b_{ij}^{(t)}] = \begin{cases} N^{-1}A^*MA - zI, & t = 1; \\ A^*A - zI, & t = 2; \\ A^{k+1} - zI, & t = 3; \\ A^2 - zI, & t = 4; \\ A - zI, & t = 5. \end{cases}$$

Let

$$F_t(z) = \mathrm{adj}(B_t(z)) = [B_{ij}^{(t)}] \quad \text{and} \quad g_t(z) = \det(B_t(z)), \tag{5.4.7}$$

where $\mathrm{adj}(B_t(z))$ is the adjoint of the matrix $B_t(z)$, whose elements $B_{ij}^{(t)}$ are the cofactors of the jth row and ith column element of $B_t(z)$, then

$$(B_t(z))^{-1} = \frac{F_t(z)}{g_t(z)}. \tag{5.4.8}$$

Theorem 5.4.3 *Let $F_t(z)$ and $g_t(z)$ be given by (5.4.7), then $F_t(z)$ and $g_t(z)$ satisfy the following ordinary differential equations*

$$\frac{dF_t}{dz} = \frac{-F_t \text{tr}(F_t) + F_t^2}{g_t}, \tag{5.4.9}$$

$$\frac{dg_t}{dz} = -\text{tr}(F_t), \tag{5.4.10}$$

where $F_t = F_t(z)$ and $g_t = g_t(z)$.

Proof Denote $B_t = B_t(z)$. Premultiplying both sides of (5.4.8) by the matrix B_t and postmultiplying both sides by $\det(B_t)$, we have

$$\det(B_t)I = B_t \text{adj}(B_t), \tag{5.4.11}$$

where I is the identity matrix. Also, by postmultiplying both sides of (5.4.8) by $B_t \det(B_t)$, we have

$$\det(B_t)I = \text{adj}(B_t)B_t. \tag{5.4.12}$$

Differentiate both sides of (5.4.11) with respect to the parameter z:

$$(B_t)_z \text{adj}(B_t) + B_t(\text{adj}(B_t))_z = (\det(B_t))_z I.$$

Premultiplying both sides of the above equation by $\text{adj}(B_t)$, we get

$$\text{adj}(B_t)(B_t)_z \text{adj}(B_t) + \text{adj}(B_t)B_t(\text{adj}(B_t))_z = \text{adj}(B_t)(\det(B_t))_z.$$

Applying (5.4.12) to the second term of the above equation, we obtain

$$\text{adj}(B_t)(B_t)_z \text{adj}(B_t) + \det(B_t)(\text{adj}(B_t))_z = \text{adj}(B_t)(\det(B_t))_z.$$

Since $\det(B_t)$ is a scalar, from the above equation we find

$$\text{adj}(B_t)_z = \frac{\text{adj}(B_t)(\det(B_t))_z - \text{adj}(B_t)(B_t)_z \text{adj}(B_t)}{\det(B_t)}. \tag{5.4.13}$$

Then differentiating $\det(B_t)$ with respect to z, we obtain

$$(\det(B_t))_z = \sum_{i,j=1}^{n} \frac{\partial \det(B_t)}{\partial b_{ij}^{(t)}} \frac{db_{ij}^{(t)}}{dz}. \tag{5.4.14}$$

However, we have

$$\frac{\partial \det(B_t)}{\partial b_{ij}^{(t)}} = B_{ij}^{(t)} \tag{5.4.15}$$

and

$$\frac{dB_t}{dz} = -I. \tag{5.4.16}$$

Substituting the above (5.4.15) and (5.4.16) into (5.4.14) gives

$$(\det(B_t))_z = \sum_{i=1}^{n} B_{ii}^{(t)} \frac{db_{ii}^{(t)}}{dz} = -\sum_{i=1}^{n} B_{ii}^{(t)} = -\text{tr}(F_t). \qquad (5.4.17)$$

By substituting (5.4.16) and (5.4.17) into the right-hand side of (5.4.13), we have

$$(\text{adj}(B_t))_z = \frac{\text{adj}(B_t)(-\text{tr}(F_t)) + (\text{adj}(B_t))^2}{\det(B_t)}.$$

Finally, substituting (5.4.7) into the above equation and (5.4.17), we obtain (5.4.9) and (5.4.10) immediately. $\qquad \square$

For a value of z suitably less than zero, $z = z_0$, we can determine the determinant and adjoint of the matrix $B_t(z_0)$ accurately by, for example, Gaussian elimination. This provides initial conditions at $z = z_0$ for the differential equations in (5.4.9) and (5.4.10), which can now be integrated numerically with z going from z_0 toward zero.

For convenience, we denote

$$A_{MN}^{\dagger} = A_{(1)}, \quad A^{\dagger} = A_{(2)}, \quad A_d = A_{(3)}, \quad A_g = A_{(4)}, \quad A^{-1} = A_{(5)}$$

and if $A \in \mathbb{C}^{m \times n}$, let

$$D_1 = N^{-1}A^*M \quad \text{and} \quad D_2 = A^*,$$

and if $A \in \mathbb{C}^{n \times n}$ and $\text{Ind}(A) = k$, let

$$D_3 = A^k, \quad D_4 = A, \quad \text{and} \quad D_5 = I.$$

Then, as z tends to zero,

$$\frac{F_t(z)}{g_t(z)} D_t$$

yields an approximation of $A_{(t)}$, for $t = 1, 2, \cdots, 5$.

Let us summarize the above in the form of a theorem.

Theorem 5.4.4 *Let the matrix F_t and the scalar g_t be determined by the differential equations:*

$$\begin{cases} \dfrac{dF_t}{dz} = \dfrac{F_t^2 - F_t \, \text{tr}(F_t)}{g_t}, \\ \dfrac{dg_t}{dz} = -\text{tr}(F_t), \end{cases} \qquad (5.4.18)$$

with the initial conditions

$$\begin{cases} F_t(z_0) = \text{adj}(D_t A - z_0 I), \\ g_t(z_0) = \det(D_t A - z_0 I), \end{cases}$$

where

$$z_0 < 0, \quad |z_0| < \min_{i \in S} |z_i|, \quad S = \{ i \, : \, z_i \neq 0 \text{ is an eigenvalue of } D_t A \}.$$

By integrating this system from z_0 to $z = 0$ and forming

$$\frac{F_t(z)}{g_t(z)} D_t, \quad t = 1, 2, \cdots, 5,$$

we obtain, in the limit, $A_{(t)}$.

5.5 Finite Algorithms

Let $A \in \mathbb{C}_n^{n \times n}$ and the characteristic polynomial of A be

$$g(\lambda) = \det(\lambda I - A) = \lambda^n + g_1 \lambda^{n-1} + \cdots + g_{n-1} \lambda + g_n,$$

where I denotes the identity matrix of order n. Let the adjoint of $\lambda I - A$ be

$$F(\lambda) = \mathrm{adj}(\lambda I - A) = F_1 \lambda^{n-1} + F_2 \lambda^{n-2} + \cdots + F_{n-1} \lambda + F_n,$$

where $F_1 = I$, then

$$(\lambda I - A)^{-1} = \frac{F(\lambda)}{g(\lambda)}. \tag{5.5.1}$$

A well-known finite algorithm attributed to Le Verrier [26] and Fadeev [27] permits simultaneous determination of the coefficients g_i and F_i by means of the formula

$$
\begin{aligned}
F_1 &= I, \quad g_1 = -\mathrm{tr}(A), \\
F_i &= A F_{i-1} + g_{i-1} I, \quad g_i = -i^{-1} \mathrm{tr}(A F_i), \quad i = 2, 3, \cdots, n.
\end{aligned}
$$

When $\lambda = 0$ in (5.5.1), then

$$A^{-1} = -\frac{F_n}{g_n}. \tag{5.5.2}$$

The above algorithm is also called Le Verrier algorithm.

This scheme finds applications in linear control theory [28]. Using an extension of this scheme, a finite algorithm for computing the Moore-Penrose inverse is given by Decell [29], and a new proof of Decell's finite algorithm for the generalized inverse is given in [30]. The finite algorithm for computing the weighted Moore-Penrose inverse $A_{MN}^\dagger$ is given in [31]. A uniform formula for the finite algorithms for computing the generalized inverses $A_{MN}^\dagger$, $A^\dagger$, A_d, A_g, and regular inverse A^{-1} using the embedding method is given in [25].

Theorem 5.5.1 *For $A \in \mathbb{C}_r^{m \times n}$, let M and N be Hermitian positive definite matrices of orders m and n respectively and denote*

$$A_{(1)} = A_{MN}^\dagger, \quad A_{(2)} = A^\dagger, \quad D_1 = N^{-1}A^*M, \quad D_2 = A^*.$$

For $A \in \mathbb{C}_r^{n \times n}$ with $\mathrm{Ind}(A) = k$, denote

$$A_{(3)} = A_d, \; A_{(4)} = A_g, \; A_{(5)} = A^{-1},$$
$$D_3 = A^k, \; D_4 = A, \quad D_5 = I,$$

and let

$$\mathrm{rank}(D_t) = r \le n, \; t = 1, 2, 3, 4, \quad \mathrm{rank}(D_5) = n,$$

and

$$F_t(\lambda) = \mathrm{adj}(D_t A - \lambda I) \tag{5.5.3}$$
$$= (-1)^{n-1}(F_1^{(t)}\lambda^{n-1} + F_2^{(t)}\lambda^{n-2} + \cdots + F_{n-1}^{(t)}\lambda + F_n^{(t)}),$$
$$g_t(\lambda) = \det(D_t A - \lambda I) \tag{5.5.4}$$
$$= (-1)^n(g_0^{(t)}\lambda^n + g_1^{(t)}\lambda^{n-1} + \cdots + g_{n-1}^{(t)}\lambda + g_n^{(t)}),$$

where $F_1^{(t)}, F_2^{(t)}, \cdots, F_n^{(t)}$ are $n \times n$ constant matrices and $g_0^{(t)} = 1$, and $g_1^{(t)}, \cdots, g_n^{(t)}$ are scalars, then

$$A_{(t)} = -\frac{F_r^{(t)}}{g_r^{(t)}}D_t, \quad t = 1, 2, 3, 4, 5.$$

Proof From (5.4.1), (5.4.2), (5.4.3), (5.4.5), (5.4.6), (5.5.3) and (5.5.4), we have

$$(D_t A - \lambda I)^{-1} = \frac{F_t(\lambda)}{g_t(\lambda)}. \tag{5.5.5}$$

Hence

$$A_{(t)} = \lim_{\lambda \to 0^-}(D_t A - \lambda I)^{-1}D_t$$
$$= \lim_{\lambda \to 0^-}\left(-\frac{F_1^{(t)}\lambda^{n-1} + F_2^{(t)}\lambda^{n-2} + \cdots + F_{n-1}^{(t)}\lambda + F_n^{(t)}}{g_0^{(t)}\lambda^n + g_1^{(t)}\lambda^{n-1} + \cdots + g_{n-1}^{(t)}\lambda + g_n^{(t)}}\right)D_t.$$

When $g_n^{(t)} \ne 0$, then

$$A_{(t)} = -\frac{F_n^{(t)}}{g_n^{(t)}}D_t.$$

Next, consider the case when $g_n^{(t)} = 0$ but $g_{n-1}^{(t)} \neq 0$. Since the above limit exists, according to Theorems 5.4.1 and 5.4.2 and Corollaries 5.4.1 and 5.4.2, we must have $F_n^{(t)} D_t = 0$, and then

$$A_{(t)} = -\frac{F_{n-1}^{(t)}}{g_{n-1}^{(t)}} D_t.$$

We know

$$\text{rank}(D_2 A) = \text{rank}(A^* A) = \text{rank}(A^*) = \text{rank}(D_2) = r.$$

Similarly, we have

$$\begin{aligned}
\text{rank}(D_1 A) &= \text{rank}(N^{-1} A^* M A) \\
&= \text{rank}(N^{1/2}(N^{-1} A^* M A) N^{-1/2}) \\
&= \text{rank}((M^{1/2} A N^{-1/2})^*(M^{1/2} A N^{-1/2})) \\
&= \text{rank}(M^{1/2} A N^{-1/2}) \\
&= \text{rank}(A) \\
&= \text{rank}(D_1) \\
&= r.
\end{aligned}$$

Since $\text{Ind}(A) = k$, we obtain

$$\begin{aligned}
\text{rank}(D_3 A) &= \text{rank}(A^{k+1}) = \text{rank}(A^k) = \text{rank}(D_3) = r, \\
\text{rank}(D_4 A) &= \text{rank}(A^2) = \text{rank}(A) = \text{rank}(D_4) = r, \\
\text{rank}(D_5 A) &= \text{rank}(A) = \text{rank}(D_5) = n.
\end{aligned}$$

So the number of the nonzero eigenvalues of $D_t A$ should be r, and we that assume $\lambda_1, \lambda_2, \cdots, \lambda_r$ are nonzero and $\lambda_{r+1} = \lambda_{r+2} = \cdots = \lambda_n = 0$. Since $g_t(\lambda)$ is the characteristic polynomial of $D_t A$, according to Vieta's relations between the roots and coefficients of a polynomial, we have

$$g_r^{(t)} \neq 0 \quad \text{and} \quad g_{r+1}^{(t)} = g_{r+2}^{(t)} = \cdots = g_n^{(t)} = 0.$$

Therefore

$$A_{(t)} = -\frac{F_r^{(t)}}{g_r^{(t)}} D_t.$$

The proof is completed. $\qquad\qquad\qquad\qquad\qquad\qquad\qquad\qquad\qquad\qquad\square$

Now we have the following finite algorithm for computing $A_{(t)}$.

Theorem 5.5.2 *The coefficients $F_i^{(t)}$ and $g_i^{(t)}$ in (5.5.3) and (5.5.4) are determined by the recurrence relations:*

$$F_{i+1}^{(t)} = D_t A F_i^{(t)} + g_i^{(t)} I, \tag{5.5.6}$$

$$g_{i+1}^{(t)} = -(i+1)^{-1} \mathrm{tr}(D_t A F_{i+1}^{(t)}), \tag{5.5.7}$$

$$i = 1, 2, \cdots, r - 1. \tag{5.5.8}$$

The initial conditions are

$$F_1^{(t)} = I,$$

$$g_1^{(t)} = -\mathrm{tr}(D_t A).$$

Proof From (5.5.5), we have

$$(F_1^{(t)} \lambda^{n-1} + F_2^{(t)} \lambda^{n-2} + \cdots + F_{n-1}^{(t)} \lambda + F_n^{(t)})(D_t A - \lambda I)$$

$$= -(\lambda^n + g_1^{(t)} \lambda^{n-1} + \cdots + g_{n-1}^{(t)} \lambda + g_n^{(t)}) I. \tag{5.5.9}$$

From Theorem 5.5.1, we have

$$g_{r+1}^{(t)} = \cdots = g_n^{(t)} = 0 \quad \text{and} \quad F_j^{(t)} D_t = O, \ j = r + 1, \cdots, n.$$

So

$$D_t A F_j^{(t)} = F_j^{(t)} D_t A = O, \quad j = r + 1, \cdots, n.$$

By comparing the identical powers of λ on both side of (5.5.9), we see that (5.5.6) holds and

$$F_{r+2}^{(t)} = F_{r+3}^{(t)} = \cdots = F_n^{(t)} = O.$$

To obtain (5.5.7), from (5.4.18) we have

$$(-1)^n (n \lambda^{n-1} + (n-1) g_1^{(t)} \lambda^{n-2} + \cdots + (n-r) g_r^{(t)} \lambda^{n-r-1})$$

$$= -(-1)^{n-1} (\lambda^{n-1} \mathrm{tr}(F_1^{(t)}) + \cdots + \lambda^{n-r} \mathrm{tr}(F_r^{(t)}) + \lambda^{n-r-1} \mathrm{tr}(F_{r+1}^{(t)})).$$

Equating the coefficients of the like power of λ, we see that

$$(n-i) g_i^{(t)} = \mathrm{tr}(F_{i+1}^{(t)}).$$

Now take the trace of both sides of (5.5.6) to obtain

$$\mathrm{tr}(F_{i+1}^{(t)}) = \mathrm{tr}(D_t A F_i^{(t)}) + n g_i^{(t)}.$$

It then follows that

$$g_{i+1}^{(t)} = -(i+1)^{-1} \mathrm{tr}(D_t A F_{i+1}^{(t)}),$$

which completes the proof. $\square$

Remarks

As for the numerical computation of the generalized inverses, the book [13] gives useful suggestions and a list of direct and iterative methods for computing the {1}-inverse, {1, 2}-inverse, {1, 4}-inverse, Moore-Penrose inverse $A^\dagger$, Drazin inverse A_d, and group inverse A_g. But no perturbation analysis is given there. This chapter is based on the computational methods for the regular inverse, we derive the numerical methods for computing the four important generalized inverses: $A^\dagger$, $A^\dagger_{MN}$, A_d and A_g. There are embedding method and finite algorithm for computing the generalized inverse $A^{(2)}_{T,S}$ [32, 33]. The finite algorithm can also be used to compute the inverse of matrix polynomial $\lambda^N I_n - \lambda^{N-1} A_1 - \cdots - \lambda A_{N-1}$ and the inverses of the singular pencils $\mu E - A$ and $\mu^2 E - \mu A_1 - A_2$ [34–36]. The Rump's method for computing the Moore-Penrose inverse is described in [37]. The gradient methods for computing the Drazin inverse are presented in [38].

The Moore-Penrose inverse and the Drazin inverse of a 2×2 block matrix, the weighted generalized inverse of a partitioned matrix, the Moore-Penrose inverse of a rank-1 modified matrix, the Drazin inverse of a modified matrix, and the Drazin inverse of a Hessenberg matrix are discussed in [8, 11, 39–44]. The algebraic perturbation method for the generalized inverse can be found in [45].

The limit representations of the generalized inverses and the alternative limit expression of the Drazin inverse are given in [46, 47].

Recursive least squares (RLS) algorithm and fast RLS algorithm for linear prediction problems are given in [48, 49]. The iterative methods are referred to Chap. 11 and [50].

It is well-known that the important generalized inverses $A^\dagger_{MN}$, $A^\dagger$, A_d, A_g, $A_{d,W}$, $A^{(-1)}_{(L)}$ and $A^{(\dagger)}_{(L)}$ can be described as the generalized inverse $A^{(2)}_{T,S}$, which has the prescribed range T and null space S, and is the outer inverse of A. A unified method for computing the generalized inverse $A^{(2)}_{T,S}$ such as the embedding method, $(T\text{-}S)$ splitting method, or iterative method can be found in [32, 51–58]. A limit representation of the outer inverse is given in [59].

The inverse and generalized inverse of a matrix polynomial often occur in the control theory (see [60]). The finite algorithm in Sect. 5.5 also can be applied to the computation of such kind of generalized inverse [61], since this kind of matrices is a special case of block-matrices whose blocks can commute each other [62].

The representations and approximations of the Drazin inverse, weighted Moore-Penrose inverse, and generalized inverse $A^{(2)}_{T,S}$ are given in [53, 63–65].

Neural networks have been used for computing regular inverse [66], Moore-Penrose inverse [67], outer inverse [68], Drazin inverse [69], weighted Drazin inverse [70], and time-varying Drazin inverse [71].

Symbolic computation of the generalized inverses using Maple is presented in [72].

References

1. B. Noble, J. Demmel, *Applied Linear Algebra*, 3rd edn. (Prentice-Hall, New Jersey, 1988)
2. G.W. Stewart, *Introduction to Matrix Computation* (Academic Press, New York, 1973)
3. G.H. Golub, C.F. Van Loan, *Matrix Computations*, 4th edn. (The Johns Hopkins University Press, Baltimore, MD, 2013)
4. C.F. Van Loan, Generalizing the singular value decomposition. SIAM J. Numer. Anal. **13**, 76–83 (1976)
5. C.R. Rao, S.K. Mitra, *Generalized Inverse of Matrices and Its Applications* (John Wiley, New York, 1971)
6. S.L. Campbell, C.D. Meyer Jr., *Generalized Inverses of Linear Transformations* (Pitman, London, 1979)
7. C.D. Meyer, N.J. Rose, The index and the Drazin inverse of block triangular matrices. SIAM J. Appl. Math. **33**, 1–7 (1977)
8. J. Miao, The Moore-Penrose inverse of a rank-*r* modified matrix. Numer. Math. J. Chinese Univ. **19**, 355–361 (1989). (in Chinese)
9. J.M. Shoaf, The Drazin inverse of a rank-one modification of a square matrix. PhD thesis, North Carolina State University, 1975
10. Y. Wei, The weighted Moore-Penrose inverse of modified matrices. Appl. Math. Comput. **122**, 1–13 (2001)
11. Y. Wei, The Drazin inverse of a modified matrix. Appl. Math. Comput. **125**, 295–301 (2002)
12. T.N.E. Greville, Some applications of pseudoinverse of a martix. SIAM Rev. **2**, 15–22 (1960)
13. A. Ben-Israel, T.N.E. Greville, *Generalized Inverses: Theory and Applications*, 2nd edn. (Springer, New York, 2003)
14. S. Dang, *Matrix Theory and its Applications in Survey and Drawing* (Survey and Drawing Press, 1980) (in Chinese)
15. X. He, W. Sun, The analysis of the Greville's method. J. Nanjing Univ. **5**, 1–10 (1988). in Chinese
16. G. Wang, A new proof of Greville method for computing the Moore-Penrose generalized inverse. J. Shanghai Normal Univ. **14**, 32–38 (1985). in Chinese
17. G. Wang, Y. Chen, A recursive algorithm for computing the W-weighted Moore-Penrose inverse $A^\dagger_{MN}$. J. Comput. Math. **4**, 74–85 (1986)
18. R.E. Cline, Representation of the generalized inverse of a partitioned matrix. J. Soc. Indust. Appl. Math. **12**, 588–600 (1964)
19. L. Mihalyffy, An alternative representation of the generalized inverse of partitioned matrices. Linear Algebra Appl. **4**, 95–100 (1971)
20. A. Ben-Israel, A note on partitioned matrices and equations. SIAM Rev. **11**, 247–250 (1969)
21. J.V. Rao, Some more representations for generalized inverse of a partitioned matrix. SIAM J. Appl. Math. **24**, 272–276 (1973)
22. J. Miao, Representations for the weighted Moore-Penrose inverse of a partitioned matrix. J. Comput. Math. **7**, 320–323 (1989)
23. J. Miao, Some results for computing the Drazin inverse of a partitioned matrix. J. Shanghai Normal Univ. **18**, 25–31 (1989). in Chinese
24. R.E. Kalaba, N. Rasakhoo, Algorithm for generalized inverse. J. Optim. Theory Appl. **48**, 427–435 (1986)
25. G. Wang, An imbedding method for computing the generalized inverse. J. Comput. Math. **8**, 353–362 (1990)
26. U.J. Le Verrier, *Memoire sur les variations séculaires des éléments des orbites, pour les sept Planetes principales Mercure, Venus, La Terre, Mars, Jupiter* (Bachelier, Saturne et Uranus, 1845)
27. D.K. Fadeev, V.N. Fadeeva, *Computational Methods of Linear Algebra* (W.H. Freeman & Co., Ltd, San Francisco, 1963)
28. M. Clique, J.G. Gille, On companion matrices and state variable feedback. Podstawy Sterowania **15**, 367–376 (1985)

29. H.P. Decell Jr., An application of the Cayley-Hamilton theorem to generalized matrix inversion. SIAM Rev. **7**(4), 526–528 (1965)
30. R.E. Kalaba et al., A new proof for Decell's finite algorithm for the generalized inverse. Appl. Math. Comput. **12**, 199–211 (1983)
31. G. Wang, A finite algorithm for computing the weighted Moore-Penrose inverse $A_{MN}^{\dagger}$. Appl. Math. Comput. **23**, 277–289 (1987)
32. G. Wang, Y. Wei, Limiting expression for generalized inverse $A_{T,S}^{(2)}$ and its corresponding projectors. Numer. Math., J. Chinese Univ. (English Series), **4**, 25–30 (1995)
33. Y. Chen, Finite algorithm for the (2)-generalized inverse $A_{T,S}^{(2)}$. Linear Multilinear Algebra **40**, 61–68 (1995)
34. G. Wang, Y. Lin, A new extension of Leverier's algorithm. Linear Algebra Appl. **180**, 227–238 (1993)
35. G. Wang, L. Qiu, Leverrier-Chebyshev algorithm for singular pencils. Linear Algebra Appl. **345**, 1–8 (2002)
36. Z. Wu, B. Zheng, G. Wang, Leverrier-Chebyshev algorithm for the matrix polynomial of degree two. Numer. Math. J. Chinese Univ. (English Series), **11**, 226–234 (2002)
37. Y. Chen, X. Shi, Y. Wei, Convergence of Rump's method for computing the Moore-Penrose inverse. Czechoslovak Math. J., **66**(141)(3), 859–879 (2016)
38. S. Miljković, M. Milandinović, P.S. Stanimirović, Y. Wei, Gradient methods for computing the Drazin-inverse solution. J. Comput. Appl. Math. **253**, 255–263 (2013)
39. J. Miao, General expressions for the Moore-Penrose inverse of a 2×2 block matrix. Linear Algebra Appl. **151**, 1–15 (1991)
40. Y. Wei, Expression for the Drazin inverse of a 2×2 block matrix. Linear Multilinear Algebra **45**, 131–146 (1998)
41. S. Dang, A new method for computing the weighted generalized inverse of partitioned matrices. J. Comput. Math. **7**, 324–326 (1989)
42. J. Miao, The Moore-Penrose inverse of a rank-1 modified matrix. J. Shanghai Normal Univ. **17**, 21–26 (1988). in Chinese
43. J. Miao, The Drazin inverse of Hessenberg matrices. J. Comput. Math. **8**, 23–29 (1990)
44. Y. Wei, Y. Cao, H. Xiang, A note on the componentwise perturbation bounds of matrix inverse and linear systems. Appl. Math. Comput. **169**, 1221–1236 (2005)
45. J. Ji, The algebraic perturbation method for generalized inverses. J. Comput. Math. **7**, 327–333 (1989)
46. P.S. Stanimirović, Limit representations of generalized inverses and related methods. Appl. Math. Comput. **103**, 51–68 (1999)
47. J. Ji, An alternative limit expression of Drazin inverse and its applications. Appl. Math. Comput. **61**, 151–156 (1994)
48. S. Qiao, Recursive least squares algorithm for linear prediction problems. SIAM J. Matrix Anal. Appl. **9**, 323–328 (1988)
49. S. Qiao, Fast adaptive RLS algorithms: a generalized inverse approach and analysis. IEEE Trans. Signal Process. **39**, 1455–1459 (1991)
50. G. Wang, Y. Wei, The iterative methods for computing the generalized inverse $A_{MN}^{\dagger}$ and $A_{d,W}$. Numer. Math. J. Chinese Univ., **16**, 366–371 (1994). (in Chinese)
51. Y. Wei, H. Wu, $(T\text{-}S)$ splitting methods for computing the generalized inverse $A_{T,S}^{(2)}$ of rectangular systems. Int. J. Comput. Math. **77**, 401–424 (2001)
52. Y. Wei, A characterization and representation of the generalized inverse $A_{T,S}^{(2)}$ and its applications. Linear Algebra Appl. **280**, 87–96 (1998)
53. Y. Wei, H. Wu, The representation and approximation for the generalized inverse $A_{T,S}^{(2)}$. Appl. Math. Comput. **135**, 263–276 (2003)
54. X. Li, Y. Wei, A note on computing the generalized inverse $A_{T,S}^{(2)}$ of a matrix A. Int. J. Math. Math. Sci. **31**, 497–507 (2002)
55. J.J. Climent, N. Thome, Y. Wei, A geometrical approach on generalized inverse by Neumann-type series. Linear Algebra Appl. **332–334**, 535–542 (2001)

56. X. Chen, W. Wang, Y. Song, Splitting based on the outer inverse of matrices. Appl. Math. Comput. **132**, 353–368 (2002)
57. X. Chen, G. Chen, A splitting method for the weighted Drazin inverse of rectangular matrices. J. East China Normal Univ. **8**, 71–78 (1993). in Chinese
58. X. Chen, R.E. Hartwig, The hyperpower iteration revisited. Linear Algebra Appl. **233**, 207–229 (1996)
59. X. Liu, Y. Yu, J. Zhong, Y. Wei, Integral and limit representations of the outer inverse in Banach space. Linear Multilinear Algebra **60**(3), 333–347 (2012)
60. E.D. Sontag, On generalized inverses of polynomial and other matrtices. IEEE Trans. Auto. Control AC **25**, 514–517 (1980)
61. J. Gao, G. Wang, Two algorithms for computing the Drazin inverse of a polynomial matrix. J. Shanghai Teach. Univ. (Natural Sciences) **31**(2), 31–38 (2002). In Chinese
62. G. Wang, An application of the block-Cayley-Hamilton theorem. J. Shanghai Normal Univ. **20**, 1–10 (1991). in Chinese
63. Y. Wei, H. Wu, The representation and approximation for the Drazin inverse. J. Comput. Appl. Math. **126**, 417–432 (2000)
64. Y. Wei, H. Wu, The representation and approximation for the weighted Moore-Penrose inverse. Appl. Math. Comput. **121**, 17–28 (2001)
65. Y. Wei, G. Wang, Approximate methods for the generalized inverse $A_{T,S}^{(2)}$. J. Fudan Univ. **38**, 233–240 (1999)
66. J. Wang, A recurrent neural networks for real-time matrix inversion. Appl. Math. Comput. **55**, 23–34 (1993)
67. J. Wang, Recurrent neural networks for computing pseudoinverse of rank-deficient matrices. SIAM J. Sci. Comput. **18**, 1479–1493 (1997)
68. P.S. Stanimirović, I.S. Živković, Y. Wei, Neural network approach to computing outer inverses based on the full rank representation. Linear Algebra Appl. **501**, 344–362 (2016)
69. P.S. Stanimirović, I.S. Živković, Y. Wei, Recurrent neural network for computing the Drazin inverse. IEEE Trans. Neural Netw. Learn. Syst. **26**(11), 2830–2843 (2015)
70. X. Wang, H. Ma, P.S. Stanimirović, Recurrent neural network for computing the W-weighted Drazin inverse. Appl. Math. Comput. **300**, 1–20 (2017)
71. S. Qiao, X. Wang, Y. Wei, Two finite-time convergent Zhang neural network models for time-varying complex matrix Drazin inverse. Linear Algebra Appl. **542**, 101–117 (2018)
72. J. Jones, N. Karampetakis, A. Pugh, The computation and application of the generalized inverse via maple. J. Symbolic Comput. **25**, 99–124 (1998)

Chapter 6
Structured Matrices and Their Generalized Inverses

A matrix is considered structured if its structure can be exploited to obtain efficient algorithms. Examples of structured matrices include Toeplitz, Hankel, circulant, Vandermonde, Cauchy, sparse. A matrix is called Toeplitz if its entries on the same diagonal are equal. For example,

$$A = \begin{bmatrix} 3 & 2 & 1 \\ 4 & 3 & 2 \\ 5 & 4 & 3 \\ 6 & 5 & 4 \end{bmatrix} \tag{6.0.1}$$

is a Toeplitz matrix. Thus an $m \times n$ Toeplitz matrix is determined by its first row and first column, total of $m + n - 1$ entries. In comparison, a general $m \times n$ matrix is determined by mn parameters. Thus, fast algorithms are expected for Toeplitz matrices and other structured matrices. This chapter includes two aspects of structured matrices and generalized inverses. One is about computing the generalized inverses of structured matrices. Particularly, we present a fast algorithm for computing the Moore-Penrose inverse of a Toeplitz matrix. The ideas can be applied to other generalized inverses of other structured matrices such as Hankel and sparse. The other aspect is about the structure of the generalized inverses of structured matrices.

6.1 Computing the Moore-Penrose Inverse of a Toeplitz Matrix

This section describes a Newton's method for computing the Moore-Penrose inverse of a Toeplitz matrix presented in [1]. How do we define the structure of a Toeplitz matrix so that it can be exploited to develop fast algorithms? The displacement struc-

© Springer Nature Singapore Pte Ltd. and Science Press 2018
G. Wang et al., *Generalized Inverses: Theory and Computations*,
Developments in Mathematics 53, https://doi.org/10.1007/978-981-13-0146-9_6

ture, defined as follows, is commonly exploited in the computation of the generalized inverses. A matrix A is said to have the displacement structure if we can find two dimensionally compatible matrices U and V such that the rank of the Sylvester displacement $AU - VA$ or the Stein displacement $A - VAU$ is much smaller than the size of A [2]. For example, denoting the $n \times n$ shift-down (or shift-left) matrix

$$Z_n = \begin{bmatrix} 0 & & & 0 \\ 1 & 0 & & \\ & \ddots & \ddots & \\ 0 & & 1 & 0 \end{bmatrix}, \tag{6.1.1}$$

for the matrix A in (6.0.1), the rank of the Sylvester displacement

$$AZ_3 - Z_4A = \begin{bmatrix} 2 & 1 & 0 \\ 3 & 2 & 0 \\ 4 & 3 & 0 \\ 5 & 4 & 0 \end{bmatrix} - \begin{bmatrix} 0 & 0 & 0 \\ 3 & 2 & 1 \\ 4 & 3 & 2 \\ 5 & 4 & 3 \end{bmatrix}$$

$$= \begin{bmatrix} 1 \\ 0 \\ 0 \\ 0 \end{bmatrix} \begin{bmatrix} 2 & 1 & 0 \end{bmatrix} - \begin{bmatrix} 0 \\ 1 \\ 2 \\ 3 \end{bmatrix} \begin{bmatrix} 0 & 0 & 1 \end{bmatrix}$$

is two, called the Sylvester displacement rank of A. Also, the rank of the Stein displacement

$$A - Z_4^T A Z_3 = \begin{bmatrix} 3 & 2 & 1 \\ 4 & 3 & 2 \\ 5 & 4 & 3 \\ 6 & 5 & 4 \end{bmatrix} - \begin{bmatrix} 3 & 2 & 0 \\ 4 & 3 & 0 \\ 5 & 4 & 0 \\ 6 & 5 & 0 \end{bmatrix}$$

$$= \begin{bmatrix} 0 \\ 0 \\ 0 \\ 1 \end{bmatrix} \begin{bmatrix} 6 & 5 & 4 \end{bmatrix} + \begin{bmatrix} 1 \\ 2 \\ 3 \\ 0 \end{bmatrix} \begin{bmatrix} 0 & 0 & 1 \end{bmatrix}$$

is two. In fact, as shown above, it can be proved that the displacement rank of a Toeplitz matrix is at most two. This low displacement rank property can be exploited to develop fast algorithms for triangular factorization, inversion, among others [2]. In [1], Wei, Cai, and Ng present a fast Newton's method for computing the Moore-Penrose inverse of a Toeplitz matrix by exploiting the displacement structure.

For an $m \times n$ Toeplitz matrix A, defining the Sylvester displacement operator

$$\Delta(A) = Z_m A - A Z_n,$$

we have $\text{rank}(\Delta(A)) \leq 2$. The following theorem shows that A can be expressed as a sum of $k \leq 2$ structured matrices.

Theorem 6.1.1 ([3]) *Suppose that* $k = \text{rank}(\Delta(A))$ *and* $\Delta(A) = \sum_{i=1}^{k} \mathbf{g}_i \mathbf{h}_i^*$, $\mathbf{g}_i \in \mathbb{R}^m$ *and* $\mathbf{h}_i \in \mathbb{R}^n$, *then* A *can be expressed as*

$$A = L(A\mathbf{e}_1) + \sum_{i=1}^{k} L(\mathbf{g}_i) U(-Z_n \mathbf{h}_i),$$

where $\mathbf{e}_1$ *is the first unit vector in* $\mathbb{R}^n$, $L(\mathbf{x})$ *is the* $m \times p$, $p = \min(m, n)$, *lower triangular Toeplitz matrix whose first column is* $\mathbf{x}$ *and* $U(\mathbf{y})$ *is the* $p \times n$, $p = \min(m, n)$, *upper triangular Toeplitz matrix whose first row is* $\mathbf{y}^{\mathrm{T}}$.

In particular, when $\Delta(A) = \sum_{i=1}^{k} \sigma_i \mathbf{u}_i \mathbf{v}_i^*$, where σ_i, $\mathbf{u}_i$, and $\mathbf{v}_i$ are respectively the ith singular value, left singular vector, and right singular vector, then

$$A = L(A\mathbf{e}_1) + \sum_{i=1}^{k} \sigma_i L(\mathbf{u}_i) U(-Z_n \mathbf{v}_i),$$

called the orthogonal displacement representation of A.

Note that $L(\mathbf{x})$ and $U(\mathbf{y})$ are Toeplitz matrices and Toeplitz matrix-vector multiplication can be efficiently computed by two FFTs, a componentwise multiplication and one inverse FFT [4]. Thus the low displacement rank combined with the fast Toeplitz-vector multiplication can be exploited to develop fast algorithms for computing the Moore-Penrose inverse of a Toeplitz matrix. The method for computing the Moore-Penrose inverse considered in [1] is the Newton's iteration:

$$X_{i+1} = 2X_i - X_i A X_i, \quad i = 0, 1, 2, \ldots$$

In [1], X_0 is initialized to $\rho^{-1} A^* A A^*$, where ρ is the spectral radius of AA^*AA^*, which can be estimated by a few power iterations [5]:

$$\mathbf{q}_i = (AA^*AA^*)\mathbf{x}_{i-1};$$
$$\mathbf{x}_i = \mathbf{q}_i / \|\mathbf{q}_i\|_2; \qquad i = 1, 2, \ldots$$
$$\lambda_i = \mathbf{x}_i^* (AA^*AA^*)\mathbf{x}_i;$$

with a random initial $\mathbf{x}_0$. It is proved in [1] that with the above initial X_0, the Newton's method converges to the Moore-Penrose inverse $A^\dagger$.

Although A can be expressed as a sum of at most three terms, the number of terms in the intermediate X_i grows quickly, as the Newton's iteration proceeds. To control the number of terms in the intermediate results, the idea of the truncated singular value decomposition is applied. Specifically, the singular value decomposition of $W_i = 2X_i - X_i A X_i$ is truncated, that is, its small singular values are set to zero, before being set to the next X_{i+1}. We refer the details to [1]. For example, for efficiency, instead of X_i, the factor Y_i in $X_i = A^* Y_i A^*$ is used. Moreover, instead of explicitly computing Y_i, the factors in the orthogonal displacement representation of Y_i are computed and updated.

Algorithm 6.1.1 ([1]) Newton's method for computing the Moore-Penrose inverse of a Toeplitz matrix using the displacement structure.
Input: The first column and first row of the Toeplitz matrix.
Output: An approximation of the Moore-Penrose inverse.

1. Compute the factors in the orthogonal displacement representation of $A^* A A^*$;
2. Estimate the spectral radius ρ of $A A^* A A^*$;
3. Set $Y_0 = \rho^{-1} A$;
   $i = 0$;
   repeat
4. Compute the factors in the truncated orthogonal displacement representation of Y_{i+1} by updating those in the orthogonal displacement representation of Y_i;
5. $X_{i+1} = A^* Y_{i+1} A^*$;
   $i = i + 1$ until X_i satisfies a predetermined tolerance.

The experimental results presented in [1] show that the method achieves high accuracy and performance. The running time grows linearly as the size of A increases. This Newton's method is applicable to any low displacement rank matrix. Example of such matrices can be found in [2]. For more on the Newton's iteration for the generalized inverses of structured matrices, see [6, 7].

6.2 Displacement Structure of the Generalized Inverses

As pointed out in the previous section, the displacement rank of a Toeplitz matrix is at most two. Is the inverse of a nonsingular Toeplitz matrix also structured with respect to the displacement rank? The answer is: Yes. Indeed, the Sylvester displacement rank of a nonsingular matrix A equals that of its inverse A^{-1}, since, for two dimensionally compatible matrices U and V, $AU - VA = A(UA^{-1} - A^{-1}V)A$. In other words, if A is structured with respect to the Sylvester displacement rank associated with (U, V), then its inverse A^{-1} is also structured with respect to the Sylvester displacement rank associated with (V, U). In particular, the Sylvester displacement rank of the inverse of a nonsingular Toeplitz matrix is also at most two. How about the generalized inverses? In [8], an upper bound for the displacement rank of the group inverse is presented.

Theorem 6.2.1 ([8]) *Let* $A, U, V \in \mathbb{C}^{n \times n}$ *and* $\text{Ind}(A) = 1$, *then, for the group inverse* A_g *of* A, *the Sylvester displacement rank is bounded by*

$$\text{rank}(A_g V - U A_g) \le \text{rank}(AU - VA) + \text{rank}(AV - UA).$$

Proof Recall that if $Ind(A) = 1$, then there exists a nonsingular R of order n such that

$$A = R \begin{bmatrix} C & O \\ O & O \end{bmatrix} R^{-1},$$

where C is a nonsingular matrix and its group inverse A_g is given by

$$A_g = R \begin{bmatrix} C^{-1} & O \\ O & O \end{bmatrix} R^{-1}.$$

Defining the two projections

$$Q = AA_g = R \begin{bmatrix} I & O \\ O & O \end{bmatrix} R^{-1} \quad \text{and} \quad P = I - AA_g = R \begin{bmatrix} O & O \\ O & I \end{bmatrix} R^{-1}, \quad (6.2.1)$$

we have

$$A_g(AU - VA)A_g = QUA_g - A_gVQ.$$

It then follows that the Sylvester displacement of A_g

$$
\begin{aligned}
A_g V &- U A_g \\
&= A_g V P + A_g V Q - P U A_g - Q U A_g \\
&= A_g V P - P U A_g - A_g(AU - VA)A_g,
\end{aligned}
$$

implying that

$$\text{rank}(A_g V - U A_g) \le \text{rank}(AU - VA) + \text{rank}(QVP) + \text{rank}(PUQ),$$

by the definitions of Q and P in (6.2.1). It then remains to show that

$$\text{rank}(QVP) + \text{rank}(PUQ) \le \text{rank}(AV - UA). \qquad (6.2.2)$$

Indeed, partitioning

$$R^{-1}UR = \begin{bmatrix} U_{11} & U_{12} \\ U_{21} & U_{22} \end{bmatrix} \quad \text{and} \quad R^{-1}VR = \begin{bmatrix} V_{11} & V_{12} \\ V_{21} & V_{22} \end{bmatrix} \qquad (6.2.3)$$

accordingly, we have

$$QVP = R \begin{bmatrix} O & V_{12} \\ O & O \end{bmatrix} R^{-1} \quad \text{and} \quad PUQ = R \begin{bmatrix} O & O \\ U_{12} & O \end{bmatrix} R^{-1},$$

implying that

$$\text{rank}(QVP) + \text{rank}(PUQ) = \text{rank}(V_{12}) + \text{rank}(U_{21}). \qquad (6.2.4)$$

On the other hand, applying the partitions (6.2.3), we have

$$\text{rank}(AV - UA) = \text{rank}\left(\begin{bmatrix} CV_{11} - U_{11}C & CV_{12} \\ -U_{21}C & O \end{bmatrix}\right).$$

Comparing the above with (6.2.4) gives (6.2.2), which completes the proof, recalling that C is nonsingular.   □

This theorem says that the Sylvester displacement rank of A_g associated with (V, U) is bounded above by the sum of the Sylvester displacement ranks of A associated with (U, V) and (V, U). In particular, when A is a Toeplitz matrix of index one, then the displacement rank of its group inverse A_g is at most four.

The above result is generalized to the Drazin inverse [9]:

$$\text{rank}(A_d V - U A_d) \leq \text{rank}(AU - VA) + \text{rank}(A^k V - U A^k),$$

where k is the index of A. In [9], the above upper bound is applied to structured matrices such as Toeplitz, close-to-Toeplitz, generalize Cauchy, among others.

An analogous upper bound for the Sylvester displacement rank of the weighted Moore-Penrose inverse is presented in [10]:

$$\text{rank}(A_{MN}^\dagger V - U A_{MN}^\dagger) \leq \text{rank}(AU - VA) + \text{rank}(AU^\# - V^\# A),$$

recalling that $U^\# = N^{-1}U^*N$ and $V^\# = M^{-1}V^*M$ are the weighted conjugate transposes of U and V respectively.

Remarks

Heinig and Hellinger [11] unified and generalized the Sylvester displacement and the Stein displacement. The upper bounds for the Sylvester displacement rank of various generalized inverses presented in Sect. 6.2 are all generalized to this generalized displacement [8–10].

As we know, the Moore-Penrose inverse $A^\dagger$ and the Drazin inverse A_d are both special cases of $A_{TS}^{(2)}$, the {2}-inverse with prescribed range T and null space S. In [12], an upper bound for the Sylvester displacement rank as well as the generalized displacement rank of $A_{TS}^{(2)}$ is established.

As shown in [9], by applying the upper bounds, the group inverse of some structured matrices, such as close-to-Toeplitz, generalized Cauchy, and Toeplitz-plus-Hankel, have low displacement ranks. Thus the Newton's method for the Moore-Penrose inverse of a Toeplitz matrix described in Sect. 6.1 can be modified for the group inverse of matrices of low displacement rank. For more on the Moore-Penrose and group inverses of a Teoplitz matrix, see [13, 14].

References

1. Y. Wei, J. Cai, M.K. Ng, Computing Moore-Penrose inverses of Toeplitz matrices by Newton's iteration. Math. Comput. Model. **40**(1–2), 181–191 (2004)
2. T. Kailath, A.H. Sayed, Displacement structure: theory and applications. SIAM Rev. **37**(3), 297–386 (1995)
3. D. Bini, V. Pan, *Polynomial and Matrix Computations: Volume 1 Fundamental Algorithms.* (Birkhäuser, Boston, 1994)
4. C.F. Van Loan, *Computational Frameworks for the Fast Fourier Transform* (SIAM, Philadelphia, 1992)
5. G.H. Golub, C.F. Van Loan, *Matrix Computations*, 4th edn. (The Johns Hopkins University Press, Baltimore, MD, 2013)
6. V. Pan, R. Schreiber, An improved Newton iteration for the generalized inverse of a matrix, with applications. SIAM J. Sci. Stat. Comput. **12**(5), 1109–1130 (1991)
7. V. Pan, Y. Rami, X. Wang, Structured matrices and Newton's iteration: unified approach. Linear Algebra Appl. **343–344**, 233–265 (2002)
8. Y. Wei, M.K. Ng. Displacement structure of group inverses. Numer. Linear Algebra Appl. **12**, 103–110 (2005)
9. H. Diao, Y. Wei, S. Qiao, Displacement rank of the Drazin inverse. J. Comput. Appl. Math. **167**(1), 147–161 (2004)
10. J. Cai, Y. Wei, Displacement structure of weighted pseudoinverses. Appl. Math. Comput. **153**(2), 317–335 (2004)
11. G. Heinig, F. Hellinger, Displacement structure of pseudoinverse. Linear Algebra Appl. **197–198**, 623–649 (1994)
12. S. Li, Displacement structure of the generalized inverse $A_{T,S}^{(2)}$. Appl. Math. Comput. **156**, 33–40 (2004)
13. P. Xie, Y. Wei, The stability of formulae of the Gohberg-Semencul-Trench type for Moore-Penrose and group inverses of toeplitz matrices. Linear Algebra Appl. **498**, 117–135 (2016)
14. Y. Wei, H. Diao, On group inverse of singular Toeplitz matrices. Linear Algebra Appl. **399**, 109–123 (2005)

References

Chapter 7
Parallel Algorithms for Computing the Generalized Inverses

The UNIVersal Automatic Computer (UNIVAC I) and the machines built in 1940s and mid 1950s are often referred to as the first generation of computers.

From 1958 to 1964, the second generation of computers was developed based on transistor technology. During this phase, IBM reengineered its 709 to use transistor technology and named it the IBM7090. It was able to calculate close to 500,000 additions per second.

In 1964, the third generation of computer was born. The new generation was based on integrated circuit (IC) technology, which was invented in 1957. An IC device is a tiny clip of silicon that hosts many transistors and other circuit components.

The Large-Scale Integration (LSI) and the Very Large-Scale Integration (VLSI) technologies have moved computers from the third to new generations. The computers developed from 1972 to 1990 are referred to as the forth generation of computers; from 1991 to present is referred to as the fifth generation.

The progress in increasing the number of transistors on single chip continues to augment the computational power of computer systems, in particular that of the small systems (personal computer and workstations). Today, multiprocessor systems are common. When a computer has multiple processors, multiple instructions can be executed on multiple processors in parallel. The processors can work independently on multiple tasks or process different parts of a same task simultaneously. Such a computer is referred to as a parallel computer.

Algorithms in which operations must be executed step by step are called serial or sequential. Algorithms in which several operations may be executed simultaneously are referred to as parallel.

Various approaches may be taken to design a parallel algorithm for a given problem. One approach is to attempt to convert a sequential algorithm to a parallel algorithm. If a sequential algorithm exists for the problem, then inherent parallelism in that algorithm may be recognized and implemented. It should be noted that exploiting inherent parallelism in a sequential algorithm might not always lead to an efficient

© Springer Nature Singapore Pte Ltd. and Science Press 2018

G. Wang et al., *Generalized Inverses: Theory and Computations*,

Developments in Mathematics 53, https://doi.org/10.1007/978-981-13-0146-9_7

parallel algorithm. Another approach is to design a totally new parallel algorithm that is more efficient than the existing one.

It follows from the unceasing progress of parallel computers, the research of the parallel algorithms for many problems get rapid development (see [1–4]).

In recent years, the research of the parallel computation of the generalized inverses is discussed in [5–9]. Some of our results are given in this chapter.

The model of parallel processors are briefly introduced as follows. The detail can be found in [3, 10–13].

7.1   The Model of Parallel Processors

The parallel processors were constructed in 1970s. According to the computer architecture systems, there are array processor, pipeline processor and multiprocessor and so on. Brief examples of them are given as follows.

7.1.1   Array Processor

The first parallel array computer Illiac IV was constructed in 1972. Illiac IV consisted of $N = 64$ fast processors, with memories of 2048 64bit words connected in an 8×8 array as illustrated in Fig. 7.1. The individual processors were controlled by a separate control unit and all processors did the same instruction (or nothing) at a given time.

7.1.2   Pipeline Processor

Pipelining is one way of improving the overall processing performance of a processor. This architectural approach allows the simultaneous execution of several instructions. The pipeline processor Cray I was constructed in 1976.

The pipeline design technique decomposes a sequential process into several subprocesses, called stages or segments. A stage performs a particular function and produces an intermediate result. It consists of an input latch, also called a register, followed by a processing circuit. The processing circuit of a given stage is connected to the input latch of the next stage (see Fig. 7.2).

An arithmetic pipeline is used for implementing complex arithmetic functions. These functions can be decomposed into consecutive subfunctions. For example the floating-point addition can be divided into three stages: mantissa alignment, mantissa addition, and result normalization.

Figure 7.3 depicts a pipeline architecture for floating-point addition of two numbers.

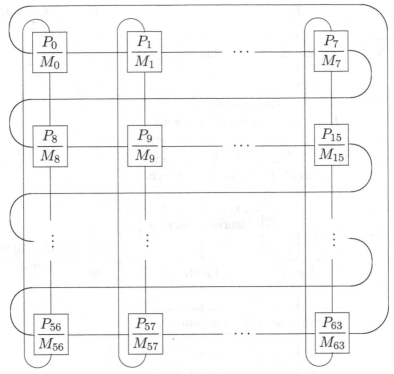

P_i: Processor M_i: Memory

Fig. 7.1 An 8 × 8 array processor

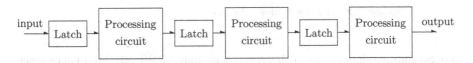

Fig. 7.2 Pipeline

7.1.3 Multiprocessor

A multiprocessor architecture has a memory system that is addressable by each processor. As such, the memory system consists of one or more memory modules whose address space is shared by all the processors.

In addition to the central memory system, each processor might also have a small cache memory. These cache also help reduce memory contention and make the system more efficient.

A multiprocessor computer has one operating system used by all processors. The operating system provides interaction between processors and their tasks at the

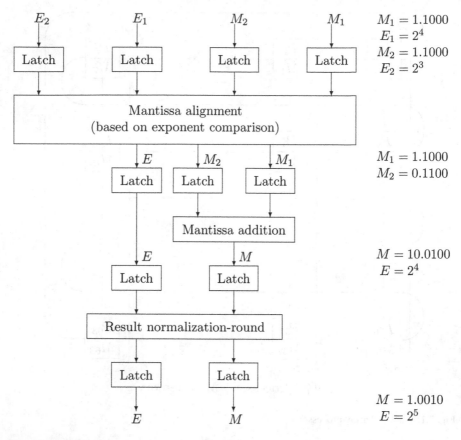

Fig. 7.3 A pipeline architecture for floating-point addition

process and data element level. (The term process may be defined as a part of program that can run on a processor.) Each processor is capable of doing a large task on its own.

A taxonomy of computer architectures was given by Flynn in 1966. He divided machines into four categories: SISD (single instruction stream/single data stream), SIMD (single instruction stream/multiple data stream), MISD (multiple instruction stream/single data stream), and MIMD (multiple instruction stream/multiple data stream).

Traditional sequential computers belong to SISD, since instructions are processed sequentially and result in the movement of data from memory to functional unit and back to memory; Array processors and pipeline processors belong to SIMD. Since each processor executes the same instruction (or no instruction) at the same time but on different data; Multiprocessors belong to MIMD, since the instructions may differ across the processors, which need not operate synchronously.

Although these categories give a helpful coarse division, the current situation is more complicated, with some architectures exhibiting aspects of more than one category. Indeed, many of today's machines are really a hybrid design.

The design of parallel algorithms faces the SIMD or MIMD type machines. The number of parallel processors p is an important parameter in a parallel system. On SIMD machines, p processors can execute the same instruction at the same time but on different data. On MIMD machines, instructions may differ across p processors, which need not operate synchronously.

The parallel algorithms facing the SIMD and MIMD machines are the algorithms based on "p processors execute the same instruction at the same time but on different data" and "instructions may differ across p processors, which need not operate synchronously" respectively.

For convenience, three ideal assumptions for the model of parallel computers are proposed as follows.

(1) The model has an arbitrary number of identical processors with independent control at any time.
(2) The model has an arbitrary large memory with unrestricted access in any time.
(3) Each processor in the model is capable of taking its operands from the memory, performing any one of the binary operations $+$, $-$, $\times$, $\div$ and storing the result in the memory in unit time. This unit time is called a step (the bookkeeping overhead is ignored). Before starting the computation, the input data is stored in the memory.

The SIMD and MIMD models can be constructed by SIMD and MIMD system in addition to the above three ideal assumptions.

7.2 Measures of the Performance of Parallel Algorithms

A mathematical problem can be solved by several parallel algorithms, it is important to analyze which algorithm is better than the others. This work is both practical and theoretical.

Next we introduce some criteria for measuring the performance of parallel algorithms.

The problem size n is the amount of computer storage required to store all the data that define the problem instance. By the time complexity of an algorithm is meant the worst case number of steps required for its execution, we assume that it is a function of n, denoted by $T(n)$. If the time required for computing n numbers using some algorithm is cn^2 steps, where c is a constant, then $T(n) = O(n^2)$. By the space complexity of an algorithm is meant the upper bound of the processors required for its execution, denoted by $P(n)$. If the time and space complexity of two different parallel algorithms for solving same problem are $T_1(n)$, $P_1(n)$ and $T_2(n)$, $P_2(n)$ respectively, and their products (cost-optimality) satisfy

$$T_1(n) \cdot P_1(n) < T_2(n) \cdot P_2(n),$$

then the former is better than the latter.

After one has obtained a parallel algorithm for the problem of size n, it is important to measure its performance in some way. The most commonly accepted measurement is speedup.

Let $T_p(n)$ denote the execution time using the parallel algorithm on p (≥ 1) processors, $T_1(n)$ denote the execution time using the fastest sequential algorithm on one processor. Then

$$S_p(n) = \frac{T_1(n)}{T_p(n)}$$

is called the speedup.

An algorithm with excellent parallel characteristics, that is, a high speedup factor S_p, still might not yield much actual improvement on p processors as S_p would indicate. Thus we have the following measurement.

The efficiency $E_p(n)$ of p-processor system is defined by

$$E_p(n) = \frac{S_p(n)}{p}.$$

The value of $E_p(n)$ expresses, in a relative sense, how busy the processors are kept. Evidently

$$0 \leq E_p(n) \leq 1.$$

If the speedup and efficiency of two different parallel algorithms for solving some problem are $S_{p_1}(n)$, $E_{p_1}(n)$ and $S_{p_2}(n)$, $E_{p_2}(n)$ respectively, and their products satisfy

$$S_{p_1}(n) \cdot E_{p_1}(n) > S_{p_2}(n) \cdot E_{p_2}(n),$$

then the former is more efficient than the latter.

7.3 Parallel Algorithms

Some basic parallel algorithms are first given before we discuss the parallel algorithms for computing the generalized inverses. In the following the logarithm $\log p$ denotes $\log_2 p$, $\lceil x \rceil$ denotes an integer such that

$$x \leq \lceil x \rceil < x + 1$$

and $\lfloor x \rfloor$ denotes an integer such that

$$x - 1 < \lfloor x \rfloor \le x.$$

7.3.1 Basic Algorithms

(1) The sum of n numbers

To compute the sum

$$S = \sum_{i=1}^{n} b_i, \tag{7.3.1}$$

a common parallel method is the binary tree method. For example, when $n = 8$, the process is shown in Fig. 7.4.

This takes $\log 8 = 3$ steps and 4 processors. In general, the model of parallel computation assumes an arbitrary number of identical processors. The binary tree method for computing the sum of (7.3.1) takes

$$T(n) = \lceil \log n \rceil$$

steps and

$$P(n) = \left\lceil \frac{n}{2} \right\rceil$$

processors.

(2) The product of two matrices

Let $A = [a_{ij}]$ be an $m \times p$ matrix and $B = [b_{ij}]$ be a $p \times n$ matrix. The product of A and B is the $m \times n$ matrix $C = [c_{ij}]$ whose elements are given by

$$c_{ij} = \sum_{k=1}^{p} a_{ik} b_{kj}. \tag{7.3.2}$$

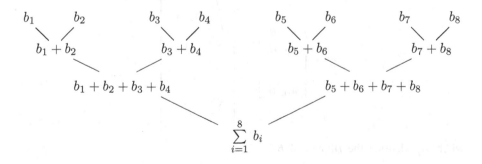

Fig. 7.4 Binary tree method

Let the ith row of A and the jth column of B be

$$\mathbf{a}_{i,:} = [a_{i1}\, a_{i2}\, \cdots\, a_{ip}] \quad \text{and} \quad \mathbf{b}_j = [b_{1j}\, b_{2j}\, \cdots\, b_{pj}]^T$$

respectively, then

$$c_{ij} = \mathbf{a}_{i,:}\mathbf{b}_j, \quad i = 1, 2, \cdots, m, \;\; j = 1, 2, \cdots, n,$$

is called the inner product of $\mathbf{a}_{i,:}$ and $\mathbf{b}_j$.

To compute $a_{ik}b_{kj}$ $(k = 1, 2, \cdots, p)$ in parallel, it takes one step and p processors. The algorithm for parallelly computing the sum (7.3.2) takes $\lceil \log p \rceil$ steps and p processors.

If using mnp processors, then the inner product algorithm for parallelly computing the product of two matrices takes $\lceil \log p \rceil + 1$ steps.

In the special case when $m = n = p$, the inner product algorithm for parallelly computing the product C of the matrices A and B of order n takes $\lceil \log n \rceil + 1 = O(\log n)$ steps and $O(n^3)$ processors.

The multiplication of two $n \times n$ matrices can be done in parallel in time $O(\log n)$ using $n^\alpha / \log n$ processors, for some real α satisfying the obvious bounds $2 \leq \alpha \leq 3$. The smallest feasible value of α is $\log 7$. The details can be found in [14].

The middle product algorithm for computing the product of $m \times p$ matrix A and $p \times n$ matrix B is described as follows. Let

$$
\begin{aligned}
C &= AB \\
&= A[\mathbf{b}_1 \;\; \mathbf{b}_2 \;\; \cdots \;\; \mathbf{b}_n] \\
&= [A\mathbf{b}_1 \;\; A\mathbf{b}_2 \;\; \cdots \;\; A\mathbf{b}_n] \\
&= \left[\sum_{j=1}^{p} b_{j1}\mathbf{a}_j \;\; \sum_{j=1}^{p} b_{j2}\mathbf{a}_j \;\; \cdots \;\; \sum_{j=1}^{p} b_{jn}\mathbf{a}_j \right],
\end{aligned}
$$

where $\mathbf{a}_j$ denotes the jth column of A.

The following is the dual-middle product algorithm for computing $C = AB$:

$$
C = \begin{bmatrix} \mathbf{a}_{1,:}B \\ \mathbf{a}_{2,:}B \\ \vdots \\ \mathbf{a}_{m,:}B \end{bmatrix} = \begin{bmatrix} \sum_{j=1}^{p} a_{1j}\mathbf{b}_{j,:} \\ \sum_{j=1}^{p} a_{2j}\mathbf{b}_{j,:} \\ \vdots \\ \sum_{j=1}^{p} a_{mj}\mathbf{b}_{j,:} \end{bmatrix},
$$

where $\mathbf{b}_{j,:}$ denotes the jth row of B.

The outer product algorithm for computing the product C of $m \times p$ matrix A and $p \times n$ matrix B is given as the following:

$$C = AB$$

$$= [\mathbf{a}_1 \ \mathbf{a}_2 \ \cdots \ \mathbf{a}_p] \begin{bmatrix} \mathbf{b}_{1,:} \\ \mathbf{b}_{2,:} \\ \vdots \\ \mathbf{b}_{p,:} \end{bmatrix}$$

$$= \sum_{i=1}^{p} \mathbf{a}_i \mathbf{b}_{i,:}$$

$$= \sum_{i=1}^{p} [b_{i1}\mathbf{a}_i \ \ b_{i2}\mathbf{a}_i \ \ \cdots, b_{in}\mathbf{a}_i].$$

The following is the dual-outer product algorithm for computing $C = AB$:

$$C = \sum_{i=1}^{p} \mathbf{a}_i \mathbf{b}_{i,:} = \sum_{i=1}^{p} \begin{bmatrix} a_{1i}\mathbf{b}_{i,:} \\ a_{2i}\mathbf{b}_{i,:} \\ \vdots \\ a_{mi}\mathbf{b}_{i,:} \end{bmatrix}.$$

The steps and the number of processors required by the above algorithms are left to the reader as an exercise.

(3) The powers of an $n \times n$ matrix

The parallel algorithm for computing the set $\{B^j \mid j = 1, 2, \cdots, w\}$ of powers of a given $n \times n$ matrix B is given in [15].

Procedure POWERS(B, w)

Input: An $n \times n$ matrix B and a positive integer w

Output: $\{B^j \mid j = 1, 2, \cdots, w\}$

```
     begin
1.       if w = 1,
2.           return B;
3.       else
4.           q ← ⌈w/2⌉;
5.           POWERS(B, q);
6.           Bⁱ = B⌊i/2⌋B⌈i/2⌉, i = q + 1, q + 2, ⋯, w;
     end.
```

We denote by $P_1(w)$ and $T_1(w)$ respectively the number of processors and steps required by POWERS(B, w).

Let

$$T = O(\log n) \quad \text{and} \quad P = \frac{n^{\alpha}}{\log n}, \quad 2 \le \alpha \le 3,$$

be the steps and the number of processors required for multiplying two $n \times n$ matrices respectively, then

$$T_1(w) = O(\log w \, \log n)$$

and

$$P_1(w) = \left\lfloor \frac{w}{2} \right\rfloor \frac{n^\alpha}{\log n}.$$

Notice that all the $\lfloor w/2 \rfloor$ matrix multiplications in step 6 can be executed in parallel, given sufficient number of processors. Therefore, since $q \approx \lfloor w/2 \rfloor$, we obtain the simple recurrence:

$$\begin{aligned}
T_1(w) &= T_1(w/2) + T \\
&= T_1(w/4) + 2T \\
&= T_1(w/4) + (\log 4)\, T \\
&= \cdots \\
&= (\log w)\, T.
\end{aligned}$$

Thus, POWERS(B, w) runs in time

$$T_1(w) = \log w \, O(\log n) = O(\log w \, \log n)$$

with

$$P_1(w) = \left\lfloor \frac{w}{2} \right\rfloor P = \left\lfloor \frac{w}{2} \right\rfloor \frac{n^\alpha}{\log n}$$

processors.

By using POWERS(B, w), an improved parallel algorithm for computing the powers of $n \times n$ matrices is given in [15].

Procedure SUPERPOWERS(B, t)

Input: An $n \times n$ matrix B and a positive integer t

Output: $\{B^j \mid j = 1, 2, \cdots, t\}$

```
    begin
1.      if t = 1
2.          return B;
3.      else
4.          a ← ⌈log t⌉;
5.          b ← ⌊t/a⌋; c ← t − ⌊t/a⌋ a;
6.          POWERS(B, b);
7.          for i ← 1 step 1 until a − 1 do
                B^{bi+j} ← B^{b(i−1)+j} B^b,   j = 1, 2, · · · , b;
8.          B^{ba+k} ← B^{b(a−1)+k} B^b,   k = 1, 2, · · · , c;
    end.
```

We denote by $P_2(t)$ the number of processors and $T_2(t)$ the number of steps required by SUPERPOWERS(B, t). It is easy to show that

$$T_2(t) \le O(\log w \, \log n)$$

and

$$P_2(t) = t \, \frac{n^\alpha}{\log t \, \log n}.$$

Notice that the steps required by SUPERPOWERS(B, t) satisfies the inequality

$$T_2(t) \le T_1(t) + aT,$$

where the first term on the right side is due to step 6 and the second is due to steps 7 and 8. Since $a \approx \log t$, $b \approx t / \log t$ and $T = O(\log n)$, we obtain

$$T_2(t) \le T_1(\frac{t}{\log t}) + \log t \, O(\log n) = O(\log t \, \log n).$$

As the number of processors, step 6 requires

$$\left\lfloor \frac{b}{2} \right\rfloor \frac{n^\alpha}{\log n}.$$

Steps 7 and 8 jointly involve the parallel execution, $a \approx \log t$ steps, of b matrix-matrix multiplications, thereby requiring

$$b \, \frac{n^\alpha}{\log n}$$

processors. It follows that

$$P_2(t) = b \, \frac{n^\alpha}{\log n}$$

$$= \left\lfloor \frac{t}{a} \right\rfloor \frac{n^\alpha}{\log n}$$

$$\approx \frac{t n^\alpha}{\log t \, \log n}.$$

(4) The solution of a lower triangular system
    We now turn to the problem of solving the lower triangular linear system

$$\mathbf{L}\mathbf{x} = \mathbf{f}, \tag{7.3.3}$$

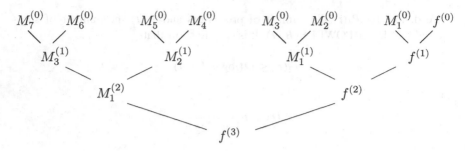

Fig. 7.5 Computation of $\mathbf{x} = M_n M_{n-1} \cdots M_1 \mathbf{f}$

where $L = [l_{ij}]$ is a nonsingular lower triangular matrix of order n. It follows from Sect. 5.1.2, the inverse of a lower triangular matrix L can be written in the form

$$L^{-1} = M_n M_{n-1} \cdots M_1,$$

where

$$M_i = \begin{bmatrix} 1 & & & & & \\ & \ddots & & & & \\ & & 1 & & & \\ & & & l_{ii}^{-1} & & \\ & & & -l_{ii}^{-1}l_{i+1,i} & 1 & \\ & & & \vdots & & \ddots \\ & & & -l_{ii}^{-1}l_{ni} & & 1 \end{bmatrix}, \quad i = 1, 2, \cdots n. \tag{7.3.4}$$

Thus

$$\mathbf{x} = M_n M_{n-1} \cdots M_1 \mathbf{f}.$$

We introduce the following notations for computing $\mathbf{x}$ in parallel. Set $s = 2^j$ and $\mu = \log n$, assuming, without loss of generality, $n = 2^\mu$. Initially, $M_i^{(0)} = M_i$, $i = 1, 2, \cdots, n-1$, and $\mathbf{f}^{(0)} = \mathbf{f}$. For example, for $n = 8$, the computation process is shown in Fig. 7.5.

It is easy to obtain a general computation process:

$$M_i^{(j+1)} = M_{2i+1}^{(j)} M_{2i}^{(j)}, \quad j = 0, 1, \cdots, \mu - 2, \ i = 1, 2, \cdots, \frac{n}{2s} - 1,$$

$$\mathbf{f}^{(j+1)} = M_1^{(j)} \mathbf{f}^{(j)}, \quad j = 0, 1, \cdots, \mu - 1,$$

$$\mathbf{x} = M_n \mathbf{f}^{(\mu)}.$$

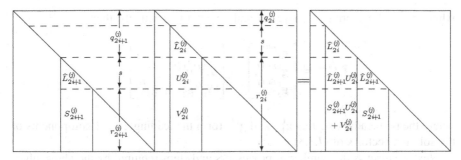

Fig. 7.6 Multiplication of $M_{2i+1}^{(j)}$ and $M_{2i}^{(j)}$

The above matrix $M_i^{(j)}$ can be written as a 3×3 block matrix of the form:

$$M_i^{(j)} = \begin{bmatrix} I_i^{(j)} & O & O \\ O & \widehat{L}_i^{(j)} & O \\ O & S_i^{(j)} & \widehat{I}_i^{(j)} \end{bmatrix},$$

where $\widehat{L}_i^{(j)}$ is a lower triangular matrix of order $s = 2^j$, $I_i^{(j)}$ and $\widehat{I}_i^{(j)}$ are two identity matrices of orders $q_i^{(j)} = is - 1$ and $r_i^{(j)} = (n+1) - (i+1)s$ respectively. It is clear that

$$\widehat{L}_i^{(0)} = \frac{1}{l_{ii}} \quad \text{and} \quad S_i^{(0)} = -\frac{1}{l_{ii}}[l_{i+1,i} \ \ l_{i+2,i} \ \ \cdots \ \ l_{n,i}]^T.$$

Partition

$$S_i^{(j)} = \begin{bmatrix} U_i^{(j)} \\ V_i^{(j)} \end{bmatrix} \in \mathbb{R}^{r_j^{(i)} \times s},$$

where $U_i^{(j)}$ is an $s \times s$ matrix.

Multiplying the two partitioned matrices $M_{2i+1}^{(j)}$ and $M_{2i}^{(j)}$, shown in Fig. 7.6, we have

$$\widehat{L}_i^{(j+1)} = \begin{bmatrix} \widehat{L}_{2i}^{(j)} & O \\ \widehat{L}_{2i+1}^{(j)} U_{2i}^{(j)} & \widehat{L}_{2i+1}^{(j)} \end{bmatrix}$$

and

$$S_i^{(j+1)} = [S_{2i+1}^{(j)} U_{2i}^{(j)} + V_{2i}^{(j)} \ \ S_{2i+1}^{(j)}].$$

Let

$$\mathbf{f}^{(j)} = \begin{bmatrix} \mathbf{g}_1^{(j)} \\ \mathbf{g}_2^{(j)} \\ \mathbf{g}_3^{(j)} \end{bmatrix},$$

where $\mathbf{g}_1^{(j)}$ and $\mathbf{g}_2^{(j)}$ are $(s-1)$-vector and s-vector respectively, then

$$
\mathbf{f}^{(j+1)} = \begin{bmatrix} \mathbf{g}_1^{(j+1)} \\ \mathbf{g}_2^{(j+1)} \\ \mathbf{g}_3^{(j+1)} \end{bmatrix} = \begin{bmatrix} \mathbf{g}_1^{(j)} \\ \widehat{L}_1^{(j)} \mathbf{g}_2^{(j)} \\ S_1^{(j)} \mathbf{g}_2^{(j)} + \mathbf{g}_3^{(j)} \end{bmatrix},
$$

where the two leading vectors $\mathbf{g}_1^{(j)}$ and $\mathbf{g}_2^{(j)}$ form the leading $2s-1$ components of the solution vector $\mathbf{x}$ of (7.3.3).

Next we discuss the number of processors and steps required by the above algorithm.

(1) Forming M_i of (7.3.4) requires 2 steps, 1 step division and 1 step subtraction, using $n-i+1$ processors. Thus forming all M_i, for $i = 1, 2, \cdots, n$, requires 2 steps and

$$
\sum_{i=1}^{n} (n-i+1) = \frac{1}{2} n(n+1)
$$

processors.

The products $\widehat{L}_{2i+1}^{(j)} U_{2i}^{(j)}$, $S_{2i+1}^{(j)} U_{2i}^{(j)}$, $\widehat{L}_1^{(j)} g_2^{(j)}$ and $S_1^{(j)} g_2^{(j)}$ can be computed in parallel, requiring $1 + \log s = 1 + j$ steps to compute the inner product of two s-vectors. The sums $S_{2i+1}^{(j)} U_{2i}^{(j)} + V_{2i}^{(j)}$ and $S_1^{(j)} g_2^{(j)} + g_3^{(j)}$ can be computed in parallel requiring one step. Therefore at the $(j-1)$st level we require

$$
\tau^{(j+1)} = 2 + j
$$

steps. In addition to the one step for computing $M_n f^{(\mu)}$, in total, we require

$$
T = 3 + \sum_{j=0}^{\mu-1} \tau^{(j+1)} = 3 + \frac{\mu(\mu+3)}{2}
$$

steps for solving the lower triangular linear system (7.3.3). Since $\mu = \log n$, we have

$$
T = O(\log^2 n).
$$

(2) First of all, we consider the processors for computing

$$
M_i^{(j+1)} = M_{2i+1}^{(j)} M_{2i}^{(j)},
$$

where $\widehat{L}_{2i+1}^{(j)} U_{2i}^{(j)}$ and $S_{2i+1}^{(j)} U_{2i}^{(j)}$ in the product can be computed in parallel. To form each column of $\widehat{L}_{2i+1}^{(j)} U_{2i}^{(j)}$, we use

$$
\sum_{k=1}^{s} k = \frac{s(s+1)}{2}
$$

processors to compute the s inner products of 1-vectors, 2-vectors, $\cdots$, and s-vectors. Thus forming all columns of $\widehat{L}_{2i+1}^{(j)} U_{2i}^{(j)}$ requires $s^2(s+1)/2$ processors.

At the same time, forming $S_{2i+1}^{(j)} U_{2i}^{(j)}$ uses $s^2 r_{2i+1}^{(j)}$ processors to compute $s r_{2i+1}^{(j)}$ inner products of s-vectors. Therefore it requires

$$P' = \frac{1}{2}s^2(s+1) + s^2 r_{2i+1}^{(j)}$$
$$= \frac{1}{2}(2n+3)s^2 - \frac{1}{2}(4i+3)s^3$$

processors.

Forming $S_{2i+1}^{(j)} U_{2i}^{(j)} + V_{2i}^{(j)}$ uses $s r_{2i+1}^{(j)}$, i.e.,

$$P'' = (n+1)s - 2(i+1)s^2$$

processors. Hence the processors required for computing $M_i^{(j+1)}$ is

$$P_i^{(j+1)} = \max\{P', \ P''\} = P'.$$

Similarly, we can show that the number of processors required for computing $\mathbf{f}^{(j+1)}$ is

$$P_0^{(j+1)} = \frac{1}{2}(2n+3)s - \frac{3}{2}s^2.$$

Hence at the $(j+1)$st level we require

$$P^{(j+1)} = \sum_{k=0}^{n/2s-1} P_k^{(j+1)}$$
$$= \frac{3}{2}s^3 - \frac{1}{4}(5n+12)s^2 + \frac{1}{4}(n^2+7n+6)s,$$

for $j = 0, 1, \cdots, \mu - 2$, processors.

Therefore the processors required for solving the lower triangular linear system (7.3.3) is

$$P = \max\left\{\max_{0 \le j \le \mu-2}(P^{(j+1)}), \ \frac{n(n+1)}{2}\right\}.$$

Set

$$f(s) = \frac{3}{2}s^3 - \frac{1}{4}(5n+12)s^2 + \frac{1}{4}(n^2+7n+6)s.$$

Since the range of $s = 2^j$, $j = 0, 1, \cdots, \mu - 2$, is $\{1, 2, \cdots, n/4\}$, it is easy to show that if $n \geq 16$, the maximum value of $f(x)$ can be reached at $s = n/8$. Thus

$$\max_{0 \leq j \leq \mu-2} P^{(j+1)} = \frac{n}{64}\left(\frac{15}{16}n^2 + 11n + 12\right)$$

and

$$P = \frac{n}{64}\left(\frac{15}{16}n^2 + 11n + 12\right) = \frac{15}{1024}n^3 + O(n^2).$$

Let us summarize the above in the following theorem.

Theorem 7.3.1 ([16]) *Let L be a nonsingular lower triangular matrix of order n, then there exists an algorithm for computing the solution of the lower triangular linear system* $Lx = f$ *in parallel requiring*

$$T = \frac{1}{2}\log^2 n + \frac{3}{2}\log n + 3$$

steps and

$$P = \frac{15}{1024}n^3 + O(n^2)$$

processors.

7.3.2 Csanky Algorithms

In 1976 Csanky proposed an algorithm for computing the inverse of an $n \times n$ matrix in time $O(\log^2 n)$ using $O(n^4)$ processors.

Algorithm 7.3.1 ([17]) Let $A \in \mathbb{R}^{n \times n}$, rank$(A) = n$. This algorithm computes the inverse of A.

(1) Parallelly compute $A^k = [a_{ij}^{(k)}]$, $k = 1, 2, \cdots, n$.
(2) Let $\lambda_1, \lambda_2, \cdots, \lambda_n$ denote the roots of the characteristic polynomial $a(\lambda)$ of A, and the trace of A^k be

$$s_k = \sum_{i=1}^{n} \lambda_i^k, \quad k = 1, 2, \cdots, n.$$

Parallelly compute

$$s_k = \text{tr}(A^k) = \sum_{i=1}^{n} a_{ii}^{(k)}, \quad k = 1, 2, \cdots, n.$$

(3) Let the characteristic polynomial of A be

$$a(\lambda) = \det(\lambda I - A)$$
$$= \prod_{i=1}^{n}(\lambda - \lambda_i)$$
$$= \lambda^n + c_1\lambda^{n-1} + \cdots + c_{n-1}\lambda + c_n.$$

From the Newton formula

$$s_k + c_1 s_{k-1} + \cdots + c_{k-1}s + kc_k = 0, \quad k = 1, 2, \cdots, n,$$

we have

$$\begin{bmatrix} 1 & & & & \\ s_1 & 2 & & & \\ s_2 & s_1 & 3 & & \\ \vdots & \vdots & \vdots & \ddots & \\ s_{n-1} & s_{n-2} & s_{n-3} & \cdots & n \end{bmatrix} \begin{bmatrix} c_1 \\ c_2 \\ c_3 \\ \vdots \\ c_n \end{bmatrix} = - \begin{bmatrix} s_1 \\ s_2 \\ s_3 \\ \vdots \\ s_n \end{bmatrix}.$$

Parallelly compute the solution of the above triangular system.

(4) It follows from the Cayley-Hamilton theorem that $a(A) = 0$ and

$$A^{-1} = -\frac{1}{c_n}(A^{n-1} + c_1 A^{n-2} + \cdots + c_{n-2}A + c_{n-1}I). \tag{7.3.5}$$

Parallelly compute A^{-1} of (7.3.5).

Theorem 7.3.2 *Let* $A \in \mathbb{R}^{n \times n}$, *rank*$(A) = n$, *then Algorithm 7.3.1 for computing* A^{-1} *can be implemented in*

$$I(n) = O(\log^2 n)$$

steps using

$$cP(n) = O\left(\frac{1}{2}n^4\right)$$

processors.

Proof Suppose that the ith stage of Algorithm 7.3.1 can be implemented in T_i steps using cP_i processors.

(1) The parallel computation of $A^k = [a_{ij}^{(k)}]$, $k = 1, 2, \cdots, n$, by the algorithm POWERS(A, n) takes $T_1 = O(\log^2 n)$ steps and

$$cP_1 = \frac{n}{2} \cdot \frac{n^\alpha}{\log n} \approx O\left(\frac{1}{2}n^4\right)$$

processors.

(2) The parallel computation of $s_k = \sum_{i=1}^n a_{ii}^{(k)}$, $k = 1, 2, \cdots, n$ by the basic algorithm for sum in Sect. 7.3.1 takes $T_2 = \log n$ steps and $cP_2 = n^2/2$ processors.

(3) The parallel computation of c_i, $i = 1, 2, \cdots, n$ by the basic algorithm for solving lower triangular systems in Sect. 7.3.1 takes $T_3 = O(\log^2 n)$ steps and $cP_3 = O(n^3)$ processors.

(4) The parallel computation of A^{-1} takes $T_4 = \log n + 2$ steps and $cP_4 = n^3/2$ processors.

Thus

$$I(n) = T_1 + T_2 + T_3 + T_4 = O(\log^2 n)$$

and

$$cP(n) = \max_{1 \leq i \leq 4} \{cP_i\} = O\left(\frac{1}{2}n^4\right),$$

which completes the proof.   □

It is easy to show that the formula (7.3.5) is the same as the finite algorithm formula (5.5.2) for computing A^{-1}. The finite algorithms for computing the generalized inverses $A_{MN}^\dagger$, $A^\dagger$, A_d and A_g are given in Chap. 5. Consequently we have the following parallel algorithms for the generalized inverses [6, 18].

Algorithm 7.3.2 Let $A \in \mathbb{R}^{m \times n}$, $\text{rank}(A) = r$. This algorithm computes the Moore-Penrose inverse $A^\dagger$ of A.

(1) Parallelly compute $B = A^T A$.

(2) Parallelly compute $B^k = [b_{ij}^{(k)}]$, $k = 1, 2, \cdots, r$.

(3) Let $\lambda_1, \lambda_2, \cdots, \lambda_n$ denote the roots of the characteristic polynomial $b(\lambda)$ of B, and the trace of B^k be

$$s_k = \sum_{i=1}^n \lambda_i^k, \quad k = 1, 2, \cdots, r.$$

Parallelly compute

$$s_k = \text{tr}(B^k) = \sum_{i=1}^n b_{ii}^{(k)}, \quad k = 1, 2, \cdots, r.$$

(4) Let the characteristic polynomial of B be

$$b(\lambda) = \det(\lambda I - B)$$

$$= \prod_{i=1}^n (\lambda - \lambda_i)$$

$$= \lambda^n + c_1 \lambda^{n-1} + \cdots + c_{n-1}\lambda + c_n.$$

Since $\text{rank}(B) = \text{rank}(A) = r$, $c_{r+1} = c_{r+2} = \cdots = c_n = 0$.

From the Newton formula we have

$$\begin{bmatrix} 1 & & & & \\ s_1 & 2 & & & \\ s_2 & s_1 & 3 & & \\ \vdots & \vdots & \vdots & \ddots & \\ s_{r-1} & s_{r-2} & s_{r-3} & \cdots & r \end{bmatrix} \begin{bmatrix} c_1 \\ c_2 \\ c_3 \\ \vdots \\ c_r \end{bmatrix} = - \begin{bmatrix} s_1 \\ s_2 \\ s_3 \\ \vdots \\ s_r \end{bmatrix} .$$

Parallelly compute the solution of the above triangular system.

(5) Parallelly compute

$$A^\dagger = -\frac{1}{c_r}((A^T A)^{r-1} + c_1(A^T A)^{r-2} + \cdots + c_{r-1}I)A^T .$$

The steps and processors required by Algorithm 7.3.2 are given in [6].

Theorem 7.3.3 *Let $A \in \mathbb{R}^{m \times n}$, $\mathrm{rank}(A) = r$. Then Algorithm 7.3.2 for computing $A^\dagger$ can be implemented in*

$$GI(m, n) = O(\log r \cdot \log n) + \log m + 2 \log n = O(f(m, n, r))$$

steps using

$$GcP(m, n) = \begin{cases} \dfrac{n^3 r}{2}, & m < \dfrac{nr}{2}, \\ mn^2, & m \geq \dfrac{nr}{2} \end{cases}$$

processors.

Proof Suppose that the ith stage of Algorithm 7.3.2 can be implemented in T_i steps using GcP_i processors.

(1) The parallel computation of $B = A^T A$ takes $T_1 = 1 + \log m$ steps, one step for multiplication, and $\log m$ steps for addition, using $GcP_1 = mn^2$ processors.

(2) The parallel computation of B^k, $k = 1, 2, \cdots, r$, takes

$$T_2 = \log r(\log n + 1) = O(\log r \cdot \log n)$$

steps using $GcP_2 = O(rn^3/2)$ processors.

(3) The parallel computation of $s_k = \sum_{i=1}^{n} b_{ii}^{(k)}$, $k = 1, 2, \cdots, r$, by the basic algorithm for sum in Sect. 7.3.1 takes $T_3 = \log n$ steps and $GcP_3 = rn/2$ processors.

(4) The parallel computation of c_k, $k = 1, 2, \cdots, r$, by the basic algorithm for solving lower triangular systems in Sect. 7.3.1 takes

$$T_4 = \frac{1}{2} \log^2 r + \frac{3}{2} \log r = O(\log^2 r)$$

steps and $GcP_4 = O(r^3)$ processors.

(5) Since $B, B^2, \cdots, B^{r-1}$ are already available, the parallel computation of

$$B^{r-1} + c_1 B^{r-2} + \cdots + c_{r-1} I$$

takes $1 + \log r$ steps and $rn^2/2$ processors. The parallel computation of $A^\dagger$ takes $\log n + 2$ steps, one step for multiplication, $\log n$ steps for addition, and one step for division, using $n^2 m$ processors. Thus

$$T_5 = \log n + \log r + 3$$

and $GcP_5 = O(n^2 m)$.

It follows that

$$GI(m, n) = \sum_{i=1}^{5} T_i$$

$$= \log r \left(\log n + \frac{7}{2} \right) + \frac{1}{2} \log^2 r + 2 \log n + \log m + 4$$

$$= O(\log r \cdot \log n) + \log m + 2 \log n$$

and

$$GcP(m, n) = \max_{1 \le i \le 5} (GcP_i) = \begin{cases} \dfrac{n^3 r}{2}, & m < \dfrac{nr}{2}, \\ mn^2, & m \ge \dfrac{nr}{2}. \end{cases}$$

The proof is completed.   □

Algorithm 7.3.3 Let $A \in \mathbb{R}^{m \times n}$, $\text{rank}(A) = r$, M and N be symmetric positive definite matrices of orders m and n respectively. This algorithm computes the weighted Moore-Penrose inverse $A^\dagger_{MN}$ of A.

(1) Parallelly compute N^{-1}.
(2) Parallelly compute $A^\# = N^{-1} A^T M$ and $B = A^\# A = (N^{-1} A^T)(MA)$.
(3) Parallelly compute $B^k = [b_{ij}^{(k)}]$, $k = 1, 2, \cdots, r$.
(4) Let $\lambda_1, \lambda_2, \cdots, \lambda_n$ denote the roots of the characteristic polynomial of B, and the trace of B^k be

$$s_k = \sum_{i=1}^{n} \lambda_i^k, \quad k = 1, 2, \cdots, r.$$

Parallelly compute

$$s_k = \text{tr}(B^k) = \sum_{i=1}^{n} b_{ii}^{(k)}, \quad k = 1, 2, \cdots, r.$$

(5) Let the characteristic polynomial $b(\lambda)$ of B be

$$b(\lambda) = \det(\lambda I - B)$$
$$= \lambda^n + c_1\lambda^{n-1} + \cdots + c_{n-1}\lambda + c_n.$$

Since $\text{rank}(B) = \text{rank}(A) = r$, $c_{r+1} = c_{r+2} = \cdots = c_n = 0$. From the Newton formula we have

$$
\begin{bmatrix}
1 & & & & \\
s_1 & 2 & & & \\
s_2 & s_1 & 3 & & \\
\vdots & \vdots & \vdots & \ddots & \\
s_{r-1} & s_{r-2} & s_{r-3} & \cdots & r
\end{bmatrix}
\begin{bmatrix}
c_1 \\
c_2 \\
c_3 \\
\vdots \\
c_r
\end{bmatrix}
= -
\begin{bmatrix}
s_1 \\
s_2 \\
s_3 \\
\vdots \\
s_r
\end{bmatrix}.
$$

Parallelly compute the solution of the above triangular system.
(6) Parallelly compute

$$A_{MN}^{\dagger} = -\frac{1}{c_r}((A^{\#}A)^{r-1} + c_1(A^{\#}A)^{r-2} + \cdots + c_{r-1}I)A^{\#}.$$

The steps and processors required for Algorithm 7.3.3 are given in [6].

Theorem 7.3.4 *Let $A \in \mathbb{R}^{m \times n}$, $\text{rank}(A) = r$, then Algorithm 7.3.3 for computing $A_{MN}^{\dagger}$ can be implemented in*

$$WGI(m, n) = GI(m, n) + O(\log^2 n + \log m)$$

steps using

$$
WGcP(m, n) =
\begin{cases}
\dfrac{n^4}{2}, & m \leq \dfrac{n}{2}(\sqrt{1 + 2n} - 1), \\
m^2n + mn^2, & m > \dfrac{n}{2}(\sqrt{1 + 2n} - 1)
\end{cases}
$$

processors.

Proof Suppose that the ith stage of Algorithm 7.3.3 can be implemented in T_i steps using WcP_i processors.
(1) It follows from Theorem 7.3.2 that the parallel computation of N^{-1} takes $T_1 = O(\log^2 n)$ steps and $WcP_1 = O(n^4/2)$ processors.
(2) First, the parallel computation of $N^{-1}A^T$ and MA takes $1 + \log m$ or $1 + \log n$ steps and $m^2n + mn^2$ processors. Then the parallel computation of $A^{\#} = (N^{-1}A^T)M$ and $B = (N^{-1}A^T)(MA)$ takes $1 + \log m$ steps and $m^2n + mn^2$ processors. Thus the parallel computation of $A^{\#}$ and B takes $T_2 = 2(1 + \log m)$ or $2 + \log m + \log n$ steps and $WcP_2 = m^2n + mn^2$ processors.

(3) The parallel computation of B^k, $k = 1, 2, \cdots, r$ takes

$$T_3 = \log r(1 + \log n) = O(\log r \cdot \log n)$$

steps and $WcP_3 = O(n^3 r/2)$ processors.

(4) The parallel computation of s_k, $k = 1, 2, \cdots, r$ takes $T_4 = \log n$ steps and $WcP_4 = rn/2$ processors.

(5) The parallel computation of c_k, $k = 1, 2, \cdots, r$ takes

$$T_5 = \frac{1}{2} \log^2 r + \frac{3}{2} \log r = O(\log^2 r)$$

steps and $WcP_5 = O(r^3)$ processors.

(6) The parallel computation of $A_{MN}^\dagger$ takes

$$T_6 = \log r + \log n + 3$$

steps and $WcP_6 = n^2 m$ processors.

Thus

$$WGI(m, n) = \sum_{i=1}^{6} T_i$$

$$= GI(m, n) + O(\log^2 n + \log m)$$

and

$$WGcP(m, n) = \max_{1 \le i \le 6} (WcP_i) = \begin{cases} \dfrac{n^4}{2}, & m \le \dfrac{n}{2}(\sqrt{1 + 2n} - 1), \\[3mm] m^2 n + mn^2, & m > \dfrac{n}{2}(\sqrt{1 + 2n} - 1). \end{cases}$$

The proof is completed.   □

Remark By using the algorithm SUPERPOWERS(B, r) and some techniques, an improved parallel algorithm for the generalized inverse $A^\dagger$ is given in [8]. It shows that, under the same assumptions as in [15], the time complexity and the number of processors using the improved parallel algorithm are

$$\widetilde{GI}(m, n) = O(\log r \cdot \log n)$$

and

$$\widetilde{GcP}(m, n) = \begin{cases} \dfrac{2r^{1/2} n^\alpha}{\log r \cdot \log n}, & \dfrac{m}{n} \le \dfrac{2r^{1/2}}{\log r}, \\[3mm] \dfrac{\lceil m/n \rceil n^\alpha}{\log n}, & \dfrac{m}{n} > \dfrac{2r^{1/2}}{\log r} \end{cases}$$

respectively, and proves the cost-optimality

$$\widetilde{GcP}(m, n)\widetilde{GI}(m, n) < GcP(m, n)GI(m, n).$$

By using the finite algorithms for the Drazin inverse and the group inverse, the corresponding parallel algorithms and their time complexities and the required numbers of processors can be obtained. It is omitted here.

7.4 Equivalence Theorem

In 1976, Csanky not only proposed the parallel algorithm for computing the inverse of an $n \times n$ matrix A but also gave an important theoretical result [17].

Let $I(n)$, $E(n)$, $D(n)$, and $P(n)$ respectively denote the parallel arithmetic complexities of inverting a matrix of order n, solving a system $A\mathbf{x} = \mathbf{b}$ of n linear equations with n unknowns, computing an order n determinant $\det(A)$, and finding the characteristic polynomial $a(\lambda) = \det(\lambda I - A)$ of an order n matrix A, then $I(n)$, $E(n)$, $D(n)$ and $P(n)$ have the same growth rate as follows.

Lemma 7.4.1 $2 \log n \leq I(n),\ E(n),\ D(n),\ P(n)$.

Proof The proof follows directly from the fact that in each case, at least one partial result is a nontrivial function of at least n^2 variables and fan in argument. □

Theorem 7.4.1

$$I(n) = O(f(n))$$
$$\Leftrightarrow E(n) = O(f(n))$$
$$\Leftrightarrow D(n) = O(f(n))$$
$$\Leftrightarrow P(n) = O(f(n)).$$

Proof (1) $D(n) \leq E(n) + \log n + O(1)$.
Let D_t denote an order t determinant and D_n be the determinant to be computed. Define

$$x_k = \frac{D_{n-k}}{D_{n-k+1}},$$

where $1 \leq k \leq n - 1$ and D_{n-k} is a properly chosen minor of D_{n-k+1}. Since $\prod_{k=1}^{n-1} x_k = D_1/D_n$,

$$D_n = \frac{D_1}{\prod_{k=1}^{n-1} x_k}.$$

Thus to compute D_n, we compute x_k for all k in $E(n - k + 1)$ parallel steps by solving the corresponding system of equations, then in $\log n + O(1)$ additional steps, we can compute D_n.

(2) $E(n) \le I(n) + 2$.

Transforming $A\mathbf{x} = \mathbf{b}$ into the form $A'\mathbf{x} = (I)_{*1}$ in two steps by row operations, where $(A)_{*j}$ denotes the jth column of A. Then invert A' in $I(n)$ steps and $\mathbf{x} = ((A')^{-1})_{*1}$.

(3) $I(n) \le P(n) + 1$.

It is well know that

$$A^{-1} = \frac{\text{adj}(A)}{\det(A)}, \quad \text{adj}(A) = [A_{ji}],$$

where A_{ji} is the algebraic cofactor of a_{ji}, the (j, i)-element of $A = [a_{ij}]$. There are n^2 determinants of order $n - 1$ and one determinant $\det(A)$ of order n for computing A^{-1}. These $n^2 + 1$ determinants can be evaluated in parallel. Let the characteristic polynomial of any square matrix B is

$$b(\lambda) = \det(B - \lambda I),$$

then $\det(B) = b(0)$.

(4) $P(n) \le D(n) + \log n + O(1)$.

Let the characteristic polynomial of A be

$$a(\lambda) = \lambda^n + a_1 \lambda_{n-1} + \cdots + a_{n-1}\lambda + a_n,$$

and ω a primitive $(n + 1)st$ root of unity. First compute $a(\omega^j)$ for all distinct ω^j in parallel by using the algorithm for computing determinants. This computation takes $D(n) + 1$ steps, including one step for computing the diagonal elements of $A - \omega^j I$. Let

$$F = \begin{bmatrix} 1 & 1 & 1 & \cdots & 1 \\ 1 & \omega & \omega^2 & \cdots & \omega^n \\ 1 & \omega^2 & \omega^4 & \cdots & \omega^{2n} \\ \vdots & \vdots & \vdots & & \vdots \\ 1 & \omega^n & \omega^{2n} & \cdots & \omega^{n^2} \end{bmatrix}, \quad \mathbf{a} = \begin{bmatrix} a_n \\ a_{n-1} \\ \vdots \\ a_1 \\ 1 \end{bmatrix}, \quad \mathbf{b} = \begin{bmatrix} a(1) \\ a(\omega) \\ a(\omega^2) \\ \vdots \\ a(\omega^n) \end{bmatrix},$$

then $F\mathbf{a} = \mathbf{b}$. The coefficients $a_1, a_2, \cdots, a_n$ of $a(\lambda)$ can be obtained by the fast Fourier transform [19]. This takes $\log n + O(1)$ steps.

From Lemma 7.4.1 and the four inequalities above, the theorem follows. For example, to prove $I(n) \Leftrightarrow E(n)$, we have

$$I(n) \le P(n) + 1 \le D(n) + \log n + O(1)$$
$$\le E(n) + 2\log n + O(1) \le c_1 E(n),$$
$$E(n) \le I(n) + 2 \le c_2 I(n),$$

where c_1 and c_2 are two nonzero constants. Hence $I(n) \Leftrightarrow E(n)$. $\qquad\square$

In the remaining of the section, we discuss the generalized matrix defined shortly, its parallel algorithm, relation with the generalized inverses, and equivalence theorems.

Let $A \in \mathbb{C}_r^{m \times n}$, and $U \in \mathbb{C}_{m-r}^{m \times (m-r)}$ and $V^* \in \mathbb{C}_{n-r}^{n \times (n-r)}$ be matrices whose columns form bases for $\mathcal{N}(A^*)$ and $\mathcal{N}(A)$ respectively. It follows from Corollary 3.1.1 that

$$A_1 = \begin{bmatrix} A & U \\ V & O \end{bmatrix}$$

is nonsingular and

$$A_1^{-1} = \begin{bmatrix} A^{\dagger} & V^{\dagger} \\ U^{\dagger} & O \end{bmatrix}.$$

For convenience, A_1 is called a generalized matrix (but not unique) of A. If A is nonsingular, we adopt the convention $A_1 = A$.

Let adj(A_1) be the common adjoint matrix of A_1. An $n \times m$ submatrix that lies in the upper left-hand corner of adj(A_1) is called a generalized adjoint matrix of A and denoted by Adj(A_1). If A is nonsingular, we adopt the convention Adj$(A_1) =$ adj(A_1). It is clear that

$$A^{\dagger} = \frac{\text{Adj}(A_1)}{\det(A_1)}. \tag{7.4.1}$$

The row echelon form of a matrix is given in Chap. 5. A form which is very closely related to the row echelon form is the Hermite echelon form. However, the Hermite echelon form is defined only for square matrices.

Definition 7.4.1 A matrix $H \in \mathbb{C}^{n \times n}$ is said to be in the Hermite echelon form if its elements h_{ij} satisfy the following conditions:

(1) H is an upper triangular matrix, i.e., $h_{ij} = 0, i > j$.
(2) h_{ii} is either 0 or 1.
(3) If $h_{ii} = 0$, then $h_{ik} = 0$ for all $k, 1 \le k \le n$.
(4) If $h_{ii} = 1$, then $h_{ki} = 0$ for all $k \ne i$.

For example, the matrix

$$H = \begin{bmatrix} 1 & 2 & 0 & 2 \\ 0 & 0 & 0 & 0 \\ 0 & 0 & 1 & 1 \\ 0 & 0 & 0 & 0 \end{bmatrix}$$

is in the Hermite echelon form. Below are some facts about the Hermite echelon form, the proofs can be found in [20].
Let $A \in \mathbb{C}^{n \times n}$, then

(1) A can always be row reduced to a Hermite echelon form. If A is reduced to its row echelon form, then a permutation of rows can always be performed to obtain a Hermite echelon form.

(2) For a given matrix A, the Hermite echelon form H_A obtained by row reducing A is unique.

(3) $H_A^2 = H_A$.

(4) $\mathcal{N}(A) = \mathcal{N}(H_A) = \mathcal{R}(I - H_A)$ and a basis for $\mathcal{N}(A)$ is the set of nonzero columns of $I - H_A$.

Algorithm 7.4.1 Let $A \in \mathbb{R}_r^{n \times n}$, this algorithm computes a generalized matrix A_1 of A.

(1) Row reduce A to its Hermite echelon form H_A.

(2) Form $I - H_A$ and select its nonzero columns $\mathbf{v}_1, \mathbf{v}_2, \cdots, \mathbf{v}_{n-r}$ to form the matrix $V = [\mathbf{v}_1 \ \mathbf{v}_2 \ \cdots \ \mathbf{v}_{n-r}]^\mathsf{T}$.

(3) Row reduce A^* to its Hermite echelon form H_{A^*}.

(4) Form $I - H_{A^*}$ and select its nonzero columns $\mathbf{u}_1, \mathbf{u}_2, \cdots, \mathbf{u}_{n-r}$ to form the matrix $U = [\mathbf{u}_1 \ \mathbf{u}_2 \ \cdots \ \mathbf{u}_{n-r}]$.

(5) Form the nonsingular matrix

$$A_1 = \begin{bmatrix} A & U \\ V & O \end{bmatrix}.$$

Although the above algorithm is stated for square matrices, it is easy to modify it for non-square ones. If $A \in \mathbb{C}^{m \times n}$, then we pad zero rows or zero columns such that

$$[A \ O] \quad \text{or} \quad \begin{bmatrix} A \\ O \end{bmatrix}$$

is square and use the identity

$$[A \ O]^\dagger = \begin{bmatrix} A^\dagger \\ O \end{bmatrix} \quad \text{or} \quad \begin{bmatrix} A \\ O \end{bmatrix}^\dagger = [A^\dagger \ O].$$

Let $F(U, V)$ denote the parallel arithmetic complexity of computing the submatrices U and V in the generalized matrix A_1 of A, then we have the following bounds.

Lemma 7.4.2 ([6]) *Let $A \in \mathbb{C}^{n \times n}$, rank$(A) = r$, and the number of processors used in Algorithm 7.4.1 be*

$$(n - 1)(2n - j_1 - j_1^*),$$

where j_1 and j_1^* are the indices of the first nonzero columns of A and A^* respectively, then

$$4r \le F(U, V) \le 2(n + r).$$

Proof The proof is left to the reader as an exercise.   □

Let $A \in \mathbb{C}^{m \times n}$, rank$(A) = r$, $\mathbf{b} \in \mathbb{C}^m$, and let $GI(m, n)$, $GE(m, n)$, $GP(m, n)$, and $GD(m, n)$ denote the parallel arithmetic complexity of computing $A^\dagger$, the minimum-norm least-squares solution of the inconsistent linear equations $A\mathbf{x} = \mathbf{b}$, and the characteristic polynomial and the determinant of order $m + n - r$ generalized matrix A_1, respectively. The the following lemmas show that $GI(m, n)$, $GE(m, n)$, $GP(m, n)$, and $GD(m, n)$ have the same growth rate.

Lemma 7.4.3 $GI(m, n) = D(m + n - r) + F(U, V) + O(1)$.

Proof From (7.4.1), there are mn order $m + n - r - 1$ determinants and one order $m + n - r$ determinant to be computed. They can be computed in parallel. □

Lemma 7.4.4 $GE(m, n) = D(m + n - r) + F(U, V) + O(1)$.

Proof From Corollary 3.2.1, the components x_j of the minimum-norm least-squares solution of the inconsistent linear equations $A\mathbf{x} = \mathbf{b}$ are given by

$$x_j = \frac{\det(A_1(j \to \widetilde{\mathbf{b}}))}{\det(A_1)}, \quad j = 1, 2, \cdots, n, \tag{7.4.2}$$

where A_1 is a generalized matrix of A, and

$$\widetilde{\mathbf{b}} = \begin{bmatrix} \mathbf{b} \\ \mathbf{0} \end{bmatrix}$$

is an $(m + n - r)$-vector. From (7.4.2), there are $n + 1$ order $m + n - r$ determinants to be computed. They can be computed in parallel. □

The following results are obvious.

Lemma 7.4.5

$$GD(m, n) = D(m + n - r) + F(U, V),$$
$$GP(m, n) = P(m + n - r) + F(U, V).$$

From Lemmas 7.4.3–7.4.5 and Theorems 7.4.1 and 7.3.4, we can immediately obtain the following important result.

Theorem 7.4.2

$$GI(m, n) = O(f(m, n, r))$$
$$\Leftrightarrow GE(m, n) = O(f(m, n, r))$$
$$\Leftrightarrow GD(m, n) = O(f(m, n, r))$$
$$\Leftrightarrow GP(m, n) = O(f(m, n, r)).$$

It follows from Theorem 3.1.1 that the matrix

$$A_2 = \begin{bmatrix} A & M^{-1}U \\ VN & O \end{bmatrix}$$

is nonsingular and

$$A_2^{-1} = \begin{bmatrix} A_{MN}^{\dagger} & V^*(VNV^*)^{-1} \\ (U^*M^{-1}U)^{-1}U^* & O \end{bmatrix}.$$

Let $WGI(m, n)$, $WGE(m, n)$, $WGP(m, n)$, and $WGD(m, n)$ denote the parallel arithmetic complexity of computing the weighted Moore-Penrose inverse $A_{MN}^{\dagger}$, the minimum-norm(N) least-squares(M) solution $\mathbf{x} = A_{MN}^{\dagger}\mathbf{b}$ of $A\mathbf{x} = \mathbf{b}$, and the characteristic polynomial and the determinant of order $m + n - r$ generalized matrix A_2, respectively. The following equivalence theorem is given in [6].

Theorem 7.4.3

$$WGI(m, n) = O(g(m, n, r))$$
$$\Leftrightarrow WGE(m, n) = O(g(m, n, r))$$
$$\Leftrightarrow WGD(m, n) = O(g(m, n, r))$$
$$\Leftrightarrow WGP(m, n) = O(g(m, n, r)).$$

Remarks

Besides Algorithm 7.4.1 for computing the generalized inverse $A^{\dagger}$, Wang [9], Wang and Wei [21, 22] presented a parallel Cramer rule (PCR) for computing $A^{\dagger}$, A_d and $A_{MN}^{\dagger}$, which is an extension of the parallel Cramer rule for computing A^{-1} by Sridhar [23].

A parallel algorithm for computing the Moore-Penrose inverse of a bidiagonal matrix based on SIMD machines is given in [7]. It uses n processors and $O(n\lceil\log n\rceil)$ iterations. Parallel (M-N)SVD algorithm on the SIMD computers can be found in [24].

Parallel successive matrix squaring algorithms for computing the Moore-Penrose inverse, the Drazin inverse, and the weighted Moore-Penrose inverse are presented in [25–27]. Recurrent neural networks for computing the regular inverse, the Moore-Penrose inverse, and the weighted Moore-Penrose inverse are given in [28–30].

An improved parallel method for computing the weighted Moore-Penrose inverse $A_{MN}^{\dagger}$ is discussed in [31].

References

1. D. Heller, A survey of parallel algorithms in numerical linear algebra. SIAM Review **20**, 740–777 (1978)
2. L. Kang et al., *Asynchronous Parallel Algorithms for Solving the Problem of Mathematical Physics* (Science Press, Beijing, 1985). in Chinese
3. J.M. Ortega, R.G. Voigt, Solution of partial differential equations on vector and parallel computer. SIAM Review **27**, 149–240 (1985)
4. J.M. Ortega, *Introduction to Parallel and Vector Solution of Linear Systems* (Plenum Press, New York, 1988)

5. O.Y. Devel, E.V. Krishnamurthy, An iterative pipelined arrary architecture for the generalized matrix inversion. Inform. Process. Lett. **26**, 263–267 (1988)
6. G. Wang, S. Lu, Fast parallel algorithm for computing the generalized inverses $A^\dagger$ and $A^\dagger_{MN}$. J. Comput. Math. **6**, 348–354 (1988)
7. G. Wang, The Moore-Penrose inverse of bidiagonal matrices and its parallel algorithm. Numer. Math. J. Chinese Univ., **12**, 14–23 (1990) (in Chinese)
8. G. Wang, An improved parallel algorithm for computing the generalized inverse $A^\dagger$. Inform. Process. Lett. **41**, 243–251 (1992)
9. G. Wang, PCR algorithm for parallel computing minimum-norm least-squares solution of inconsistent linear equations. Numer. Math. J. Chinese Univ. English Series **2**, 1–10 (1993)
10. J. Chen, *Parallel Numerical Methods* (Qinghua Univ. Press, Beijing, 1983). in Chinese
11. M.R. Zargham, *Computer Architecture* (Prentice-Hall Inc, Single and Paralled Systems, 1996)
12. J.J. Dongarra, I.S. Duff, D.C. Sorensen, van der Vorst, A. Henk, *Solving Linear Systems on Vector and Shared Memory Computers*. (Society for Industrial and Applied Mathematics (SIAM), Philadelphia, PA, 1991)
13. B. Zhang et al., *Principles and Methods of Numerical Parallel Computation* (National Defence Science and Technology Press, Beijing, 1999). in Chinese
14. A.K. Chandra, Maximal parallelism in matrix multiplication. Technical Report RC-6139 (I.B.M. Watson Research Center, Yorktown Heights, N.Y., 1976)
15. E.P. Preparata, D.V. Sarwate, An improved parallel processor bound in fast matrix inversion. Inform. Process. Lett. **7**, 148–150 (1978)
16. A.H. Sameh, R.P. Brent, Solving triangular systems on a parallel computer. SIAM J. Numer. Anal. **14**, 1101–1113 (1977)
17. L. Csanky, Fast parallel matrix inversion algorithm. SIAM J. Comput. **5**, 618–623 (1976)
18. G. Wang, S. Lu, Fast parallel algorithm for computing generalized inverses $A^\dagger$ and $A^\dagger_{MN}$. J. Shanghai Normal Univ. **16**, 17–22 (1987). in Chinese
19. L. Zhang et al., *Design and Analysis of Parallel Algorithms* (Hunan Science and Technology Press, Hunan, 1984). in Chinese
20. B. Noble, J. Demmel, *Applied Linear Algebra*, 3rd edn. (Prentice-Hall, New Jersey, 1988)
21. G. Wang, Y. Wei, PCR algorithm for the parallel computation of the solution of a class of singular equations. J. Shanghai Normal Univ. **23**, 1–8 (1994)
22. Y. Wei, G. Wang, PCR algorithm for parallel computing minimum-norm (T) least-squares (S) solution of inconsistent linear equations. Appl. Math. Comput. **133**, 547–557 (2002)
23. M.K. Sridhar, A new algorithm for parallel solution of linear equations. Inform. Process. Lett. **24**, 407–412 (1987)
24. G. Wang, Y. Wei, Parallel (M-N)SVD algorithm on the SIMD computers. Wuhan Univ. J. Natural Sci., vol. 1, pp. 541–546. International Parallel Computing Conference (1996)
25. L. Chen, E.V. Krishnamurthy, I. Madeod, Generalized matrix inversion and rank computation by successive matrix powering. Parallel Comput. **20**, 297–311 (1994)
26. Y. Wei, Successive matrix squaring for computing the Drazin inverse. Appl. Math. Comput. **108**, 67–75 (2000)
27. Y. Wei, H. Wu, J. Wei, Successive matrix squaring algorithm for parallel computing the weighted generalized inverse $A^\dagger_{MN}$. Appl. Math. Comput. **116**, 289–296 (2000)
28. J. Wang, A recurrent neural networks for real-time matrix inversion. Appl. Math. Comput. **55**, 23–34 (1993)
29. J. Wang, Recurrent neural networks for computing pseudoinverse of rank-deficient matrices. SIAM J. Sci. Comput. **18**, 1479–1493 (1997)
30. Y. Wei, Recurrent neural networks for computing weighted Moore-Penrose inverse. Appl. Math. Comput. **116**, 279–287 (2000)
31. G. Wang, Y. Wei, An improved parallel algorithm for computing the weighted Moore-Penrose inverse $A^\dagger_{MN}$. J. Shanghai Normal Univ. **29**, 12–20 (2000)

Chapter 8
Perturbation Analysis of the Moore-Penrose Inverse and the Weighted Moore-Penrose Inverse

Let A be a given matrix. When computing a generalized inverse of A, due to rounding error, we actually obtain the generalized inverse of a perturbed matrix $B = A + E$ of A. It is natural to ask if the generalized inverse of B is close to that of A when the perturbation E is sufficiently small. Thus, it becomes an important subject to study the perturbation analysis of the generalized inverses and find ways to reduce the effect of the perturbation.

8.1 Perturbation Bounds

Recall the perturbation bound of a nonsingular matrix A [1]. We start with the identity matrix.

Theorem 8.1.1 Let $P \in \mathbb{C}^{n \times n}$ and $\|P\| < 1$, then $I - P$ is nonsingular and

$$\|I - (I - P)^{-1}\| \leq \frac{\|P\|}{1 - \|P\|}. \tag{8.1.1}$$

The inequality (8.1.1) tells us that for an identity matrix I, if the perturbation P is small, then the error in the inverse $(I - P)^{-1}$ is approximately of the size $\|P\|$. In other words, the error in the inverse of a perturbed identity matrix is about the same as the perturbation. Now, for a general nonsingular matrix we have the following theorem.

Theorem 8.1.2 (1) Let A, $B = A + E \in \mathbb{C}_n^{n \times n}$, then

$$\frac{\|B^{-1} - A^{-1}\|}{\|A^{-1}\|} \leq \|B^{-1}\| \, \|E\|.$$

© Springer Nature Singapore Pte Ltd. and Science Press 2018
G. Wang et al., *Generalized Inverses: Theory and Computations*,
Developments in Mathematics 53, https://doi.org/10.1007/978-981-13-0146-9_8

(2) *If A is nonsingular, $B = A + E$ and $\Delta = \|A^{-1}\|\,\|E\| < 1$, then B is also nonsingular and*

$$\|B^{-1}\| \le \frac{\|A^{-1}\|}{1 - \Delta}$$

and

$$\frac{\|B^{-1} - A^{-1}\|}{\|A^{-1}\|} \le \frac{\Delta}{1 - \Delta}$$
$$= \frac{\kappa(A)\|E\|/\|A\|}{1 - \kappa(A)\|E\|/\|A\|}, \tag{8.1.2}$$

where $\kappa(A) = \|A\|\,\|A^{-1}\|$.

The left side of the inequality (8.1.2) is the relative error in B^{-1}. If E is sufficiently small, then the right side of the inequality (8.1.2) is about $\kappa(A)\|E\|/\|A\|$. Since $\|E\|/\|A\|$ is the relative error in $B = A + E$, the inequality (8.1.2) implies that the relative error in B may be magnified by a factor of $\kappa(A)$ in the relative error in B^{-1}. We call $\kappa(A)$ the condition number with respect to the inversion of the matrix A. If $\kappa(A)$ is large, then the problem of inverting A is sensitive to the perturbation in A. It is called ill-conditioned. Since

$$\kappa(A) = \|A\|\,\|A^{-1}\| \ge \|AA^{-1}\| = \|I\| = 1,$$

$\kappa(A)$ is indeed an enlargement constant.

It follows from (8.1.2) that $(A + E)^{-1}$ is close to A^{-1} when the inversion of the matrix A is not ill-conditioned and the perturbation E is small enough. But it does not hold for the Moore-Penrose inverse. For example, let

$$A = \begin{bmatrix} 1 & 0 \\ 0 & 0 \end{bmatrix}, \quad F = \begin{bmatrix} 0 & 1 \\ 0 & 0 \end{bmatrix}, \quad \text{and} \quad G = \begin{bmatrix} 0 & 0 \\ 0 & 1 \end{bmatrix},$$

then

$$A^\dagger = A, \quad (A + \varepsilon F)^\dagger = \frac{1}{1 + \varepsilon^2} \begin{bmatrix} 1 & 0 \\ \varepsilon & 0 \end{bmatrix}, \quad (A + \varepsilon G)^\dagger = \begin{bmatrix} 1 & 0 \\ 0 & \varepsilon^{-1} \end{bmatrix}.$$

Clearly, $\lim_{\varepsilon \to 0}(A + \varepsilon F)^\dagger = A^\dagger$, however $\lim_{\varepsilon \to 0}(A + \varepsilon G)^\dagger$ does not exist.

The above example shows that the perturbation εF has little effect on the generalized inversion of A, but εG affects a big effect. Through direct observation, the possible reason is that $\text{rank}(A + \varepsilon F) = \text{rank}(A)$, whereas $\text{rank}(A + \varepsilon G) > \text{rank}(A)$. We will discuss the effect of the change of the rank on the Moore-Penrose inverse and the weighted Moore-Penrose inverse. In the following discussion, M and N represent Hermitian positive definite matrices of orders m and n, respectively. First, we consider the weighted Moore-Penrose inverse.

Theorem 8.1.3 *Let* $A, E \in \mathbb{C}^{m \times n}$, $B = A + E$, *and* $\operatorname{rank}(A) = r$.

(1) *If* $\operatorname{rank}(B) > \operatorname{rank}(A)$, *then*

$$\|B_{MN}^{\dagger}\|_{NM} \geq \frac{1}{\|E\|_{MN}}. \tag{8.1.3}$$

(2) *If* $\|A_{MN}^{\dagger}\|_{NM} \|E\|_{MN} < 1$, *then*

$$\operatorname{rank}(B) \geq \operatorname{rank}(A). \tag{8.1.4}$$

Proof (1) Suppose $\operatorname{rank}(B) = p > r$, it follows from Lemma 5.2.2 that

$$\mu_p(A + E) \leq \|E\|_{MN} + \mu_p(A).$$

Since $\operatorname{rank}(A) = r$, we have $\mu_p(A) = 0$. Using (5.2.5), we get

$$\|(A + E)_{MN}^{\dagger}\|_{NM} = \frac{1}{\mu_p(A + E)},$$

implying that (8.1.3) is true.

(2) If $\|A_{MN}^{\dagger}\|_{NM} \|E\|_{MN} < 1$, then

$$\frac{1}{\|A_{MN}^{\dagger}\|_{NM}} - \|E\|_{MN} > 0,$$

and

$$\mu_r(A) - \|E\|_{MN} > 0.$$

From the (M, N) singular value decomposition in Chap. 5,

$$\mu_r(B) \geq \mu_r(A) - \|E\|_{MN} > 0.$$

Consequently, (8.1.4) holds. □

The Moore-Penrose inverse is a special case of the above theorem.

Corollary 8.1.1 *Let* $A \in \mathbb{C}_r^{m \times n}$ *and* $B = A + E$.

(1) *If* $\operatorname{rank}(B) > \operatorname{rank}(A)$, *then*

$$\|B^{\dagger}\|_2 \geq \frac{1}{\|E\|_2}.$$

(2) *If* $\|A^\dagger\|_2 \|E\|_2 < 1$, *then*

$$\text{rank}(B) \geq \text{rank}(A).$$

Theorem 8.1.3 states that (1) if $\text{rank}(B) > \text{rank}(A)$, even for $B = A + E$ close to A, that is, the perturbation E is small, $\|B_{MN}^\dagger\|_{NM}$ can be large and $B_{MN}^\dagger$ can be totally different from $A_{MN}^\dagger$; (2) if $\|E\|_{NM}$ is small enough, and $\|A_{MN}^\dagger\|_{NM} \|E\|_{MN} < 1$, then the rank of the perturbed matrix $B = A + E$ will not decrease. If $\text{rank}(B) > \text{rank}(A)$, the above example tells us that $B^\dagger$ may not be close to $A^\dagger$, it may not even exist. Thus, in the following discussion, we assume that $\text{rank}(B) = \text{rank}(A)$, that is, the perturbation maintains the rank.

For convenience, the following conditions are called Conditions I and II.

Condition I   Let $A, E \in \mathbb{C}^{m \times n}$, $B = A + E$, $\text{rank}(B) = \text{rank}(A) = r$, and $\Delta_1 = \|A_{MN}^\dagger\|_{NM} \|E\|_{MN} < 1$.

Condition II   Let $A, E \in \mathbb{C}^{m \times n}$, $B = A + E$, $\text{rank}(B) = \text{rank}(A) = r$, and $\Delta_2 = \|A^\dagger\|_2 \|E\|_2 < 1$.

Theorem 8.1.4 *If* Condition I *holds, then*

$$\|B_{MN}^\dagger\|_{NM} \leq \frac{\|A_{MN}^\dagger\|_{NM}}{1 - \Delta_1}. \tag{8.1.5}$$

Proof Suppose the non-zero (M, N) singular values of A and $B = A + E$ are $\mu_1(A) \geq \mu_2(A) \geq \cdots \geq \mu_r(A) > 0$ and $\mu_1(A + E) \geq \mu_2(A + E) \geq \cdots \geq \mu_r(A + E) > 0$, respectively. It follows from Theorem 5.2.2 and Lemma 5.2.2 that

$$\begin{aligned}
\|B_{MN}^\dagger\|_{NM}^{-1} &= \mu_r(B) \\
&\geq \mu_r(A) - \|E\|_{MN} \\
&= \|A_{MN}^\dagger\|_{NM}^{-1} - \|E\|_{MN},
\end{aligned}$$

implying that (8.1.5) holds.   □

The Condition II is a special case of Condition I.

Corollary 8.1.2 *If* Condition II *holds, then*

$$\|B^\dagger\|_2 \leq \frac{\|A^\dagger\|_2}{1 - \Delta_2}.$$

In order to derive a bound for the relative error

$$\frac{\|B_{MN}^\dagger - A_{MN}^\dagger\|_{NM}}{\|A_{MN}^\dagger\|_{NM}},$$

we give a decomposition of $B_{MN}^\dagger - A_{MN}^\dagger$.

Lemma 8.1.1 *Let $A, E \in \mathbb{C}^{m \times n}$, and $B = A + E$, then*

$$B_{MN}^{\dagger} - A_{MN}^{\dagger} = -B_{MN}^{\dagger} E A_{MN}^{\dagger} + B_{MN}^{\dagger}(B_{MN}^{\dagger})^{\#} E^{\#}(I - A A_{MN}^{\dagger})$$
$$- (I - B_{MN}^{\dagger} B) E^{\#}(A_{MN}^{\dagger})^{\#} A_{MN}^{\dagger}. \tag{8.1.6}$$

Proof From the left side of (8.1.6),

$$B_{MN}^{\dagger} - A_{MN}^{\dagger}$$
$$= -B_{MN}^{\dagger} E A_{MN}^{\dagger} + (B_{MN}^{\dagger} - A_{MN}^{\dagger}) + B_{MN}^{\dagger}(B - A) A_{MN}^{\dagger}$$
$$= -B_{MN}^{\dagger} E A_{MN}^{\dagger} + B_{MN}^{\dagger}(I - A A_{MN}^{\dagger}) - (I - B_{MN}^{\dagger} B) A_{MN}^{\dagger}. \tag{8.1.7}$$

Using

$$A^* M(I - A A_{MN}^{\dagger}) = A^* M - A^*(A_{MN}^{\dagger})^* A^* M = O,$$

we get

$$B_{MN}^{\dagger}(I - A A_{MN}^{\dagger}) = B_{MN}^{\dagger} B B_{MN}^{\dagger}(I - A A_{MN}^{\dagger})$$
$$= B_{MN}^{\dagger} M^{-1}(M B B_{MN}^{\dagger})^*(I - A A_{MN}^{\dagger})$$
$$= B_{MN}^{\dagger} M^{-1}(B_{MN}^{\dagger})^*(A + E)^* M(I - A A_{MN}^{\dagger})$$
$$= B_{MN}^{\dagger} M^{-1}(B_{MN}^{\dagger})^* E^* M(I - A A_{MN}^{\dagger})$$
$$= B_{MN}^{\dagger} M^{-1}(B_{MN}^{\dagger})^* N N^{-1} E^* M(I - A A_{MN}^{\dagger})$$
$$= B_{MN}^{\dagger}(B_{MN}^{\dagger})^{\#} E^{\#}(I - A A_{MN}^{\dagger}).$$

Similarly, from

$$(I - B_{MN}^{\dagger} B) N^{-1} B^* = O,$$

we get

$$(I - B_{MN}^{\dagger} B) A_{MN}^{\dagger} = -(I - B_{MN}^{\dagger} B) E^{\#}(A_{MN}^{\dagger})^{\#} A_{MN}^{\dagger}. \tag{8.1.8}$$

Thus, (8.1.6) follows. □

Lemma 8.1.2 *If $O \neq P \in \mathbb{C}^{n \times n}$ and $P^2 = P = P^{\#}$, then*

$$\|P\|_{NN} = 1.$$

Proof It follows from (1.4.11) that

$$\|P\|_{NN}^2 = \|P^{\#} P\|_{NN} = \|P^2\|_{NN} = \|P\|_{NN},$$

thus

$$\|P\|_{NN}(\|P\|_{NN} - 1) = O.$$

Since $P \neq O$, we get $\|P\|_{NN} = 1$. $\qquad\qquad\qquad\qquad\qquad\qquad\qquad\qquad$ □

Now, we have a bound for the relative error in the weighted Moore-Penrose inverse.

Theorem 8.1.5 *If* Condition I *holds, then*

$$\frac{\|B_{MN}^{\dagger} - A_{MN}^{\dagger}\|_{NM}}{\|A_{MN}^{\dagger}\|_{NM}} \leq \left(1 + \frac{1}{1 - \Delta_1} + \frac{1}{(1 - \Delta_1)^2}\right) \Delta_1. \qquad (8.1.9)$$

Proof Since

$$(I - AA_{MN}^{\dagger})^2 = I - AA_{MN}^{\dagger} = (I - AA_{MN}^{\dagger})^{\#}$$

and

$$(I - B_{MN}^{\dagger}B)^2 = I - B_{MN}^{\dagger}B = (I - B_{MN}^{\dagger}B)^{\#},$$

from Lemma 8.1.2, we have

$$\|I - AA_{MN}^{\dagger}\|_{MM} = 1 \quad \text{and} \quad \|I - B_{MN}^{\dagger}B\|_{NN} = 1.$$

It then follows from Lemma 8.1.1, (1.4.10), and (1.4.11) that

$$\|B_{MN}^{\dagger} - A_{MN}^{\dagger}\|_{NM}$$
$$\leq (\|A_{MN}^{\dagger}\|_{NM}\|B_{MN}^{\dagger}\|_{NM} + \|B_{MN}^{\dagger}\|_{NM}^2 + \|A_{MN}^{\dagger}\|_{NM}^2)\|E\|_{MN}.$$

From Theorem 8.1.4, we obtain

$$\|B_{MN}^{\dagger} - A_{MN}^{\dagger}\|_{NM}$$
$$\leq \left(\frac{\|A_{MN}^{\dagger}\|_{NM}^2}{1 - \Delta_1} + \frac{\|A_{MN}^{\dagger}\|_{NM}^2}{(1 - \Delta_1)^2} + \|A_{MN}^{\dagger}\|_{NM}^2\right)\|E\|_{MN},$$

which implies (8.1.9). $\qquad\qquad\qquad\qquad\qquad\qquad\qquad\qquad\qquad\qquad$ □

The above theorem says that if A has a small perturbation E and $\text{rank}(A + E) = \text{rank}(A)$, then the perturbation on $A_{MN}^{\dagger}$ is small.

Corollary 8.1.3 *If* Condition II *holds, then*

$$\frac{\|B^{\dagger} - A^{\dagger}\|_2}{\|A^{\dagger}\|_2} \leq \left(1 + \frac{1}{1 - \Delta_2} + \frac{1}{(1 - \Delta_2)^2}\right) \Delta_2.$$

In this special case, if A has a small perturbation E, and $\text{rank}(A + E) = \text{rank}(A)$, then the perturbation on $A^{\dagger}$ is small.

In the following, we derive another bound for the relative error of the weighted Moore-Penrose inverse.

Lemma 8.1.3 *If* Condition I *holds, then*

$$\|BB^{\dagger}_{MN}(I - AA^{\dagger}_{MN})\|_{MM} = \|AA^{\dagger}_{MN}(I - BB^{\dagger}_{MN})\|_{MM}. \tag{8.1.10}$$

Proof Suppose that the (M, N) singular value decompositions of A and B are respectively:

$$A = U_1 \begin{bmatrix} D_1 & O \\ O & O \end{bmatrix} V_1^*, \quad \text{and} \quad B = U_2 \begin{bmatrix} D_2 & O \\ O & O \end{bmatrix} V_2^*,$$

where $D_1 = \text{diag}(\mu_1(A), \ldots, \mu_r(A))$, $D_2 = \text{diag}(\mu_1(B), \ldots, \mu_r(B))$, and

$$U_1^* M U_1 = I_m \; V_1^* N^{-1} V_1 = I_n,$$
$$U_2^* M U_2 = I_m \; V_2^* N^{-1} V_2 = I_n.$$

Thus, we have

$$A^{\dagger}_{MN} = N^{-1} V_1 \begin{bmatrix} D_1^{-1} & O \\ O & O \end{bmatrix} U_1^* M,$$

$$B^{\dagger}_{MN} = N^{-1} V_2 \begin{bmatrix} D_2^{-1} & O \\ O & O \end{bmatrix} U_2^* M,$$

$$AA^{\dagger}_{MN} = U_1 \begin{bmatrix} I_r & O \\ O & O \end{bmatrix} U_1^* M = U_1 \begin{bmatrix} I_r & O \\ O & O \end{bmatrix} U_1^{-1},$$

$$I - AA^{\dagger}_{MN} = U_1 \begin{bmatrix} O & O \\ O & I_{m-r} \end{bmatrix} U_1^{-1} = U_1 \begin{bmatrix} O & O \\ O & I_{m-r} \end{bmatrix} U_1^* M,$$

$$BB^{\dagger}_{MN} = U_2 \begin{bmatrix} I_r & O \\ O & O \end{bmatrix} U_2^* M = U_2 \begin{bmatrix} I_r & O \\ O & O \end{bmatrix} U_2^{-1},$$

$$I - BB^{\dagger}_{MN} = U_2 \begin{bmatrix} O & O \\ O & I_{m-r} \end{bmatrix} U_2^{-1} = U_2 \begin{bmatrix} O & O \\ O & I_{m-r} \end{bmatrix} U_2^* M.$$

Let $\widehat{U}_1 = M^{1/2} U_1$ and $\widehat{U}_2 = M^{1/2} U_2$, then $(\widehat{U}_1)^* \widehat{U}_1 = I_m$ and $(\widehat{U}_2)^* \widehat{U}_2 = I_m$. Thus $\widehat{U}_1$ and $\widehat{U}_2$ are unitary matrices, and

$$U_2^* M U_1 = (\widehat{U}_2)^* \widehat{U}_1 \equiv W$$

is also unitary. Partitioning

$$W = \begin{bmatrix} W_{11} & W_{12} \\ W_{21} & W_{22} \end{bmatrix} \begin{matrix} r \\ m-r \end{matrix}$$
$$\quad\quad\; r \quad m-r$$

we have

$$\|BB_{MN}^{\dagger}(I - AA_{MN}^{\dagger})\|_{MM}$$

$$= \left\| U_2 \begin{bmatrix} I_r & O \\ O & O \end{bmatrix} U_2^* M U_1 \begin{bmatrix} O & O \\ O & I_{m-r} \end{bmatrix} U_1^* M \right\|_{MM}$$

$$= \left\| M^{1/2} U_2 \begin{bmatrix} I_r & O \\ O & O \end{bmatrix} \begin{bmatrix} W_{11} & W_{12} \\ W_{21} & W_{22} \end{bmatrix} \begin{bmatrix} O & O \\ O & I_{m-r} \end{bmatrix} U_1^* M^{1/2} \right\|_2$$

$$= \left\| \begin{bmatrix} I_r & O \\ O & O \end{bmatrix} \begin{bmatrix} W_{11} & W_{12} \\ W_{21} & W_{22} \end{bmatrix} \begin{bmatrix} O & O \\ O & I_{m-r} \end{bmatrix} \right\|_2$$

$$= \left\| \begin{bmatrix} O & W_{12} \\ O & O \end{bmatrix} \right\|_2$$

$$= \|W_{12}\|_2.$$

Similarly, we obtain

$$\|AA_{MN}^{\dagger}(I - BB_{MN}^{\dagger})\|_{MM} = \|W_{21}\|_2.$$

To derive (8.1.10), it remains to show that $\|W_{21}\|_2 = \|W_{12}\|_2$. For any $\mathbf{x} \in \mathbb{C}^{m-r}$ with $\|\mathbf{x}\|_2 = 1$, we set

$$\mathbf{y} = \begin{bmatrix} \mathbf{0} \\ \mathbf{x} \end{bmatrix} \begin{matrix} r \\ m-r \end{matrix},$$

then, since W is unitary, we obtain

$$\|\mathbf{x}\|_2^2 = \|\mathbf{y}\|_2^2 = \|W\mathbf{y}\|_2^2 = \|W_{12}\mathbf{x}\|_2^2 + \|W_{22}\mathbf{x}\|_2^2.$$

Thus $\|W_{12}\mathbf{x}\|_2^2 = \|\mathbf{x}\|_2^2 - \|W_{22}\mathbf{x}\|_2^2$, consequently,

$$\|W_{12}\|_2^2 = \max_{\|\mathbf{x}\|_2=1} \|W_{12}\mathbf{x}\|_2^2$$

$$= 1 - \min_{\|\mathbf{x}\|_2=1} \|W_{22}\mathbf{x}\|_2^2$$

$$= 1 - \sigma_{m-r}^2(W_{22}),$$

where $\sigma_{m-r}(W_{22})$ is the smallest singular value of W_{22}.

Similarly, it follows from $\|\mathbf{x}\|_2^2 = \|\mathbf{y}\|_2^2 = \|W^*\mathbf{y}\|_2^2 = \|W_{21}^*\mathbf{x}\|_2^2 + \|W_{22}^*\mathbf{x}\|_2^2$ that

$$\|W_{21}\|_2^2 = \|W_{21}^*\|_2^2 = 1 - \min_{\|\mathbf{x}\|_2=1} \|W_{22}^*\mathbf{x}\|_2^2 = 1 - \sigma_{m-r}^2(W_{22}^*).$$

Since $\sigma_{m-r}(W_{22}) = \sigma_{m-r}(W_{22}^*)$, we have $\|W_{12}\|_2 = \|W_{21}\|_2$.   □

Lemma 8.1.4 *If* Condition I *holds, denoting* $G = B_{MN}^{\dagger}(I - AA_{MN}^{\dagger})$, *then*

$$\|G\|_{NM} \leq \|B_{MN}^{\dagger}\|_{NM} \|A_{MN}^{\dagger}\|_{NM} \|E\|_{MN}. \tag{8.1.11}$$

Proof It follows from $G = B_{MN}^\dagger B B_{MN}^\dagger (I - A A_{MN}^\dagger)$ that

$$\|G\|_{NM} \leq \|B_{MN}^\dagger\|_{NM} \|B B_{MN}^\dagger (I - A A_{MN}^\dagger)\|_{MM}.$$

Notice that

$$B^* M (I - B B_{MN}^\dagger) = O$$

and

$$(I - B B_{MN}^\dagger)^2 = I - B B_{MN}^\dagger = (I - B B_{MN}^\dagger)^\#.$$

From Lemma 8.1.2, we have $\|I - B B_{MN}^\dagger\|_{MM} = 1$. Based on Lemma 8.1.3, we get

$$
\begin{aligned}
&\|B B_{MN}^\dagger (I - A A_{MN}^\dagger)\|_{MM} \\
&= \|A A_{MN}^\dagger (I - B B_{MN}^\dagger)\|_{MM} \\
&= \|M^{1/2} A A_{MN}^\dagger (I - B B_{MN}^\dagger) M^{-1/2}\|_2 \\
&= \|M^{-1/2} (M A A_{MN}^\dagger)^* (I - B B_{MN}^\dagger) M^{-1/2}\|_2 \\
&= \|M^{-1/2} {A_{MN}^\dagger}^* E^* M^{1/2} M^{1/2} (I - B B_{MN}^\dagger) M^{-1/2}\|_2 \\
&\leq \|M^{-1/2} {A_{MN}^\dagger}^* E^* M^{1/2}\|_2 \\
&= \|M^{1/2} E A_{MN}^\dagger M^{-1/2}\|_2 \\
&= \|M^{1/2} E N^{-1/2} N^{1/2} A_{MN}^\dagger M^{-1/2}\|_2 \\
&\leq \|M^{1/2} E N^{-1/2}\|_2 \|N^{1/2} A_{MN}^\dagger M^{-1/2}\|_2 \\
&= \|E\|_{MN} \|A_{MN}^\dagger\|_{NM}.
\end{aligned}
$$

Thus (8.1.11) follows.   □

From (8.1.7), (8.1.8), Theorem 8.1.4, and Lemma 8.1.4, we have the following relative perturbation bounds:

$$\frac{\|B_{MN}^\dagger - A_{MN}^\dagger\|_{NM}}{\|A_{MN}^\dagger\|_{NM}} \leq \left(1 + \frac{2}{1 - \Delta_1}\right) \Delta_1 \leq \frac{3\Delta_1}{1 - \Delta_1}$$

and

$$\frac{\|B^\dagger - A^\dagger\|_2}{\|A^\dagger\|_2} \leq \left(1 + \frac{2}{1 - \Delta_2}\right) \Delta_2 \leq \frac{3\Delta_2}{1 - \Delta_2}.$$

If we study the perturbation analysis carefully, we can get some special results.

Theorem 8.1.6 *If Condition I holds, then*

$$\frac{\|B_{MN}^\dagger - A_{MN}^\dagger\|_{NM}}{\|A_{MN}^\dagger\|_{NM}} \leq C \frac{\Delta_1}{1 - \Delta_1}, \tag{8.1.12}$$

where

$$C = \begin{cases} (1 + \sqrt{5})/2, & if \operatorname{rank}(A) < \min\{m, n\}, \\ \sqrt{2}, & if \operatorname{rank}(A) = \min\{m, n\}, \ (m \neq n), \\ 1, & if \operatorname{rank}(A) = m = n. \end{cases} \tag{8.1.13}$$

Denote the three terms in (8.1.7) as

$$F = -B_{MN}^{\dagger} E A_{MN}^{\dagger},$$
$$G = B_{MN}^{\dagger}(I - A A_{MN}^{\dagger}),$$
$$H = -(I - B_{MN}^{\dagger} B) A_{MN}^{\dagger}.$$

It follows from Theorem 8.1.4, Lemma 8.1.4, and (8.1.8) that

$$\|F\|_{NM}, \ \|G\|_{NM} \leq \frac{\Delta_1}{1 - \Delta_1} \|A_{MN}^{\dagger}\|_{NM},$$

$$\|H\|_{NM} \leq \Delta_1 \|A_{MN}^{\dagger}\|_{NM},$$

where

$$\Delta_1 = \|A_{MN}^{\dagger}\|_{NM} \|E\|_{MN} = \|(A_{MN}^{\dagger})^{\#}\|_{MN} \|E^{\#}\|_{NM}.$$

Setting

$$\alpha = \frac{\Delta_1}{1 - \Delta_1},$$

we have

$$\|F\|_{NM}, \ \|G\|_{NM}, \ \|H\|_{NM} \leq \alpha \|A_{MN}^{\dagger}\|_{NM}.$$

Proof (1) For any $\mathbf{x} \in \mathbb{C}^m$ and $\|\mathbf{x}\|_M = 1$, $\mathbf{x}$ can be decomposed as $\mathbf{x} = \mathbf{x}_1 + \mathbf{x}_2$, where $\mathbf{x}_1 = A A_{MN}^{\dagger} \mathbf{x}$ and $\mathbf{x}_2 = (I - A A_{MN}^{\dagger}) \mathbf{x}$. clearly, $\mathbf{x}_1$ and $\mathbf{x}_2$ are M-orthogonal. Thus $1 = \|\mathbf{x}\|_M^2 = \|\mathbf{x}_1\|_M^2 + \|\mathbf{x}_2\|_M^2$ and there exists a φ such that $\cos \varphi = \|\mathbf{x}_1\|_M$ and $\sin \varphi = \|\mathbf{x}_2\|_M$. From (8.1.7), we have

$$(B_{MN}^{\dagger} - A_{MN}^{\dagger})\mathbf{x} = F\mathbf{x}_1 + G\mathbf{x}_2 + H\mathbf{x}_1 \equiv \mathbf{y}_1 + \mathbf{y}_2 + \mathbf{y}_3.$$

Using $(I - B_{MN}^{\dagger} B)^* N B_{MN}^{\dagger} = O$, it is easy to verify that $\mathbf{y}_3$ is N-orthogonal to both $\mathbf{y}_1$ and $\mathbf{y}_2$. So,

$$\|(B_{MN}^\dagger - A_{MN}^\dagger)\mathbf{x}\|_N^2$$
$$= \|\mathbf{y}_1 + \mathbf{y}_2\|_N^2 + \|\mathbf{y}_3\|_N^2$$
$$\leq \alpha^2 \|A_{MN}^\dagger\|_{NM}^2 ((\|\mathbf{x}_1\|_M + \|\mathbf{x}_2\|_M)^2 + \|\mathbf{x}_1\|_M^2)$$
$$= \alpha^2 \|A_{MN}^\dagger\|_{NM}^2 ((\cos\varphi + \sin\varphi)^2 + \cos^2\varphi)$$
$$= \alpha^2 \|A_{MN}^\dagger\|_{NM}^2 (3 + 2\sin 2\varphi + \cos 2\varphi)/2$$
$$\leq \alpha^2 \|A_{MN}^\dagger\|_{NM}^2 (3 + \sqrt{5})/2. \tag{8.1.14}$$

Thus

$$\|B_{MN}^\dagger - A_{MN}^\dagger\|_{NM} = \max_{\|\mathbf{x}\|_M = 1} \|(B_{MN}^\dagger - A_{MN}^\dagger)\mathbf{x}\|_N$$

$$\leq \alpha \|A_{MN}^\dagger\|_{NM}(1 + \sqrt{5})/2.$$

(2) When $\mathrm{rank}(B) = \mathrm{rank}(A) = n < m$, by $B_{MN}^\dagger = (B^* M B)^{-1} B^* M$, we have $I - B_{MN}^\dagger B = O$. Thus $H = O$ and $\mathbf{y}_3 = \mathbf{0}$.

When $\mathrm{rank}(B) = \mathrm{rank}(A) = m < n$, by $A_{MN}^\dagger = N^{-1} A^* (A N^{-1} A^*)^{-1}$, we have $I - A A_{MN}^\dagger = O$. Thus $G = O$ and $\mathbf{y}_2 = \mathbf{0}$.

It follows from (8.1.14), where either $\mathbf{y}_2$ or $\mathbf{y}_3$ is zero, that

$$\|(B_{MN}^\dagger - A_{MN}^\dagger)\mathbf{x}\|_N^2 \leq 2\alpha^2 \|A_{MN}^\dagger\|_{NM}^2.$$

Thus

$$\|B_{MN}^\dagger - A_{MN}^\dagger\|_{NM} = \max_{\|\mathbf{x}\|_M = 1} \|(B_{MN}^\dagger - A_{MN}^\dagger)\mathbf{x}\|_N$$

$$\leq \sqrt{2}\alpha \|A_{MN}^\dagger\|_{NM}.$$

(3) When $\mathrm{rank}(A) = m = n$ and $\mathrm{rank}(B) = m = n$. From the proof in (2), we know that $G = O$ and $H = O$. Thus the third case in (8.1.13) follows immediately. $\square$

Finally, the Moore-Penrose inverse is a special case of the weighted Moore-Penrose inverse.

Corollary 8.1.4 *If Condition II holds, then*

$$\frac{\|B^\dagger - A^\dagger\|_2}{\|A^\dagger\|_2} \leq C \frac{\Delta_2}{1 - \Delta_2},$$

where

$$C = \begin{cases} (1 + \sqrt{5})/2 & \textit{if } \mathrm{rank}(A) < \min\{m, n\}, \\ \sqrt{2} & \textit{if } \mathrm{rank}(A) = \min\{m, n\}, \ (m \neq n), \\ 1 & \textit{if } \mathrm{rank}(A) = m = n. \end{cases}$$

The perturbation bound for the Moore-Penrose inverse can be found in [2, 3], the bound for the weighted Moore-Penrose inverse can be found in [4], on which this section is based.

8.2 Continuity

It follows from Theorem 8.1.2 that if $E \to O$, i.e., $B \to A$, we have $\|B^{-1} - A^{-1}\|_2 \to 0$, i.e., $B^{-1} \to A^{-1}$. This implies that the inverse of a nonsingular matrix is a continuous function of the elements of the matrix. If we replace matrix B with sequence $\{A_k\}$, we have the following conclusion.

Theorem 8.2.1 *Let* $A \in \mathbb{C}_n^{n \times n}$, $\lim_{k \to \infty} A_k = A$, *then for sufficiently large* k, A_k *is nonsingular and*

$$\lim_{k \to \infty} A_k^{-1} = A^{-1}. \tag{8.2.1}$$

Proof Suppose $E_k = A_k - A$, then $\lim_{k \to \infty} \|E_k\| = 0$. Thus for sufficiently large k, $\|A^{-1}\| \|E_k\| < 1$. It follows from Theorem 8.1.2 that A_k is nonsingular and $\lim_{k \to \infty} \|A_k^{-1} - A^{-1}\| = 0$, thus (8.2.1) holds. $\square$

Just as the continuity of the inversion of nonsingular matrices, the Moore-Penrose inverse and the weighted Moore-Penrose inverse of full column rank or full row rank matrices are also continuous.

Theorem 8.2.2 *Let* $A \in \mathbb{C}_n^{m \times n}$, M *and* N *be Hermitian positive definite matrices of orders* m *and* n, *respectively,* $\lim_{k \to \infty} A_k = A$, *then for sufficiently large* k, A_k *is of full column rank and*

$$\lim_{k \to \infty} (A_k)_{MN}^{\dagger} = A_{MN}^{\dagger}. \tag{8.2.2}$$

Proof Suppose $E_k = A_k - A$, then $\lim_{k \to \infty} \|E_k\| = 0$. Since $A \in \mathbb{C}_n^{m \times n}$, A^*A is nonsingular and

$$A_k^* A_k = (A + E_k)^*(A + E_k)$$
$$= A^*A \left(I + (A^*A)^{-1}((A + E_k)^* E_k + E_k^* A)\right).$$

Since $\|E_k\| \to 0$, we have $\|(A^*A)^{-1}((A + E_k)^* E_k + E_k^* A)\| < 1$, thus

$$I + (A^*A)^{-1}((A + E_k)^* E_k + E_k^* A)$$

is nonsingular, and so is $A_k^* A_k$, i.e., A_k is of full column rank. Also, it can be derived that $\widetilde{A}_k = M^{1/2} A_k N^{-1/2}$ is of full column rank. Denote

$$\widetilde{A}_k = M^{1/2} A_k N^{-1/2}$$
$$= M^{1/2} A N^{-1/2} + M^{1/2} E_k N^{-1/2}$$
$$\equiv \widetilde{A} + \widetilde{E}_k.$$

Notice that $\widetilde{E}_k \to O$, as $k \to \infty$,

$$\lim_{k \to \infty} (\widetilde{A} + \widetilde{E}_k)^* = \widetilde{A}^*,$$
$$\lim_{k \to \infty} ((\widetilde{A} + \widetilde{E}_k)^*(\widetilde{A} + \widetilde{E}_k))^{-1} = (\widetilde{A}^* \widetilde{A})^{-1}.$$

Furthermore,

$$\lim_{k \to \infty} \widetilde{A}_k^\dagger = \lim_{k \to \infty} ((\widetilde{A} + \widetilde{E}_k)^*(\widetilde{A} + \widetilde{E}_k))^{-1}(\widetilde{A} + \widetilde{E}_k)^*$$
$$= (\widetilde{A}^* \widetilde{A})^{-1} \widetilde{A}^*$$
$$= \widetilde{A}^\dagger$$
$$= (M^{1/2} A N^{-1/2})^\dagger.$$

Thus

$$(A_k)_{MN}^\dagger = N^{-1/2}(M^{1/2} A_k N^{-1/2})^\dagger M^{1/2}$$
$$= N^{-1/2} \widetilde{A}_k^\dagger M^{1/2}$$
$$\to N^{-1/2}(M^{1/2} A N^{-1/2})^\dagger M^{1/2}$$
$$= A_{MN}^\dagger,$$

as $k \to \infty$.   □

Corollary 8.2.1 *Let $A \in \mathbb{C}_n^{m \times n}$, $\lim_{k \to \infty} A_k = A$, then for sufficiently large k, A_k is of full column rank and*

$$\lim_{k \to \infty} A_k^\dagger = A^\dagger.$$

Next we discuss the continuity of the Moore-Penrose inverse and the weighted Moore-Penrose inverse of a rank-deficient matrix.

Theorem 8.2.3 *Let $A \in \mathbb{C}_r^{m \times n}$, M and N be Hermitian positive definite matrices of orders m and n, respectively. Suppose $\{A_k\}$ is an $m \times n$ matrix sequence and $A_k \to A$, then the necessary and sufficient condition for $(A_k)_{MN}^\dagger \to A_{MN}^\dagger$ is*

$$\text{rank}(A_k) = \text{rank}(A). \tag{8.2.3}$$

for the sufficiently large k.

Proof Let $E_k = A_k - A$. If $A_k \to A$, then $E_k \to O$.
Sufficiency: If $\text{rank}(A_k) = \text{rank}(A)$, then Condition I holds. It follows from Theorem 8.1.6 and $E_k \to O$ that $\|(A_k)_{MN}^\dagger - A_{MN}^\dagger\|_{NM} \to 0$, thus $(A_k)_{MN}^\dagger \to A_{MN}^\dagger$.

Necessity: Suppose $(A_k)^\dagger_{MN} \to A^\dagger_{MN}$. If $\text{rank}(A_k) \neq \text{rank}(A)$, then we have $\text{rank}(A)$ $< \text{rank}(A + E_k) = \text{rank}(A_k)$. It follows from Theorem 8.1.3 that $\|(A_k)^\dagger_{MN}\|_{NM} \geq \|E_k\|^{-1}_{MN}$, which shows that if $E_k \to O$, then $(A_k)^\dagger_{MN}$ does not exist.   □

Corollary 8.2.2 *Suppose $A \in \mathbb{C}^{m \times n}_r$, $\{A_k\}$ is an $m \times n$ matrix sequence. If $A_k \to A$, then the necessary and sufficient condition for $A^\dagger_k \to A^\dagger$ is*

$$\text{rank}(A_k) = \text{rank}(A)$$

for the sufficiently large k.

The continuity of the Moore-Penrose inverse can be found in [1, 5], that of the weighted Moore-Penrose inverse can be found in [4], on which this section is based.

8.3   Rank-Preserving Modification

From the discussion in the previous section, we know that after the matrix A is modified to $B = A + E$, if $\text{rank}(B) > \text{rank}(A)$, it will cause the discontinuity of the Moore-Penrose inverse or the weighted Moore-Penrose inverse. The modification E can be computational errors. Therefore the computed result can be far from the true solution, which is an serious issue in scientific computing. So, we are interested in rank-preserving modifications. The rank-preserving modification of a rank deficient matrix is discussed in [4, 5], which can overcome the discontinuity problem.

The basic idea behind rank-preserving modification is to deal with $A + E$ suitably, specifically, to construct its rank-preserving matrix A_E such that when $\|E\| \to 0$,

(1) $A_E \to A$;
(2) $\text{rank}(A_E) = \text{rank}(A)$.

The tool used is the full-rank decomposition. Let $A \in \mathbb{R}^{m \times n}_r$. Without loss of generality, suppose the first r columns of A are linearly independent. Let the full-rank factorization of A be

$$A = QU,$$

where Q is a product of orthogonal (or elementary lower triangular) matrices of order m, U is an $m \times n$ upper echelon matrix.

Set

$$Q = [Q_1 \ Q_2] \quad \text{and} \quad U = \begin{bmatrix} R & S \\ O & O \end{bmatrix},$$

where $Q_1 \in \mathbb{R}^{m \times r}$, $Q_2 \in \mathbb{R}^{m \times (m-r)}$, $R \in \mathbb{R}^{r \times r}$ an upper triangular matrix, and $S \in \mathbb{R}^{r \times (n-r)}$, thus

$$A = Q_1[R \ S].$$

Suppose that the full-rank decomposition of $B = A + E = [\mathbf{a}_1 + \mathbf{e}_1 \cdots \mathbf{a}_n + \mathbf{e}_n]$ is

$$B = \widetilde{Q}\widetilde{U},$$

where $\widetilde{Q}$ is a product of orthogonal (or elementary lower triangular) matrices of order m and $\widetilde{U}$ is an $m \times n$ upper echelon matrix. Set

$$\widetilde{Q} = [\widetilde{Q}_1 \ \widetilde{Q}_2] \quad \text{and} \quad \widetilde{U} = \begin{bmatrix} \widetilde{R} & \widetilde{S} \\ O & \widehat{S} \end{bmatrix},$$

where $\widetilde{Q}_1 \in \mathbb{R}^{m \times r}$, $\widetilde{Q}_2 \in \mathbb{R}^{m \times (m-r)}$, $\widetilde{R} \in \mathbb{R}^{r \times r}$ an upper triangular matrix, $\widetilde{S} \in \mathbb{R}^{r \times (n-r)}$, and $\widehat{S} \in \mathbb{R}^{(m-r) \times (n-r)}$. Construct

$$A_E = \widetilde{Q}_1[\widetilde{R} \ \widetilde{S}]. \tag{8.3.1}$$

Obviously, when $\|E\| \to 0$, $\widetilde{Q}_1 \to Q_1$ and $[\widetilde{R} \ \widetilde{S}] \to [R \ S]$. Therefore $A_E \to A$. From the construction of the full-rank decomposition, when E is sufficiently small, $\mathrm{rank}(A_E) = \mathrm{rank}(A) = r$. We call the definition (8.3.1) of A_E the rank-preserving modification of $A + E$, that is, A_E is a modification of $A + E$ that preserves the rank of A. From Theorem 8.2.3 and Corollary 8.2.2, we have

$$\lim_{\|E\| \to 0} (A_E)^\dagger_{MN} = A^\dagger_{MN} \quad \text{and} \quad \lim_{\|E\| \to 0} A^\dagger_E = A^\dagger$$

respectively.

Example Let

$$A = \begin{bmatrix} 1 & 1 \\ 1 & 1 \end{bmatrix}, \quad E = \begin{bmatrix} \epsilon & 0 \\ 0 & 0 \end{bmatrix}, \quad M = \begin{bmatrix} 1 & 0 \\ 0 & 4 \end{bmatrix}, \quad \text{and} \quad N = \begin{bmatrix} 4 & 0 \\ 0 & 1 \end{bmatrix},$$

then

$$A^\dagger_{MN} = N^{1/2}(M^{1/2}AN^{-1/2})^\dagger M^{1/2} = \frac{1}{25}\begin{bmatrix} 1 & 4 \\ 4 & 16 \end{bmatrix}$$

and

$$(A + E)^\dagger_{MN} = \frac{1}{\epsilon}\begin{bmatrix} 1 & -1 \\ -1 & 1 + \epsilon \end{bmatrix}.$$

When $\epsilon \to 0$, $B = A + E \to A$, however, $B^\dagger_{MN} = (A + E)^\dagger_{MN}$ does not exist. Now consider the full-rank decomposition

$$A + E = \begin{bmatrix} 1 + \epsilon & 1 \\ 1 & 1 \end{bmatrix} = \begin{bmatrix} 1 & 0 \\ \dfrac{1}{1 + \epsilon} & 1 \end{bmatrix}\begin{bmatrix} 1 + \epsilon & 1 \\ 0 & \dfrac{\epsilon}{1 + \epsilon} \end{bmatrix}$$

and modify it into the full-rank decomposition

$$A_E = \begin{bmatrix} 1 \\ 1 \\ 1+\epsilon \end{bmatrix} [1+\epsilon \ \ 1] = FG$$

to reserve the rank of A. Then when $\epsilon \to 0$, we have

$$(A_E)_{MN}^{\dagger} = N^{-1}G^T(F^T M A_E N^{-1}G^T)^{-1}F^T M$$

$$= \frac{4(1+\epsilon)^2}{(4+(1+\epsilon)^2)^2} \begin{bmatrix} \dfrac{1+\epsilon}{4} & 1 \\ 1 & \dfrac{4}{1+\epsilon} \end{bmatrix}$$

$$\to \frac{1}{25} \begin{bmatrix} 1 & 4 \\ 4 & 16 \end{bmatrix}$$

$$= A_{MN}^{\dagger}.$$

8.4 Condition Numbers

Based on the perturbation bounds presented in Sect. 8.1, this section gives condition numbers. As before, we start with the nonsingular case. When $A \in \mathbb{C}_n^{n \times n}$, from (8.1.2), the perturbation bound for the inversion is

$$\frac{\|(A+E)^{-1} - A^{-1}\|}{\|A^{-1}\|} \leq \frac{\kappa(A)\|E\|/\|A\|}{1 - \kappa(A)\|E\|/\|A\|},$$

where $\kappa(A) = \|A\| \, \|A^{-1}\|$ is the condition number with respect to the inversion of matrix A.

When $A \in \mathbb{C}_r^{m \times n}$, from (8.1.12), the perturbation bound for the weighted Moore-Penrose inverse is

$$\frac{\|(A+E)_{MN}^{\dagger} - A_{MN}^{\dagger}\|_{NM}}{\|A_{MN}^{\dagger}\|_{NM}} \leq C \frac{\|A_{MN}^{\dagger}\|_{NM}\|E\|_{MN}}{1 - \|A_{MN}^{\dagger}\|_{NM}\|E\|_{MN}}.$$

Denoting

$$\kappa_{MN}(A) = \|A\|_{MN}\|A_{MN}^{\dagger}\|_{NM}, \tag{8.4.1}$$

we have the following result.

Theorem 8.4.1 *If* Condition I *holds, then*

$$\frac{\|(A+E)_{MN}^{\dagger} - A_{MN}^{\dagger}\|_{NM}}{\|A_{MN}^{\dagger}\|_{NM}} \leq C \frac{\kappa_{MN}(A)\|E\|_{MN}/\|A\|_{MN}}{1 - \kappa_{MN}(A)\|E\|_{MN}/\|A\|_{MN}}.$$

This implies that if $\kappa_{MN}(A)$ is small, then the effect of the perturbation E on $A_{MN}^{\dagger}$ is also small; if $\kappa_{MN}(A)$ is large, E may have large effect on $A_{MN}^{\dagger}$. Therefore, $\kappa_{MN}(A)$ is called the condition number with respect to the weighted Moore-Penrose inverse.

It follows from Lemma 8.1.4 that the perturbation bound for the Moore-Penrose inverse is

$$\frac{\|(A+E)^{\dagger} - A^{\dagger}\|_2}{\|A^{\dagger}\|_2} \leq C \frac{\|A^{\dagger}\|_2 \|E\|_2}{1 - \|A^{\dagger}\|_2 \|E\|_2}.$$

Denoting

$$\kappa_2(A) = \|A\|_2 \|A^{\dagger}\|_2,$$

we have the following result:

Corollary 8.4.1 *If* Condition II *holds, then*

$$\frac{\|(A+E)^{\dagger} - A^{\dagger}\|_2}{\|A^{\dagger}\|_2} \leq C \frac{\kappa_2(A)\|E\|_2/\|A\|_2}{1 - \kappa_2(A)\|E\|_2/\|A\|_2},$$

and $\kappa_2(A)$ *is called the condition number with respect to the Moore-Penrose inverse.*

The classical normwise relative condition number measures the sensitivity of matrix inversion. In this section, we will discuss the normwise relative condition number for the weighted Moore-Penrose inverse and the Moore-Penrose inverse. First, for a nonsingular $A \in \mathbb{C}_n^{n \times n}$ with matrix norm $\| \cdot \|$, this normwise relative condition number is defined by

$$\text{cond}(A) = \lim_{\epsilon \to 0^+} \sup_{\|E\| \leq \epsilon \|A\|} \frac{\|(A+E)^{-1} - A^{-1}\|}{\epsilon \|A^{-1}\|}.$$

That is, we look at an upper bound for the relative change in A^{-1} compared with a relative change in A of size ϵ. We take the limit as $\epsilon \to 0^+$. Hence a condition number records the worst case sensitivity to small perturbations. When the matrix norm is induced by a vector norm, $\text{cond}(A)$ can be expressed by

$$\text{cond}(A) = \kappa(A) = \|A\| \|A^{-1}\|.$$

For more discussion of the condition numbers of nonsingular matrices, see [6].

Now, we give the normwise relative condition numbers for the weighted Moore-Penrose inverse and the Moore-Penrose inverse.

Theorem 8.4.2 *The condition number of the weighted Moore-Penrose inverse defined by*

$$\text{cond}(A) = \lim_{\epsilon \to 0^+} \sup_{\substack{\|E\|_{MN} \leq \epsilon \|A\|_{MN} \\ \mathcal{R}(E) \subset \mathcal{R}(A) \\ \mathcal{R}(E^*) \subset \mathcal{R}(A^*)}} \frac{\|(A+E)_{MN}^{\dagger} - A_{MN}^{\dagger}\|_{NM}}{\epsilon \|A_{MN}^{\dagger}\|_{NM}}$$

can be expressed by $\operatorname{cond}(A) = \|A\|_{MN} \|A^\dagger_{MN}\|_{NM}$.

Proof From [7], the conditions $\|E\|_{MN} \leq \epsilon\|A\|_{MN}$, $\mathcal{R}(E) \subset \mathcal{R}(A)$, and $\mathcal{R}(E^*) \subset \mathcal{R}(A^*)$, neglecting $O(\epsilon^2)$ terms in a standard expansion, we get

$$(A + E)^\dagger_{MN} - A^\dagger_{MN} = -A^\dagger_{MN} E A^\dagger_{MN}.$$

Let $E = \epsilon\|A\|_{MN}\widehat{E}$, then we have $\|\widehat{E}\|_{MN} \leq 1$. Since

$$\|A^\dagger_{MN}\widehat{E}A^\dagger_{MN}\|_{NM} \leq \|A^\dagger_{MN}\|_{NM} \|\widehat{E}\|_{MN} \|A^\dagger_{MN}\|_{NM}$$
$$\leq \|A^\dagger_{MN}\|^2_{NM}.$$

the result follows if we can show that

$$\sup_{\substack{\|\widehat{E}\|_{MN} \leq 1 \\ \mathcal{R}(\widehat{E}) \subset \mathcal{R}(A) \\ \mathcal{R}(\widehat{E}^*) \subset \mathcal{R}(A^*)}} \|A^\dagger_{MN}\widehat{E}A^\dagger_{MN}\|_{NM} = \|A^\dagger_{MN}\|^2_{NM}. \tag{8.4.2}$$

From Theorem 5.2.2,

$$A = U \begin{bmatrix} D & O \\ O & O \end{bmatrix} V^*, \quad A^\dagger_{MN} = N^{-1}V \begin{bmatrix} D^{-1} & O \\ O & O \end{bmatrix} U^*M,$$

and $M^{1/2}U$, $N^{-1/2}V$, $U^*M^{1/2}$, and $V^*N^{-1/2}$ are all unitary matrices. Let $\mathbf{e}_i$ be the ith column of the identity matrix and $\widehat{E} = U\mathbf{e}_r\mathbf{e}_r^*V^*$, then

$$\|\widehat{E}\|_{MN} = \|M^{1/2}U\mathbf{e}_r\mathbf{e}_r^*V^*N^{-1/2}\|_2 = \|\mathbf{e}_r\mathbf{e}_r^*\|_2 = 1$$

and

$$\|A^\dagger_{MN}\widehat{E}A^\dagger_{MN}\|_{NM}$$
$$= \left\|N^{-1}V \begin{bmatrix} D^{-1} & O \\ O & O \end{bmatrix} \mathbf{e}_r\mathbf{e}_r^* \begin{bmatrix} D^{-1} & O \\ O & O \end{bmatrix} U^*M\right\|_{NM}$$
$$= \left\|N^{-1/2}V \begin{bmatrix} D^{-1} & O \\ O & O \end{bmatrix} \mathbf{e}_r\mathbf{e}_r^* \begin{bmatrix} D^{-1} & O \\ O & O \end{bmatrix} U^*M^{1/2}\right\|_2$$
$$= \left\|\begin{bmatrix} D^{-1} & O \\ O & O \end{bmatrix} \mathbf{e}_r\mathbf{e}_r^* \begin{bmatrix} D^{-1} & O \\ O & O \end{bmatrix}\right\|_2$$
$$= \mu_r^{-2}$$
$$= \|A^\dagger_{MN}\|^2_{NM}.$$

It is easy to see that $\mathcal{R}(\widehat{E}) \subset \mathcal{R}(A)$ and $\mathcal{R}(\widehat{E}^*) \subset \mathcal{R}(A^*)$ from the relations of A and $\widehat{E}$. Thus, (8.4.2) is proved. $\qquad\square$

In the special case when $M = I \in \mathbb{C}^{m \times m}$ and $N = I \in \mathbb{C}^{n \times n}$, we have the following corollary.

Corollary 8.4.2 *The condition number of the Moore-Penrose inverse defined by*

$$\text{cond}(A) = \lim_{\epsilon \to 0^+} \sup_{\substack{\|E\|_2 \leq \epsilon \|A\|_2 \\ \mathcal{R}(E) \subset \mathcal{R}(A) \\ \mathcal{R}(E^*) \subset \mathcal{R}(A^*)}} \frac{\|(A + E)^\dagger - A^\dagger\|_2}{\epsilon \|A^\dagger\|_2}$$

satisfies $\text{cond}(A) = \|A\|_2 \|A^\dagger\|_2$.

Now we consider another weighted matrix norm:

$$\|A\|_{MN}^{(F)} = \|M^{1/2} A N^{-1/2}\|_F, \quad \text{for } A \in \mathbb{C}^{m \times n}$$

and

$$\|B\|_{NM}^{(F)} = \|N^{1/2} B M^{-1/2}\|_F, \quad \text{for } B \in \mathbb{C}^{n \times m}.$$

Then we have the following condition numbers.

Theorem 8.4.3 *The condition number of the weighted Moore-Penrose inverse defined by*

$$\text{cond}_F(A) = \lim_{\epsilon \to 0^+} \sup_{\substack{\|E\|_{MN}^{(F)} \leq \epsilon \|A\|_{MN}^{(F)} \\ \mathcal{R}(E) \subset \mathcal{R}(A) \\ \mathcal{R}(E^*) \subset \mathcal{R}(A^*)}} \frac{\|(A + E)_{MN}^\dagger - A_{MN}^\dagger\|_{NM}^{(F)}}{\epsilon \|A_{MN}^\dagger\|_{NM}^{(F)}}$$

can be given by

$$\text{cond}_F(A) = \frac{\|A\|_{MN}^{(F)} \|A_{MN}^\dagger\|_{NM}^2}{\|A_{MN}^\dagger\|_{NM}^{(F)}}.$$

Proof Analogous to the proof of Theorem 8.4.2, we need to show that

$$\sup_{\substack{\|\widehat{E}\|_{MN}^{(F)} \leq 1 \\ \mathcal{R}(\widehat{E}) \subset \mathcal{R}(A) \\ \mathcal{R}(\widehat{E}^*) \subset \mathcal{R}(A^*)}} \|A_{MN}^\dagger \widehat{E} A_{MN}^\dagger\|_{NM}^{(F)} = \|A_{MN}^\dagger\|_{NM}^2. \tag{8.4.3}$$

The inequality $\|\widehat{E}\|_{MN}^{(F)} \leq 1$ in (8.4.3) implies

$$\|A_{MN}^\dagger \widehat{E} A_{MN}^\dagger\|_{NM}^{(F)} = \|N^{1/2} A_{MN}^\dagger \widehat{E} A_{MN}^\dagger M^{-1/2}\|_F$$
$$= \|N^{1/2} A_{MN}^\dagger M^{-1/2} M^{1/2} \widehat{E} N^{-1/2} N^{1/2} A_{MN}^\dagger M^{-1/2}\|_F$$
$$\leq \|A_{MN}^\dagger\|_{NM} \|\widehat{E}\|_{MN}^{(F)} \|A_{MN}^\dagger\|_{NM}$$
$$\leq \|A_{MN}^\dagger\|_{NM}^2,$$

where the inequalities $\|BC\|_F \leq \|B\|_2 \|C\|_F$ and $\|BC\|_F \leq \|B\|_F \|C\|_2$ are used.

Let $\widehat{E} = U e_r e_r^* V^*$, we have

$$\|\widehat{E}\|_{MN}^{(F)} = 1$$

and

$$\|A_{MN}^\dagger \widehat{E} A_{MN}^\dagger\|_{NM}^{(F)} = \|A_{MN}^\dagger\|_{NM}^2.$$

It is easy to see that $\mathcal{R}(\widehat{E}) \subset \mathcal{R}(A)$ and $\mathcal{R}(\widehat{E}^*) \subset \mathcal{R}(A^*)$. So (8.4.3) is proved. $\square$

As a special case of Theorem 8.4.3, we have:

Corollary 8.4.3 *The condition number of the Moore-Penrose inverse defined by*

$$\text{cond}_F(A) = \lim_{\epsilon \to 0^+} \sup_{\substack{\|E\|_F \leq \epsilon \|A\|_F \\ \mathcal{R}(E) \subset \mathcal{R}(A) \\ \mathcal{R}(E^*) \subset \mathcal{R}(A^*)}} \frac{\|(A + E)^\dagger - A^\dagger\|_F}{\epsilon \|A^\dagger\|_F}$$

satisfies

$$\text{cond}_F(A) = \frac{\|A\|_F \|A^\dagger\|_2^2}{\|A^\dagger\|_F}.$$

Note that from Corollary 8.4.3 $\text{cond}_F(A) \neq \|A\|_F \|A^\dagger\|_F$.
This section is based on [8].

8.5 Expression for the Perturbation of Weighted Moore-Penrose Inverse

In this section, we consider a perturbation formula for the weighted Moore-Penrose inverse of a rectangular matrix and give an explicit expression for the weighted Moore-Penrose inverse of a perturbed matrix under the weakest rank condition.

Let $B = A + E \in \mathbb{C}^{m \times n}$. We know that $\text{rank}(B) = \text{rank}(A)$ is the necessary and sufficient condition for the continuity of the weighted Moore-Penrose inverse. We first present a condition that is equivalent to $\text{rank}(B) = \text{rank}(A)$.

Lemma 8.5.1 *Let $A \in \mathbb{C}^{m \times n}$ with $\text{rank}(A) = r$ and $B = A + E$ such that $I + A_{MN}^\dagger E$ is nonsingular, then $\text{rank}(B) = \text{rank}(A)$ is equivalent to*

$$(I - AA_{MN}^\dagger)E(I + A_{MN}^\dagger E)^{-1}(I - A_{MN}^\dagger A) = O \qquad (8.5.1)$$

or

$$(I - AA_{MN}^\dagger)(I + EA_{MN}^\dagger)^{-1}E(I - A_{MN}^\dagger A) = O. \qquad (8.5.2)$$

Proof Denote $L = U^* M B N^{-1} V \in \mathbb{C}^{m \times n}$. It follows from the (M, N) singular value decomposition Theorem 5.2.2 that L can be written as

$$L = \begin{bmatrix} L_{11} & L_{12} \\ L_{21} & L_{22} \end{bmatrix} = \begin{bmatrix} D + P_1 Q_1 & P_1 Q_2 \\ P_2 Q_1 & P_2 Q_2 \end{bmatrix} \begin{matrix} r \\ m-r \end{matrix}, \qquad (8.5.3)$$
$$\begin{matrix} r & n-r \end{matrix}$$

where $P_i = U_i^* M E$ and $Q_i = N^{-1} V_i, i = 1, 2$.

It is easy to show

$$I + A_{MN}^\dagger E = I + N^{-1} V_1 D^{-1} U_1^* M E = I + Q_1 D^{-1} P_1.$$

Thus, if $I + A_{MN}^\dagger E$ is nonsingular, then so is $D + P_1 Q_1$. Therefore, rank$(B) = $ rank(A) is equivalent to

$$\text{rank}(L) = \text{rank}(D + P_1 Q_1) = \text{rank}(L_{11}). \qquad (8.5.4)$$

From Lemma 4.1.3, we have

$$\text{rank}(L) = \text{rank}(D + P_1 Q_1) + \text{rank}(P_2 Q_2 - P_2 Q_1 (D + P_1 Q_1)^{-1} P_1 Q_2).$$

Combining (8.5.4) and the above equation, we see that rank$(B) = $ rank(A) is equivalent to

$$P_2 Q_2 - P_2 Q_1 (D + P_1 Q_1)^{-1} P_1 Q_2 = O. \qquad (8.5.5)$$

Multiplying (8.5.5) with U_2 on the left and V_2^* on the right, (8.5.5) is equivalent to

$$(I - A A_{MN}^\dagger) E (I - N^{-1} V_1 (D + U_1^* M E N^{-1} V_1)^{-1} U_1^* M E)(I - A_{MN}^\dagger A)$$
$$= O. \qquad (8.5.6)$$

Using the Scherman-Morrison-Woodburg formula, we have

$$N^{-1} V_1 (D + U_1^* M E N^{-1} V_1)^{-1} U_1^* M$$
$$= N^{-1} V_1 (D^{-1} - D^{-1} U_1^* M E (I + N^{-1} V_1 D^{-1} U_1^* M E)^{-1} N^{-1} V_1 D^{-1})$$
$$U_1^* M$$
$$= A_{MN}^\dagger - A_{MN}^\dagger E (I + A_{MN}^\dagger E)^{-1} A_{MN}^\dagger$$
$$= (I + A_{MN}^\dagger E)^{-1} A_{MN}^\dagger. \qquad (8.5.7)$$

Substituting (8.5.7) into (8.5.6), we get

$$(I - A A_{MN}^\dagger) E (I - (I + A_{MN}^\dagger E)^{-1} A_{MN}^\dagger E)(I - A_{MN}^\dagger A) = O,$$

that is,

$$(I - A A_{MN}^\dagger) E (I + A_{MN}^\dagger E)^{-1} (I - A_{MN}^\dagger A) = O.$$

Notice that $E(I + A_{MN}^{\dagger}E) = (I + EA_{MN}^{\dagger})E$, the equivalence of (8.5.1) and (8.5.2) is obvious and the proof is completed. $\square$

Now, we are ready to give an expression of $B_{MN}^{\dagger}$, the main topic of this section.

Theorem 8.5.1 *Let* $B = A + E \in \mathbb{C}^{m \times n}$ *with* $\text{rank}(B) = \text{rank}(A) = r$, *M and N be Hermitian positive definite matrices of orders m and n respectively. Assuming that* $I + A_{MN}^{\dagger}E$ *is nonsingular, we have*

$$
\begin{aligned}
& B_{MN}^{\dagger} \\
&= (A_{MN}^{\dagger}A + N^{-1}X^*N)(A_{MN}^{\dagger}A - X(N + X^*NX)^{-1}X^*N) \\
&\quad (I + A_{MN}^{\dagger}E)^{-1}A_{MN}^{\dagger}(AA_{MN}^{\dagger} - M^{-1}Y^*(M^{-1} + YM^{-1}Y^*)^{-1}Y) \\
&\quad (AA_{MN}^{\dagger} + M^{-1}Y^*M),
\end{aligned}
\tag{8.5.8}
$$

where

$$
X = (I + A_{MN}^{\dagger}E)^{-1}A_{MN}^{\dagger}E(I - A_{MN}^{\dagger}A)
$$

and

$$
Y = (I - AA_{MN}^{\dagger})EA_{MN}^{\dagger}(I + EA_{MN}^{\dagger})^{-1}.
$$

Proof It can be easily verified that $B_{MN}^{\dagger} = N^{-1}VL^{\dagger}U^*M$, where L is given in (8.5.3). From [9, p. 34], we have

$$
\begin{aligned}
& L^{\dagger} \\
&= \begin{bmatrix} L_{11}^* \\ L_{12}^* \end{bmatrix} (L_{11}^* + L_{11}^{-1}L_{12}L_{12}^*)^{-1}L_{11}^{-1}(L_{11}^* + L_{21}^*L_{21}L_{11}^{-1})^{-1}[L_{11}^* \ L_{21}^*] \\
&= \begin{bmatrix} I \\ (L_{11}^{-1}L_{12})^* \end{bmatrix} (I + L_{11}^{-1}L_{12}(L_{11}^{-1}L_{12})^*)^{-1}L_{11}^{-1} \\
&\quad (I + (L_{21}L_{11}^{-1})^*L_{21}L_{11}^{-1})^{-1}[I \ (L_{21}L_{11}^{-1})^*].
\end{aligned}
$$

Denote $F = L_{11}^{-1}L_{12}$ and $G = L_{21}L_{11}^{-1}$. Then

$$
L^{\dagger} = \begin{bmatrix} I \\ F^* \end{bmatrix} (I + FF^*)^{-1}L_{11}^{-1}(I + G^*G)^{-1}[I \ G^*].
$$

Hence, from $V_1^*N^{-1}V_1 = U_1^*MU_1 = I$, we have

$$B_{MN}^{\dagger}$$
$$= (N^{-1}V_1 + N^{-1}V_2 F^*)(I + FF^*)^{-1}L_{11}^{-1}(I + G^*G)^{-1}(U_1^* M + G^* U_2^* M)$$
$$= (N^{-1}V_1 + N^{-1}V_2 F^*)V_1^* N^{-1}V_1(I + FF^*)^{-1}V_1^* N^{-1}V_1 L_{11}^{-1}$$
$$\quad U_1^* M U_1(I + G^*G)^{-1}U_1^* M U_1(U_1^* M + G^* U_2^* M)$$
$$= (A_{MN}^{\dagger}A + N^{-1}V_2 F^* V_1^*)N^{-1}V_1(I + FF^*)^{-1}V_1^* N^{-1}V_1 L_{11}^{-1}$$
$$\quad U_1^* M U_1(I + G^*G)^{-1}U_1^* M(AA_{MN}^{\dagger} + U_1 G^* U_2^* M)$$
$$= (A_{MN}^{\dagger}A + B_2)B_4 B_1 B_5(AA_{MN}^{\dagger} + B_3), \tag{8.5.9}$$

where

$$B_1 = N^{-1}V_1 L_{11}^{-1}U_1^* M,$$
$$B_2 = N^{-1}V_2 F^* V_1^*,$$
$$B_3 = U_1 G^* U_2^* M,$$
$$B_4 = N^{-1}V_1(I + FF^*)^{-1}V_1^*,$$

and

$$B_5 = U_1(I + G^*G)^{-1}U_1^* M.$$

We now compute B_1 to B_5 individually. By (8.5.7), we get

$$B_1 = (I + A_{MN}^{\dagger}E)^{-1}A_{MN}^{\dagger} = A_{MN}^{\dagger}(I + EA_{MN}^{\dagger})^{-1}. \tag{8.5.10}$$

Since

$$F = L_{11}^{-1}L_{12}$$
$$= V_1^* N^{-1}V_1 L_{11}^{-1}U_1^* M U_1 U_1^* M E N^{-1}V_2$$
$$= V_1^*(I + A_{MN}^{\dagger}E)^{-1}A_{MN}^{\dagger}AA_{MN}^{\dagger}EN^{-1}V_2$$
$$= V_1^*(I + A_{MN}^{\dagger}E)^{-1}A_{MN}^{\dagger}EN^{-1}V_2. \tag{8.5.11}$$

Similarly,

$$G = U_2^* M E A_{MN}^{\dagger}(I + EA_{MN}^{\dagger})^{-1}U_1.$$

For B_2 and B_3, we have

$$B_2 = N^{-1}V_2 V_2^* N^{-1}(A_{MN}^{\dagger}E)^*(I + (A_{MN}^{\dagger}E)^*)^{-1}V_1 V_1^*$$
$$= N^{-1}(I - A_{MN}^{\dagger}A)^*(A_{MN}^{\dagger}E)^*(I + (A_{MN}^{\dagger}E)^*)^{-1}(A_{MN}^{\dagger}A)^* N$$
$$= N^{-1}(I - A_{MN}^{\dagger}A)^*(I + (A_{MN}^{\dagger}E)^*)^{-1}(A_{MN}^{\dagger}E)^*(A_{MN}^{\dagger}A)^* N$$
$$= N^{-1}(I - A_{MN}^{\dagger}A)^*(A_{MN}^{\dagger}E)^*(I + (A_{MN}^{\dagger}E)^*)^{-1}N$$
$$= N^{-1}X^* N, \tag{8.5.12}$$

where

$$X = (I + A_{MN}^{\dagger} E)^{-1} A_{MN}^{\dagger} E (I - A_{MN}^{\dagger} A)$$

and

$$B_3 = M^{-1}((I - AA_{MN}^{\dagger})EA_{MN}^{\dagger}(I + EA_{MN}^{\dagger})^{-1})^* M$$
$$= M^{-1} Y^* M, \tag{8.5.13}$$

where

$$Y = (I - AA_{MN}^{\dagger})EA_{MN}^{\dagger}(I + EA_{MN}^{\dagger})^{-1}.$$

From (8.5.11), we have

$$
\begin{aligned}
& FF^* \\
&= V_1^*(I + A_{MN}^{\dagger} E)^{-1} A_{MN}^{\dagger} E N^{-1} V_2 V_2^* N^{-1}((I + A_{MN}^{\dagger} E)^{-1} A_{MN}^{\dagger} E)^* V_1 \\
&= V_1^*(I + A_{MN}^{\dagger} E)^{-1} A_{MN}^{\dagger} E (I - A_{MN}^{\dagger} A) N^{-1}((I + A_{MN}^{\dagger} E)^{-1} A_{MN}^{\dagger} E)^* V_1 \\
&= V_1^*(I + A_{MN}^{\dagger} E)^{-1} A_{MN}^{\dagger} E (I - A_{MN}^{\dagger} A) N^{-1}(I - A_{MN}^{\dagger} A)^* \\
& \quad ((I + A_{MN}^{\dagger} E)^{-1} A_{MN}^{\dagger} E)^* V_1 \\
&= V_1^* X N^{-1} X^* V_1.
\end{aligned}
$$

Likewise,

$$G^* G = U_1^* Y_1^* M Y U_1,$$

using the Sherman-Morrison-Woodburg formula again, we have

$$
\begin{aligned}
(I + FF^*)^{-1} &= (I + V_1^* X N^{-1} X^* V_1)^{-1} \\
&= I - V_1^* X N^{-1}(I + X^* V_1 V_1^* X N^{-1})^{-1} X^* V_1 \\
&= I - V_1^* X (N + X^* N X)^{-1} X^* V_1.
\end{aligned}
$$

As for B_4, we obtain

$$B_4 = A_{MN}^{\dagger} A - X(N + X^* N X)^{-1} X^* N. \tag{8.5.14}$$

Finally,

$$B_5 = AA_{MN}^{\dagger} - M^{-1} Y^*(M^{-1} + YM^{-1}Y^*)^{-1} Y.$$

Substituting Eqs. (8.5.10)–(8.5.14) into (8.5.9) leads to (8.5.8).   □

From Theorem 8.5.1, we can immediately obtain the following corollaries.

Corollary 8.5.1 *Let $B = A + E \in \mathbb{C}^{m \times n}$ with $\mathrm{rank}(B) = \mathrm{rank}(A) = r$. If $I + A^{\dagger} E$ is nonsingular, then*

$$B^\dagger = (A^\dagger A + X^*)(A^\dagger A - X(I + X^*X)^{-1}X^*)(I + A^\dagger E)^{-1}A^\dagger$$
$$(AA^\dagger - Y^*(I + YY^*)^{-1}Y)(AA^\dagger + Y^*),$$

where

$$X = (I + A^\dagger E)^{-1}A^\dagger E(I - A^\dagger A)$$

and

$$Y = (I - AA^\dagger)EA^\dagger(I + EA^\dagger)^{-1}.$$

Corollary 8.5.2 *Let* $B = A + E \in \mathbb{C}^{m \times n}$ *and* $I + A^\dagger_{MN}E$ *be invertible.*
(1) *If* $\mathcal{R}(E^*) \subset \mathcal{R}(A^*)$, *then*

$$B^\dagger_{MN}$$
$$= (I + A^\dagger_{MN}E)^{-1}A^\dagger_{MN}(AA^\dagger_{MN} - M^{-1}Y^*(M^{-1} + YM^{-1}Y^*)^{-1}Y)$$
$$(AA^\dagger_{MN} + M^{-1}Y^*M).$$

(2) *If* $\mathcal{R}(E) \subset \mathcal{R}(A)$, *then*

$$B^\dagger_{MN}$$
$$= (A^\dagger_{MN}A + N^{-1}X^*N)(A^\dagger_{MN}A - X(N + X^*NX)^{-1}X^*N)$$
$$(I + A^\dagger_{MN}E)^{-1}A^\dagger_{MN}.$$

(3) *[7] If* $\mathcal{R}(E) \subset \mathcal{R}(A)$ *and* $\mathcal{R}(E^*) \subset \mathcal{R}(A^*)$, *then*

$$B^\dagger_{MN} = (I + A^\dagger_{MN}E)^{-1}A^\dagger_{MN} = A^\dagger_{MN}(I + EA^\dagger_{MN})^{-1}.$$

This section is based on [10].

Remarks

It is difficult to compute the condition numbers $\kappa(A)$, $\kappa_2(A)$ and $\kappa_{MN}(A)$. They involve A^{-1}, $A^\dagger$, and $A^\dagger_{MN}$ or the eigenvalues $\lambda_1(A)$ and $\lambda_n(A)$, singular values $\sigma_1(A)$ and $\sigma_r(A)$, (M, N) singular values $\mu_1(A)$ and $\mu_r(A)$. Many researchers have tried to alleviate the difficulty by defining new condition numbers that are related to the known condition numbers whereas easy to compute under certain circumstances. Readers interested in this topic are referred to [6, 11–13]. Other researchers also have investigated the minimal problem of condition numbers [7, 14, 15].

The condition numbers in this chapter are given in matrix norms. For more on the condition numbers for the Moore-Penrose inverse, see [16] for Frobenius normwise condition numbers, [17–19] for mixed and componentwise condition numbers, [20] for condition numbers involving Kronecker products, and [21] for optimal perturbation bounds.

The bounds for the difference between the generalized inverses $B^\dagger$ and $A^\dagger$ is discussed in [2, 22, 23] when B is an acute perturbation of A ($\mathcal{R}(A)$ and $\mathcal{R}(B)$ are acute, so are $\mathcal{R}(A^*)$ and $\mathcal{R}(B^*)$). More recent results on the acute perturbation of the weighted M-P inverse can be found in [24].

The perturbation bounds of weighted pseudoinverse and weighted least squares problem can be found in [25–27].

The continuity of the generalized inverse $A_{T,S}^{(2)}$ is given in [28] and the continuity of {1} inverse is given in [29].

The condition numbers for the generalized inversion of structured matrices, such as symmetric, circulant, and Hankel, can be found in [30]. Wei and Zhang [31] presented a condition number related to the generalized inverse $A_{T,S}^{(2)}$. Condition numbers for the outer inverse can be found in [32].

There are more types of perturbation analysis, for example, smoothed analysis [33, 34], weighted acute perturbation [35], multiplicative perturbation [36], stable perturbation [37, 38], and level-2 condition number [39]. A study of null space perturbation with applications in presented in [40]. There is a book on the condition number for PDE [41].

References

1. G.W. Stewart, On the continuity of generalized inverse. SIAM J. Appl. Math. **17**, 33–45 (1969)
2. G.W. Stewart, On the perturbation of pseudo-inverse, projections and linear squares problems. SIAM Rev. **19**, 634–662 (1977)
3. P.Å. Wedin, Perturbation theory for pseudo-inverse. BIT **13**, 217–232 (1973)
4. G. Wang, Perturbation theory for weighted Moore-Penrose inverse. Comm. Appl. Math. Comput. **1**, 48–60 (1987). in Chinese
5. X. He, On the continuity of generalized inverses application of the theory of numerical dependence. Numer. Math. J. Chinese Univ. **1**, 168–172 (in Chinese, 1979)
6. D.J. Higham, Condition numbers and their condition numbers. Linear Algebra Appl. **214**, 193–213 (1995)
7. G. Chen, The minimizing property of the weighted condition number in the problem of matrix perturbation. J. East China Norm. Univ. **7**, 1–7 (1992). in Chinese
8. Y. Wei, D. Wang, Condition numbers and perturbation of the weighted Moore-Penrose inverse and weighted linear least squares problem. Appl. Math. Comput. **145**, 45–58 (2003)
9. R.E. Hartwig, Singular value decomposition and the Moore-Penrose inverse of bordered matices. SIAM J. Appl. Math. **31**, 31–41 (1976)
10. Y. Wei, H. Wu, Expression for the perturbation of the weighted Moore-Penrose inverse. Comput. Math. Appl. **39**, 13–18 (2000)
11. W. Kahan, Numerical linear algebra. Can. Math. Bull. **9**, 756–801 (1966)
12. G. Wang, J. Kuang, On a new measure of degree of ill-condition for matrix. Numer. Math. J. Chinese Univ. **1**, 20–30 (in Chinese, 1979)
13. J. Rohn, A new condition number for matrices and linear systems. Computing **41**, 167–169 (1989)
14. G. Chen, Y. Wei, Y. Xue, The generalized condition numbers of bounded linear operators in Banach spaces. J. Aust. Math. Soc. **76**, 281–290 (2004)
15. G. Wang, Some necessary and sufficient condition for minimizing the condition number of a matrix. J. Shanghai Norm. Univ. **15**, 10–14 (1986). in Chinese

16. H. Diao, Y. Wei, On Frobenius normwise condition numbers for Moore-Penrose inverse and linear least-squares problems. Numer. Linear Algebra Appl. **14**, 603–610 (2007)
17. Y. Wei, W. Xu, S. Qiao, H. Diao, Componentwise condition numbers for generalized matrix inversion and linear least squares. Numer. Math. J. Chinese Univ. (Engl. Ser.) **14**, 277–286 (2005)
18. Y. Wei, Y. Cao, H. Xiang, A note on the componentwise perturbation bounds of matrix inverse and linear systems. Appl. Math. Comput. **169**, 1221–1236 (2005)
19. F. Cucker, H. Diao, Y. Wei, On mixed and componentwise condition numbers for Moore-Penrose inverse and linear least squares problems. Math. Comp. **76**, 947–963 (2007)
20. H. Diao, W. Wang, Y. Wei, S. Qiao, On condition numbers for Moore-Penrose inverse and linear least squares problem involving Kronecker products. Numer. Linear Algebra Appl. **20**, 44–59 (2013)
21. L. Meng, B. Zheng, The optimal perturbation bounds of the Moore-Penrose inverse under the Frobenius norm. Linear Algebra Appl. **432**, 956–963 (2010)
22. G.W. Stewart, J. Sun, *Matrix Perturbation Theory* (Academic Press, New York, 1990)
23. J. Sun, *Matrix Perturbation Analysis*, 2nd edn. (Science Press, Beijing, in Chinese, 2001)
24. H. Ma, Acute perturbation bounds of weighted Moore-Penrose inverse. Int. J. Comput. Math. **95**, 710–720 (2018)
25. M.E. Gulliksson, X. Jin, Y. Wei, Perturbation bound for constrained and weighted least squares problem. Linear Algebra Appl. **349**, 221–232 (2002)
26. M.E. Gulliksson, P.Å. Wedin, Y. Wei, Perturbation identities for regularized Tikhonov inverses and weighted pseudoinverse. BIT **40**, 513–523 (2000)
27. M. Wei, *Supremum and Stability of Weighted Pseudoinverses and Weighted Least Squares Problems Analysis and Computations* (Nova Science Publisher Inc., Huntington, 2001)
28. Y. Wei, G. Wang, On continuity of the generalized inverse $A_{T,S}^{(2)}$. Appl. Math. Comput. **136**, 289–295 (2003)
29. X. Liu, W. Wang, Y. Wei, Continuity properties of the {1}-inverse and perturbation bounds for the Drazin inverse. Linear Algebra Appl. **429**, 1026–1037 (2008)
30. W. Xu, Y. Wei, S. Qiao, Condition numbers for structured least squares. BIT **46**, 203–225 (2006)
31. Y. Wei, N. Zhang, Condition number related with generalized inverse $A_{T,S}^{(2)}$ and constrained linear systems. J. Comput. Appl. Math. **157**, 57–72 (2003)
32. H. Diao, M. Qin, Y. Wei, Condition numbers for the outer inverse and constrained singular linear system. Appl. Math. Comput. **174**, 588–612 (2006)
33. F. Cucker, H. Diao, Y. Wei, Smoothed analysis of some condition numbers. Numer. Linear Algebra Appl. **13**, 71–84 (2006)
34. P. Burgisser, F. Cucker, Smoothed analysis of Moore-Penrose inversion. SIAM J. Matrix Anal. Appl. **31**, 2769–2783 (2010)
35. Z. Xu, C. Gu, B. Feng, Weighted acute perturbation for two matrices. Arab. J. Sci. Eng. ASJE. Math. **35**(1D), 129–143 (2010)
36. X. Zhang, X. Fang, C. song, Q. Xu, Representations and norm estimations for the Moore-Penrose inverse of multiplicative perturbations of matrices. Linear Multilinear Algebra **65**(3), 555–571 (2017)
37. Z. Li, Q. Xu, Y. Wei, A note on stable perturbations of Moore-Penrose inverses. Numer. Linear Alegbra Appl. **20**, 18–26 (2013)
38. Q. Xu, C. Song, Y. Wei, The stable perturbation of the Drazin inverse of the square matrices. SIAM J. Matrix Anal. Appl. **31**(3), 1507–1520 (2009)
39. L. Lin, T.-T. Lu, Y. Wei, On level-2 condition number for the weighted Moore-Penrose inverse. Comput. Math. Appl. **55**, 788–800 (2008)
40. K. Avrachenkov, M. Haviv, Perturbation of null spaces with application to the eigenvalue problem and generalized inverses. Linear Algebra Appl. **369**, 1–25 (2003)
41. Z.-C. Li, H.-T. Huang, Y. Wei, A.H.-D. Cheng, *Effective Condition Number for Numerical Partial Differential Equations*, 2nd edn. (Science Press and Alpha Science International Ltd., Beijing, 2015)

Chapter 9
Perturbation Analysis of the Drazin Inverse and the Group Inverse

Having studied the perturbation of the M-P inverse and the weighted M-P inverse, we now turn to the perturbation analysis of the Drazin and group inverses. Let $A \in \mathbb{C}^{n \times n}$ with $\mathrm{Ind}(A) = k$. When $B = A + E$ and E is small, we discuss whether the Drazin inverse of B is close to that of A and how to reduce the effect of the perturbation.

9.1 Perturbation Bound for the Drazin Inverse

When we derived the perturbation bounds for the M-P inverse and the weighted M-P inverse, we had $B^{\dagger} \to A^{\dagger}$ and $B_{MN}^{\dagger} \to A_{MN}^{\dagger}$ provided that $B \to A$ and $\mathrm{rank}(B) = \mathrm{rank}(A)$. In this section, we study the perturbation analysis of the Drazin inverse. First, let us observe the following example. Let

$$A = \begin{bmatrix} 0 & 1 & 0 & 0 \\ 0 & 0 & 0 & 0 \\ 0 & 0 & 0 & 1 \\ 0 & 0 & 0 & 0 \end{bmatrix}, \quad E = \begin{bmatrix} \epsilon & 0 & 0 & 0 \\ 0 & 0 & 0 & 0 \\ 0 & 0 & 0 & 0 \\ 0 & 0 & 0 & 0 \end{bmatrix},$$

and

$$B = A + E = \begin{bmatrix} \epsilon & 1 & 0 & 0 \\ 0 & 0 & 0 & 0 \\ 0 & 0 & 0 & 1 \\ 0 & 0 & 0 & 0 \end{bmatrix},$$

then $B \to A$ as $\epsilon \to 0$. It can be verified that

© Springer Nature Singapore Pte Ltd. and Science Press 2018
G. Wang et al., *Generalized Inverses: Theory and Computations*,
Developments in Mathematics 53, https://doi.org/10.1007/978-981-13-0146-9_9

$$A_d = \begin{bmatrix} 0\,0\,0\,0 \\ 0\,0\,0\,0 \\ 0\,0\,0\,0 \\ 0\,0\,0\,0 \end{bmatrix} \quad \text{and} \quad B_d = (A+E)_d = \begin{bmatrix} \epsilon^{-1}\ \epsilon^{-2}\ 0\ 0 \\ 0\ \ \ 0\ \ 0\ 0 \\ 0\ \ \ 0\ \ 0\ 0 \\ 0\ \ \ 0\ \ 0\ 0 \end{bmatrix}.$$

This shows that although $\text{rank}(B) = \text{rank}(A) = 2$, $B_d \not\to A_d$ as $\epsilon \to 0$.

Now we give a perturbation bound for the Drazin inverse. For convenience, we denote the following condition as Condition (W).

Condition (W) $B = A + E$ with $\text{Ind}(A) = k$, $E = AA_d E AA_d$, and $\Delta = \|A_d E\| < 1$.

Lemma 9.1.1 *If* Condition (W) *is satisfied, then* $\mathcal{R}(B^k) = \mathcal{R}(A^k)$ *and* $\mathcal{N}(B^k) = \mathcal{N}(A^k)$, *where* $\text{Ind}(B) = k$.

Proof From the condition, we have

$$B = A + E = A + AA_d E = A(I + A_d E) \tag{9.1.1}$$
$$= A + EA_d A = (I + EA_d)A, \tag{9.1.2}$$

in which, since $\Delta = \|A_d E\| < 1$, $I + A_d E$ is nonsingular. Also, the eigenvalues of $A_d E$ are less than unity in absolute value, hence so are those of EA_d. Thus $I + EA_d$ is also nonsingular.

It follows from (9.1.1) and (9.1.2) that $\mathcal{R}(B) = \mathcal{R}(A)$ and $\mathcal{N}(B) = \mathcal{N}(A)$. It then remains to show that

$$\text{rank}(B^i) = \text{rank}(A^i), \quad i = 1, \ldots, k,$$

which follows by induction.

Likewise it follows by induction that

$$\mathcal{R}(B^i) \subset \mathcal{R}(A^i), \quad i = 1, \ldots, k,$$

from which we arrive at

$$\mathcal{R}(B^i) = \mathcal{R}(A^i), \quad i = 1, \ldots, k.$$

Similarly, we can obtain

$$\mathcal{N}(B^j) = \mathcal{N}(A^j), \quad j = 1, \ldots, k.$$

Thus

$$\mathcal{R}(B^k) \oplus \mathcal{N}(B^k) = \mathcal{R}(A^k) \oplus \mathcal{N}(A^k) = \mathbb{C}^n,$$

and
$$AA_d = BB_d,$$

from which the conclusion follows.   □

We are now ready to prove a theorem on a bound for $\|B_d - A_d\|/\|A_d\|$.

Theorem 9.1.1 *Suppose* $E = AA_dEAA_d$ *and* $\Delta = \|A_dE\| < 1$, *then*

$$B_d = (I + A_dE)^{-1}A_d = A_d(I + EA_d)^{-1}$$

and
$$\frac{\|B_d - A_d\|}{\|A_d\|} \leq \frac{\Delta}{1 - \Delta}. \tag{9.1.3}$$

Proof First, we have

$$\begin{aligned}
B_d - A_d &= -A_dEB_d + B_d - A_d + A_d(B - A)B_d \\
&= -A_dEB_d + (B_d - A_dAB_d) + (A_dBB_d - A_d) \\
&= -A_dEB_d.
\end{aligned}$$

The last equality follows from Lemma 9.1.1. Thus,

$$(I + A_dE)B_d = A_d.$$

Similarly, we can prove $B_d - A_d = -B_dEA_d$ and $B_d(I + EA_d) = A_d$. Since $\|A_dE\| < 1$, both $I + A_dE$ and $I + EA_d$ are nonsingular and

$$B_d = (I + A_dE)^{-1}A_d = A_d(I + EA_d)^{-1}.$$

Consequently,
$$\|B_d\| \leq \frac{\|A_d\|}{1 - \|A_dE\|}.$$

Next, applying the above inequality to

$$\|B_d - A_d\| \leq \|A_dE\|\|B_d\|,$$

which can be obtained from $B_d - A_d = -A_dEB_d$, we get the inequality (9.1.3). □

Corollary 9.1.1 *If* Condition (W) *holds, then*

$$\frac{\|A_d\|}{1 + \Delta} \leq \|B_d\| \leq \frac{\|A_d\|}{1 - \Delta}.$$

Corollary 9.1.2 *If, in addition to* Condition (W), $\|A_d\| \|E\| < 1$, *then*

$$\frac{\|B_d - A_d\|}{\|A_d\|} \le \frac{\kappa_d(A)\|E\|/\|A\|}{1 - \kappa_d(A)\|E\|/\|A\|},$$

where $\kappa_d(A) = \|A\| \|A_d\|$ *is defined as the condition number with respect to the Drazin inverse.*

This section is based on [1].

9.2   Continuity of the Drazin Inverse

The continuity of the Drazin inverse is discussed in [2]. Let $A \in \mathbb{C}^{n \times n}$ with $\mathrm{Ind}(A) = k$. From Chap. 2, the core-nilpotent decomposition

$$A = P \begin{bmatrix} C & O \\ O & N \end{bmatrix} P^{-1}$$

can be written as

$$A = C_A + N_A, \tag{9.2.1}$$

where

$$C_A = P \begin{bmatrix} C & O \\ O & O \end{bmatrix} P^{-1}, \quad N_A = P \begin{bmatrix} O & O \\ O & N \end{bmatrix} P^{-1}, \tag{9.2.2}$$

P and C are nonsingular matrices of orders n and r, respectively, and N is nilpotent matrix and $N^k = O$. It is easy to see that $\mathrm{rank}(A^k) = \mathrm{rank}(C_A) = \mathrm{rank}(C^k)$, i.e., the rank of A^k is the same as the rank of C_A, we call $\mathrm{rank}(A^k)$ as the core-rank of A and denote it by Core-rank(A). We will show how the continuity of A_d is related to the core-rank of A.

First, we need the following two lemmas.

Lemma 9.2.1 *Let* $A \in \mathbb{C}_r^{n \times n}$ *with* $\mathrm{Ind}(A) = k$, *then*

$$\mathrm{rank}(A A_d) = Core\text{-}rank(A). \tag{9.2.3}$$

Proof It follows from the core-nilpotent decomposition (9.2.1)–(9.2.2) that

$$A^k = P \begin{bmatrix} C^k & O \\ O & O \end{bmatrix} P^{-1}.$$

Thus

$$Core\text{-}rank(A) = \mathrm{rank}(A^k) = \mathrm{rank}(C^k) = r.$$

Since

$$A_d = P \begin{bmatrix} C^{-1} & O \\ O & O \end{bmatrix} P^{-1} \quad \text{and} \quad AA_d = P \begin{bmatrix} I_r & O \\ O & O \end{bmatrix} P^{-1},$$

where I_r is the identity matrix of order r,

$$\text{rank}(AA_d) = \text{rank}(I_r) = r.$$

Hence (9.2.3) holds. □

Lemma 9.2.2 *Suppose that the P_j and P are projectors, not necessarily orthogonal, on $\mathbb{C}^n$, and $P_j \to P$, then there exists a j_0 such that* $\text{rank}(P_j) = \text{rank}(P)$ *for $j \geq j_0$.*

Proof Suppose that $P_j \to P$, where $P_j^2 = P_j$ and $P^2 = P$. Since $P_j \to P$, we have $\text{rank}(P_j) \geq \text{rank}(P)$ for large j from Corollary 8.1.1. Let $E_j = P_j - P$, then $P_j = P + E_j$ and $E_j \to O$, as $j \to \infty$. Suppose that there does not exist a j_0 such that $\text{rank}(P_j) = \text{rank}(P)$ for $j \geq j_0$. Then there is a subsequence $P_{j_k} = P + E_{j_k}$ such that $\text{rank}(P_{j_k}) > \text{rank}(P)$, i.e., $\dim(\mathcal{R}(P_{j_k})) > \dim(\mathcal{R}(P))$. But $\mathcal{R}(P)$ is complementary to $\mathcal{N}(P)$. Hence for every j_k, there exists a vector $\mathbf{u}_{j_k} \neq \mathbf{0}$ such that $\mathbf{u}_{j_k} \in \mathcal{R}(P_{j_k})$ and $\mathbf{u}_{j_k} \in \mathcal{N}(P)$, thus

$$\mathbf{u}_{j_k} = P_{j_k} \mathbf{u}_{j_k} = (P + E_{j_k}) \mathbf{u}_{j_k} = E_{j_k} \mathbf{u}_{j_k}.$$

Let $\| \cdot \|$ denote an operator norm on $\mathbb{C}^{n \times n}$, then $\|E_{j_k}\| \geq 1$ for all j_k, which contradicts the assumption that $E_{j_k} \to 0$. Thus the required j_0 exists. □

Next we present a theorem on the continuity of the Drazin inverse.

Theorem 9.2.1 *Let $A \in \mathbb{C}^{n \times n}$, $\text{Ind}(A) = k$, and $\{A_j\}$ be a square matrix sequence and $\lim_{j \to \infty} A_j = A$, then $(A_j)_d \to A_d$ if and only if there exists an integer j_0 such that*

$$\text{Core-rank}(A_j) = \text{Core-rank}(A) \quad \text{for all} \quad j \geq j_0. \tag{9.2.4}$$

Proof Sufficiency: Suppose that $A_j = A + E_j$ and $E_j \to O$. If (9.2.4) holds, then the condition of Theorem 2 in [3] also holds. Thus if $\|E_j\| \to 0$, we have $\|(A_j)_d - A_d\| \to 0$, therefore $(A_j)_d \to A_d$.
Necessity: Suppose that $(A_j)_d \to A_d$, then $(A_j)(A_j)_d \to AA_d$. Let $\text{Ind}(A) = k$ and $\text{Ind}(A_j) = k_j$, then

$$P_j = (A_j)(A_j)_d = P_{\mathcal{R}(A_j^{k_j}), \mathcal{N}(A_j^{k_j})} \quad \text{and} \quad P = AA_d = P_{\mathcal{R}(A^k), \mathcal{N}(A^k)}.$$

It follows Lemma 9.2.1 that

$$\text{rank}((A_j)(A_j)_d) = \text{Core-rank}(A_j), \quad \text{and} \quad \text{rank}(AA_d) = \text{Core-rank}(A),$$

which implies $\text{Core-rank}(A_j) = \text{Core-rank}(A)$ for large j from Lemma 9.2.2. □

9.3   Core-Rank Preserving Modification of Drazin Inverse

It follows from the discussion in the previous section that when a matrix A is perturbed to $A + E$, if Core-rank$(A + E) >$ Core-rank(A), discontinuity occurs and the computed solution can be far from the true solution, which is a serious problem. A core-rank preserving method is presented in [4], which alleviates the discontinuity problem.

First, we state that if $A_j \to A$, $j \to \infty$, then for sufficiently large j, we have

$$\text{Core-rank}(A_j) \geq \text{Core-rank}(A)$$

and

$$\text{Core-rank}(A) = \text{the number of the nonzero eigenvalues of } A,$$

where repeated eigenvalues are counted repeatedly.

Suppose $\{A_j\}$ is a sequence of matrices of order n, $A_j \to A$, as $j \to \infty$, Core-rank$(A) = l$, and

$$W_j A_j W_j^{-1} = \begin{bmatrix} B_j & C_j \\ O & N_j \end{bmatrix}, \tag{9.3.1}$$

where, for each j, W_j is a product of orthogonal or elementary lower triangular matrices, N_j is a strictly upper triangular matrix, B_j is a nonsingular matrix of order l_j. Let Core-rank$(A_j) = l_j$, and $l_j \geq l$ for sufficiently large j. Applying similarity transformation to B_j

$$B_j = Q_j R_j Q_j^{-1},$$

where R_j is upper triangular and setting the $l_j - l$ small (in modular) diagonal elements in R_j to zero, we can get another matrix $\widetilde{R}_j$. Thus B_j is modified into $\widetilde{B}_j = Q_j \widetilde{R}_j Q_j^{-1}$. Denote the diagonal matrix $D_j = R_j - \widetilde{R}_j$, then we have

$$\lim_{j \to \infty} D_j = O. \tag{9.3.2}$$

In fact, let $\lambda_i(A)$, $i = 1, ..., n$ be the eigenvalues of A and

$$|\lambda_1(A)| \geq |\lambda_2(A)| \geq \cdots \geq |\lambda_l(A)| > |\lambda_{l+1}(A)| = \cdots = |\lambda_n(A)| = 0.$$

By the assumptions $A_j \to A$ and (9.3.1), when $j \to \infty$, we have

$$\lambda_i(B_j) = \lambda_i(A_j) \to 0, \quad i = l+1, \ldots l_j.$$

Thus (9.3.2) holds.

We change A_j to

$$\widetilde{A}_j = W_j^{-1} \begin{bmatrix} \widetilde{B}_j & C_j \\ O & N_j \end{bmatrix} W_j. \tag{9.3.3}$$

Next we will prove $A_j \to A$, as $j \to \infty$, then

$$\widetilde{A}_j \to A. \tag{9.3.4}$$

Indeed, from (9.3.1)–(9.3.3), we know that as $j \to \infty$,

$$\widetilde{A}_j - A_j = W_j^{-1} \begin{bmatrix} \widetilde{B}_j - B_j & O \\ O & O \end{bmatrix} W_j \to O$$

and

$$\widetilde{A}_j - A = (\widetilde{A}_j - A_j) + (A_j - A) \to O,$$

that is, (9.3.4) holds.

It is easy to show that Core-rank$(\widetilde{A}_j) =$ Core-rank(A) for sufficiently large j. Thus $\widetilde{A}_j$ is called the core-rank preserving modification of A_j.

According to the continuity of the Drazin inverse, Theorem 9.2.1, we have

$$\lim_{j \to \infty} (\widetilde{A}_j)_d = A_d.$$

It implies that the Drazin inverse is continuous after the core-rank preserving modification of A_j.

Example Let

$$A = \begin{bmatrix} 1 & 2 & 4 & 0 \\ 0 & -1 & 1 & 1 \\ 0 & -1 & 1 & 0 \\ 0 & 0 & 0 & 0 \end{bmatrix} \quad \text{and} \quad A + E = \begin{bmatrix} 1 & 2 & 4 & 0 \\ 0 & -1+\epsilon & 1+\epsilon & 1 \\ 0 & -1-\epsilon & 1+3\epsilon & 0 \\ 0 & 0 & 0 & 0 \end{bmatrix},$$

then

$$A_d = \begin{bmatrix} 1 & -4 & 10 & -4 \\ 0 & 0 & 0 & 0 \\ 0 & 0 & 0 & 0 \\ 0 & 0 & 0 & 0 \end{bmatrix}$$

and $A + E \to A$, as $\epsilon \to 0$. But $(A+E)_d$ does not exist when $\epsilon \to 0$. Since Core-rank$(A + E) = 3 >$ Core-rank$(A) = 1$, we now make the core-rank preserving modification of $A + E$. Denote

$$Q = \begin{bmatrix} 1 & 0 & 0 & 0 \\ 0 & \dfrac{\sqrt{2}}{2} & -\dfrac{\sqrt{2}}{2} & 0 \\ 0 & \dfrac{\sqrt{2}}{2} & \dfrac{\sqrt{2}}{2} & 0 \\ 0 & 0 & 0 & 1 \end{bmatrix},$$

then

$$Q^T(A+E)Q = \begin{bmatrix} 1 & 3\sqrt{2} & \sqrt{2} & 0 \\ 0 & 2\epsilon & 2+2\epsilon & \dfrac{\sqrt{2}}{2} \\ 0 & 0 & 2\epsilon & -\dfrac{\sqrt{2}}{2} \\ 0 & 0 & 0 & 0 \end{bmatrix}.$$

Set

$$\widetilde{A+E} = Q \begin{bmatrix} 1 & 3\sqrt{2} & \sqrt{2} & 0 \\ 0 & 0 & 2+2\epsilon & \dfrac{\sqrt{2}}{2} \\ 0 & 0 & 0 & -\dfrac{\sqrt{2}}{2} \\ 0 & 0 & 0 & 0 \end{bmatrix} Q^T.$$

Hence

$$(\widetilde{A+E})_d = \begin{bmatrix} 1 & -4-6\epsilon & 10+6\epsilon & -4-6\epsilon \\ 0 & 0 & 0 & 0 \\ 0 & 0 & 0 & 0 \\ 0 & 0 & 0 & 0 \end{bmatrix} \to A_d, \quad \text{as } \epsilon \to 0.$$

9.4   Condition Number of the Drazin Inverse

The following condition number of the Drazin inverse A_d:

$$C(A) = \left(2\sum_{i=0}^{k-1} \|A_d\|_2^{i+1}\|A^i\|_2(1+\|A\|_2\|A_d\|_2) + \|A_d\|_2 \right)\|A\|_2 \qquad (9.4.1)$$

and the perturbation bound

$$\frac{\|B_d - A_d\|_2}{\|A_d\|_2} \le C(A)\frac{\|E\|_2}{\|A\|_2} + O(\|E\|_2^2)$$

can be found in [3]. This reflects that if the condition number $C(A)$ is small, the perturbation E of A has little effect on A_d; if the condition number $C(A)$ is large, the perturbation E may have large effect on A_d.

In this section, we discuss the normwise relative condition number of the Drazin inverse. We assume that

$$A = P \begin{bmatrix} C & O \\ O & N \end{bmatrix} P^{-1}$$

is the Jordan canonical form of A and define

$$\|A\|_P = \|P^{-1}AP\|_2 \quad \text{and} \quad \|A\|_P^{(F)} = \|P^{-1}AP\|_F.$$

Theorem 9.4.1 *Let $A, E \in \mathbb{C}^{n \times n}$ with $\mathrm{Ind}(A) = k$. The condition number defined by*

$$\mathrm{cond}_P(A) = \lim_{\epsilon \to 0^+} \sup_{\substack{\|E\|_P \le \epsilon \|A\|_P \\ \mathcal{R}(E) \subset \mathcal{R}(A^k) \\ \mathcal{R}(E^*) \subset \mathcal{R}(A^{k*})}} \frac{\|(A+E)_d - A_d\|_P}{\epsilon \|A_d\|_P}$$

can be given by $\mathrm{cond}_P(A) = \|A\|_P \|A_d\|_P$.

Proof Following Theorem 9.1.1 and neglecting $O(\epsilon^2)$ and higher order terms in a standard expansion, we have

$$(A+E)_d - A_d = -A_d E A_d.$$

Let $E = \epsilon \|A\|_P \widehat{E}$, where $\|\widehat{E}\|_P \le 1$, then

$$\|A_d \widehat{E} A_d\|_P \le \|A_d\|_P \|\widehat{E}\|_P \|A_d\|_P \le \|A_d\|_P^2.$$

It then remains to show that

$$\sup_{\substack{\|\widehat{E}\|_P \le 1 \\ \mathcal{R}(\widehat{E}) \subset \mathcal{R}(A^k) \\ \mathcal{R}(\widehat{E}^*) \subset \mathcal{R}(A^{k*})}} \|A_d \widehat{E} A_d\|_P = \|A_d\|_P^2.$$

Indeed, set

$$\widehat{E} = P \begin{bmatrix} \mathbf{y} \\ \mathbf{0} \end{bmatrix} [\mathbf{x}^* \ \mathbf{0}^T] P^{-1},$$

where $\|C^{-1}\mathbf{y}\|_2 = \|\mathbf{x}^* C^{-1}\|_2 = \|C^{-1}\|_2$ and $\|\mathbf{x}\|_2 = \|\mathbf{y}\|_2 = 1$. Thus

$$\begin{aligned}
&\|A_d \widehat{E} A_d\|_P \\
&= \left\| P \begin{bmatrix} C^{-1} & O \\ O & O \end{bmatrix} P^{-1} P \begin{bmatrix} \mathbf{y} \\ \mathbf{0} \end{bmatrix} [\mathbf{x}^* \ \mathbf{0}^T] P^{-1} P \begin{bmatrix} C^{-1} & O \\ O & O \end{bmatrix} P^{-1} \right\|_P \\
&= \left\| \begin{bmatrix} C^{-1} & O \\ O & O \end{bmatrix} \begin{bmatrix} \mathbf{y} \\ \mathbf{0} \end{bmatrix} [\mathbf{x}^* \ \mathbf{0}^T] \begin{bmatrix} C^{-1} & O \\ O & O \end{bmatrix} \right\|_2 \\
&= \|C^{-1}\mathbf{y}\|_2 \|\mathbf{x}^* C^{-1}\|_2 \\
&= \|C^{-1}\|_2^2 \\
&= \|A_d\|_P^2.
\end{aligned}$$

Since

$$\widehat{E} = P \begin{bmatrix} \mathbf{y} \\ \mathbf{0} \end{bmatrix} [\mathbf{x}^* \ \mathbf{0}^T] P^{-1} = P \begin{bmatrix} \mathbf{yx}^* & O \\ O & O \end{bmatrix} P^{-1},$$

it is easy to check that $\mathcal{R}(\widehat{E}) \subset \mathcal{R}(A^k)$ and $\mathcal{R}(\widehat{E}^*) \subset \mathcal{R}(A^{k*})$. The proof is completed. $\qquad\square$

Next we characterize the condition number of the F-norm.

Theorem 9.4.2 *Let* $A, E \in \mathbb{C}^{n \times n}$ *with* $\mathrm{Ind}(A) = k$. *The condition number defined by*

$$\mathrm{cond}_P^{(F)}(A) = \lim_{\epsilon \to 0^+} \sup_{\substack{\|E\|_P^{(F)} \leq \epsilon \|A\|_P^{(F)} \\ \mathcal{R}(E) \subset \mathcal{R}(A^k) \\ \mathcal{R}(E^*) \subset \mathcal{R}(A^{k*})}} \frac{\|(A + E)_d - A_d\|_P^{(F)}}{\epsilon \|A_d\|_P^{(F)}}$$

can be given by

$$\mathrm{cond}_P^{(F)}(A) = \frac{\|A\|_P^{(F)} \|A_d\|_P^2}{\|A_d\|_P^{(F)}}.$$

Proof The proof is similar to that of Theorem 9.4.1. $\qquad\square$

This section is adopted from [5].

9.5   Perturbation Bound for the Group Inverse

In this section, we present a perturbation bound for the group inverse. The details can be found in [6, 7].

Lemma 9.5.1 *Suppose that* $B = A + E$ *with* $\mathrm{Ind}(A) \leq 1$ *and* $\mathrm{rank}(B) = \mathrm{rank}(A)$. *If*

$$\|A_g\| \|E\| < \frac{1}{1 + \mathrm{Ind}(A)\sqrt{\|AA_g\|}}, \tag{9.5.1}$$

then

$$\mathrm{Ind}(B) = \mathrm{Ind}(A) \quad and \quad \|Y\| < 1,$$

where

$$Y = A_g(I + EA_g)^{-1}E(I - AA_g)E(I + A_gE)^{-1}A_g. \tag{9.5.2}$$

Proof Without loss of generality, we assume that $\mathrm{Ind}(A) = 1$ because $\mathrm{Ind}(A) = 0$ implies that $A + E$ is nonsingular and $Y = O$.

It follows from (9.5.1) that $\|A_g E\| \le \|A_g\| \|E\| < 1$ and

$$\|A_g\| \|E\| \operatorname{Ind}(A)\sqrt{\|AA_g\|} < 1 - \|A_g\| \|E\|.$$

Thus

$$\|Y\| \le \frac{\|A_g\|^2 \|E\|^2 \operatorname{Ind}(A)\|AA_g\|}{(1 - \|A_g\| \|E\|)^2} < 1.$$

Next we will show that $\operatorname{Ind}(A + E) = 1$. Since $\operatorname{Ind}(A) = 1$, there is a nonsingular matrix P such that

$$A = P \begin{bmatrix} D & O \\ O & O \end{bmatrix} P^{-1},$$

where D is a nonsingular matrix. Partition

$$P = [P_1 \ P_2] \quad \text{and} \quad P^{-1} = \begin{bmatrix} Q_1 \\ Q_2 \end{bmatrix},$$

where P_1 and Q_1^* have the same column dimensions as D. Let

$$P^{-1} E P = \begin{bmatrix} F_{11} & F_{12} \\ F_{21} & F_{22} \end{bmatrix}.$$

As shown in [6, Theorem 4.1], $D + F_{11}$ is nonsingular and therefore we have

$$P^{-1}(A + E)P = \begin{bmatrix} D + F_{11} & F_{12} \\ F_{21} & F_{22} \end{bmatrix} = \begin{bmatrix} I \\ S \end{bmatrix} (D + F_{11})[I \ T], \tag{9.5.3}$$

where T and S are defined and expressed as

$$T = (D + F_{11})^{-1} F_{12} = Q_1 A_g E(I + A_g E)^{-1} P_2, \tag{9.5.4}$$

$$S = F_{21}(D + F_{11})^{-1} = Q_2(I + E A_g)^{-1} E A_g P_1. \tag{9.5.5}$$

It follows that $TS = Q_1 Y P_1$ and

$$\rho(TS) = \rho(Y P_1 Q_1) \le \rho(Y) \le \|Y\| < 1,$$

which implies that $I + TS$ is nonsingular. It follows from (9.5.3) and [2, Corollary 7.7.5] that

$$\operatorname{Ind}(B) = \operatorname{Ind}(P^{-1}(A + E)P) = 1,$$

which completes the proof. □

The following theorem gives a new general upper bound for the relative error $\|B_g - A_g\|/\|A_g\|$.

Theorem 9.5.1 *Let* $B = A + E$ *such that* $\text{Ind}(A) \leq 1$ *and* $\text{rank}(B) = \text{rank}(A)$. *If* (9.5.1) *holds, then*

$$
\frac{\|B_g - A_g\|}{\|A_g\|}
$$
$$
\leq \frac{(1 - \|A_g\| \|E\|)(1 - \|A_g\| \|E\| + \|A_g\| \|(I - AA_g)E\|)}{((1 - \|A_g\| \|E\|)^2 - \|A_g\|^2 \|E(I - AA_g)E\|)^2}
$$
$$
\cdot (1 - \|A_g\| \|E\| + \|A_g\| \|E(I - AA_g)\|) - 1. \tag{9.5.6}
$$

Proof We give an outline of the proof and refer the details to [6, 7].
From (9.5.3), we have

$$
B_g = P \begin{bmatrix} I \\ S \end{bmatrix} (I + TS)^{-1} (D + F_{11})^{-1} (I + TS)^{-1} \begin{bmatrix} I & T \end{bmatrix} P^{-1}. \tag{9.5.7}
$$

By the Eqs. (9.5.4) and (9.5.5) and the definition (9.5.2) of Y, we get $I + TS = Q_1(1 + Y)P_1$. Consequently,

$$
(I + TS)^{-1} = Q_1(1 + Y)^{-1}P_1.
$$

Applying the above equation and

$$
(D + F_{11})^{-1} = Q_1 A_g (I + EA_g)^{-1} P_1 = Q_1(I + A_g E)^{-1} A_g P_1
$$

to (9.5.7), we obtain

$$
B_g
$$
$$
= P \begin{bmatrix} I \\ S \end{bmatrix} Q_1(1 + Y)^{-1} P_1 Q_1 A_g (I + EA_g)^{-1} P_1 Q_1(1 + Y)^{-1} P_1 \begin{bmatrix} I & T \end{bmatrix} P^{-1}
$$
$$
= (P_1 + P_2 S) Q_1(1 + Y)^{-1} A_g (I + EA_g)^{-1} (1 + Y)^{-1} P_1 (Q_1 + T Q_2).
$$

Expanding the right-hand side of the above equation, we get

$$
B_g = (I + Y)^{-1}(I + A_g E)^{-1} A_g (I + Y)^{-1}
$$
$$
+ (I - AA_g)(I + EA_g)^{-1} EA_g (I + Y)^{-1}(I + A_g E)^{-1} A_g (I + Y)^{-1}
$$
$$
+ (I + Y)^{-1} A_g (I + EA_g)^{-1}(I + Y)^{-1} A_g E (I + A_g E)^{-1}(I - AA_g)
$$
$$
+ (I - AA_g)(I + EA_g)^{-1} EA_g (I + Y)^{-1} A_g (I + EA_g)^{-1}(I + Y)^{-1}
$$
$$
\cdot A_g E (I + A_g E)^{-1}(I - AA_g). \tag{9.5.8}
$$

where Y is given in (9.5.2).

Let the four terms in the summation on the right-hand side of (9.5.8) be denoted by t_1, t_2, t_3 and t_4. Then

$$B_g - A_g = t_1 - A_g + t_2 + t_3 + t_4. \tag{9.5.9}$$

Now,

$$t_1 - A_g = (I + Y)^{-1}((I + A_g E)^{-1} - I)A_g(I + Y)^{-1}$$
$$- (I + Y)^{-1}A_g Y(I + Y)^{-1} - (I + Y)^{-1} Y A_g,$$

then

$$\frac{\|t_1 - A_g\|}{\|A_g\|} \leq \frac{\|A_g\| \|E\|}{(1 - \|Y\|)^2(1 - \|A_g\| \|E\|)} + \frac{2\|Y\| - \|Y\|^2}{(1 - \|Y\|)^2}$$
$$= \frac{1}{(1 - \|Y\|)^2(1 - \|A_g\| \|E\|)} - 1.$$

Similarly, we have

$$\frac{\|t_2\|}{\|A_g\|} \leq \frac{\|A_g\| \|(I - AA_g)E\|}{(1 - \|Y\|^2(1 - \|A_g\| \|E\|)^2}$$

$$\frac{\|t_3\|}{\|A_g\|} \leq \frac{\|A_g\| \|E(I - AA_g)\|}{(1 - \|Y\|)^2(1 - \|A_g\| \|E\|)^2}$$

$$\frac{\|t_3\|}{\|A_g\|} \leq \frac{\|A_g\|^2 \|(I - AA_g)E\| \|E(I - AA_g)\|}{(1 - \|Y\|)^2(1 - \|A_g\| \|E\|)^3}.$$

Applying the above four inequalities to (9.5.9), we have

$$\frac{\|B_g - A_g\|}{\|A_g\|}$$

$$\leq \frac{1}{(1 - \|Y\|)^2(1 - \|A_g\| \|E\|)} + \frac{\|A_g\| \|(I - AA_g)E\|}{(1 - \|Y\|^2(1 - \|A_g\| \|E\|)^2}$$

$$+ \frac{\|A_g\| \|E(I - AA_g)\|}{(1 - \|Y\|)^2(1 - \|A_g\| \|E\|)^2}$$

$$+ \frac{\|A_g\|^2 \|(I - AA_g)E\| \|E(I - AA_g)\|}{(1 - \|Y\|)^2(1 - \|A_g\| \|E\|)^3} - 1.$$

Simplifying the right-hand side of the above inequality, we obtain

$$\frac{\| B_g - A_g \|}{\| A_g \|}$$
$$\leq \frac{(1 - \| A_g \| \, \| E \| + \| A_g \| \, \| (I - AA_g)E \|)}{(1 - \| Y \|)^2 (1 - \| A_g \| \, \| E \|)^3}$$
$$(1 - \| A_g \| \, \| E \| + \| A_g \| \, \| E(I - AA_g) \|) - 1. \qquad (9.5.10)$$

It is shown in [6] that

$$\frac{1}{1 - \| Y \|} \leq \frac{(1 - \| A_g E \|)(1 - \| E A_g \|)}{(1 - \| A_g \| \, \| E \|)^2 - \| A_g \|^2 \, \| E(I - AA_g)E \|}.$$

The upper bound (9.5.6) then follows from the above inequality and (9.5.10).   □

Remarks

The condition number $C(A)$ in (9.4.1) is more complicated than those of regular inverse, the M-P inverse, and the weighted M-P inverse: $\| A \| \, \| A^{-1} \|$, $\| A \|_2 \, \| A^\dagger \|_2$, and $\| A \|_{MN} \| A^\dagger_{MN} \|_{NM}$. In 1979, Campbell and Meyer discussed this problem in [2] and pointed out:

If $A = PJP^{-1}$ is the Jordan canonical form of A, P is nonsingular, and $\text{Ind}(A) = k$, and we compute A_d by $A_d = A^k(A^{2k+1})^\dagger A^k$, then $\| A \|(\| A_d \| + 1)$ or

$$C(A) = \| P \| \, \| P^{-1} \|(\| J \|^k + \| J^\dagger \|^k)$$

can be regarded as the condition number.

More results on the continuity and perturbation analysis of the matrix Drazin inverse and W-weighted Drazin inverse can be found in [8–17]. The sign analysis of Drazin and group inverses is presented in [18, 19]. Some additive properties of the Drazin inverse are given in [20, 21]. The stable perturbation of the Drazin inverse is discussed in [16, 22] and acute perturbation of the group inverse in [23]. In [24], perturbation bounds are derived by the separation of simple invariant subspaces.

Condition numbers of the Bott-Duffin inverse and their condition numbers are presented in [25].

The perturbation theories of the Bott-Duffin inverse, the generalized Bott-Duffin inverse, the W-weighted Drazin inverse and the generalized inverse $A^{(2)}_{T,S}$ are presented in [26–30].

Index splitting for the Drazin inverse and the singular linear system can be found in [31–38].

The perturbation and subproper splitting for the generalized inverse $A^{(2)}_{T,S}$ are discussed in [39, 40].

References

1. Y. Wei, G. Wang, The perturbation theory for the Drazin inverse and its applications. Linear Algebra Appl. **258**, 179–186 (1997)
2. S.L. Campbell, C.D. Meyer Jr., *Generalized Inverses of Linear Transformations* (Pitman, London, 1979)
3. G. Rong, The error bound of the perturbation of Drazin inverse. Linear Algebra Appl. **47**, 159–168 (1982)
4. X. Li, On computation of Drazin inverses of a matrix. J. Shanghai Norm. Univ. **11**, 9–16 (1982). in Chinese
5. Y. Wei, G. Wang, D. Wang, Condition number of Drazin inverse and their condition numbers of singular linear systems. Appl. Math. Comput. **146**, 455–467 (2003)
6. Y. Wei, On the perturbation of the group inverse and oblique projection. Appl. Math. Comput. **98**, 29–42 (1999)
7. X. Li, Y. Wei, An improvement on the perturbation of the group inverse and oblique projection. Linear Algebra Appl. **338**, 53–66 (2001)
8. X. Chen, G. Chen, On the continuity and perturbation of the weighted Drazin inverses. J. East China Norm. Univ. **7**, 21–26 (1992). in Chinese
9. N.C. González, J.J. Koliha, Y. Wei, Perturbation of the Drazin inverse for matrices with equal eigenprojections at zero. Linear Algebra Appl. **312**, 181–189 (2000)
10. R.E. Hartwig, G. Wang, Y. Wei, Some additive results on the Drazin inverse. Linear Algebra Appl. **322**, 207–217 (2001)
11. Y. Wei, Perturbation bound of the Drazin inverse. Appl. Math. Comput. **125**, 231–244 (2002)
12. Y. Wei, The Drazin inverse of updating of a square matrix with application to perturbation formula. Appl. Math. Comput. **108**, 77–83 (2000)
13. Y. Wei, H. Wu, On the perturbation of the Drazin inverse and oblique projection. Appl. Math. Lett. **13**, 77–83 (2000)
14. Y. Wei, H. Wu, Challenging problems on the perturbation of Drazin inverse. Ann. Oper. Res. **103**, 371–378 (2001)
15. Y. Wei, H. Diao, Condition number for the Drazin inverse and the Drazin-inverse solution of singular linear system with their condition numbers. J. Comput. Appl. Math. **182**, 270–289 (2005)
16. N. Castro González, J. Robles, J.Y. Vélez-Cerrada, Characterizations of a class of matrices and perturbation of the Drazin inverse. SIAM J. Matrix Anal. Appl. **30**, 882–897 (2008)
17. X. Liu, W. Wang, Y. Wei, Continuity properties of the {1}-inverse and perturbation bounds for the Drazin inverse. Linear Algebra Appl. **429**, 1026–1037 (2008)
18. J. Zhou, C. Bu, Y. Wei, Some block matrices with signed Drazin inverses. Linear Algebra Appl. **437**, 1779–1792 (2012)
19. J. Zhou, C. Bu, Y. Wei, Group inverse for block matrices and some related sign analysis. Linear Multilinear Algebra **60**, 669–681 (2012)
20. G. Zhuang, J. Chen, D. Cvetković-Ilić, Y. Wei, Additive property of Drazin invertibility of elements in a ring. Linear Multilinear Algebra **60**, 903–910 (2012)
21. Y. Wei, C. Deng, A note on additive results for the Drazin inverse. Linear Multilinear Algebra **59**, 1319–1329 (2011)
22. Q. Xu, C. Song, Y. Wei, The stable perturbation of the Drazin inverse of the square matrices. SIAM J. Matrix Anal. Appl. **31**(3), 1507–1520 (2009)
23. Y. Wei, Acute perturbation of the group inverse. Linear Algebra Appl. **534**, 135–157 (2017)
24. Y. Wei, X. Li, F. Bu, A perturbation bound of the Drazin inverse of a matrix by separation of simple invariant subspaces. SIAM J. Matrix Anal. Appl. **27**, 72–81 (2005)
25. Y. Wei, W. Xu, Condition number of Bott-Duffin inverse and their condition numbers. Appl. Math. Comput. **142**, 79–97 (2003)
26. G. Wang, Y. Wei, Perturbation theory for the Bott-Duffin inverse and its applications. J. Shanghai Norm. Univ. **22**, 1–6 (1993)

27. G. Wang, Y. Wei, The perturbation analysis of doubly perturbed constrained systems. J. Shanghai Norm. Univ. **25**, 9–14 (1996)
28. G. Chen, G. Liu, Y. Xue, Perturbation theory for the generalized Bott-Duffin inverse and its applications. Appl. Math. Comput. **129** (2002)
29. G. Wang, Y. Wei, Perturbation theory for the W-weighted Drazin inverse and its applications. J. Shanghai Norm. Univ. **26**, 9–15 (1997)
30. Y. Wei, G. Wang, Perturbation theory for the generalized inverse $A_{T,S}^{(2)}$. J. Fudan Univ. **39**, 482–488 (2000)
31. Y. Wei, Index splitting for Drazin inverse and the singular linear system. Appl. Math. Comput. **95**, 115–124 (1998)
32. C. Zhu, G. Chen, Index splitting for the Drazin inverse of linear operator in Banach space. Appl. Math. Comput. **135**, 201–209 (2003)
33. Y. Wei, H. Wu, Additional results on index splittings for Drazin inverse solutions of singular systems. Electron. J. Linear Algebra **8**, 83–93 (2001)
34. Y. Wei, Perturbation analysis of singular linear systems with index one. Int. J. Comput. Math. **74**, 483–491 (2000)
35. Y. Wei, H. Wu, Convergence properties of Krylov subspace methods for singular linear systems with arbitrary index. J. Comput. Appl. Math. **114**, 305–318 (2000)
36. Y. Wei, H. Wu, On the use of incomplete semiiterative methods for singular systems and applications in Markov chain modeling. Appl. Math. Comput. **125**, 245–259 (2002)
37. Z. Cao, Polynomial acceleration methods for solving singular systems of linear equations. J. Comput. Math. **9**, 378–387 (1991)
38. Z. Cao, On the convergence of iterative methods for solving singular linear systems. J. Comput. Appl. Math. **145**, 1–9 (2002)
39. Y. Wei, H. Wu, On the perturbation and subproper splittings for the generalized inverse $A_{T,S}^{(2)}$ of rectangular matrix A. J. Comput. Appl. Math. **137**, 317–329 (2001)
40. G. Wang, Y. Wei, Proper splittings for restricted linear equations and the generalized inverse $A_{T,S}^{(2)}$. Numer. Math. J. Chin. Univ. (Ser. B) **7**, 1–13 (1998)

Chapter 10
Generalized Inverses of Polynomial Matrices

A polynomial matrix is a matrix whose entries are polynomials. Equivalently, a polynomial matrix can be expressed as a polynomial with matrix coefficients. Formally speaking, in the univariable case, $(\mathbb{R}[x])^{m \times n}$ and $(\mathbb{R}^{m \times n})[x]$ are isomorphic. In other words, extending the entries of matrices to polynomials is the same as extending the coefficients of polynomials to matrices. An example of a 3×2 polynomial matrix of degree 2:

$$\begin{bmatrix} 1 & x^2 \\ x & 0 \\ x+1 & x^2-1 \end{bmatrix} = \begin{bmatrix} 0 & 1 \\ 0 & 0 \\ 0 & 1 \end{bmatrix} x^2 + \begin{bmatrix} 0 & 0 \\ 1 & 0 \\ 1 & 0 \end{bmatrix} x + \begin{bmatrix} 1 & 0 \\ 0 & 0 \\ 1 & -1 \end{bmatrix}.$$

In this chapter, we study the Moore-Penrose and Drazin inverses of a polynomial matrix and algorithms for computing the generalized inverses.

10.1 Introduction

We start with the scalar nonsingular case. Let $A \in \mathbb{R}^{n \times n}$ be nonsingular and

$$p(\lambda) = \det(\lambda I - A) = c_0 \lambda^n + c_1 \lambda^{n-1} + \cdots + c_{n-1}\lambda + c_n, \qquad (10.1.1)$$

where $c_0 = 1$, be the characteristic polynomial of A. The Cayley-Hamilton theorem says

$$c_0 A^n + c_1 A^{n-1} + \cdots + c_{n-1}A + c_n I = O.$$

Thus, we have

$$A^{-1} = -c_n^{-1} P_n = (-1)^{n-1} \frac{P_n}{\det(A)},$$

© Springer Nature Singapore Pte Ltd. and Science Press 2018
G. Wang et al., *Generalized Inverses: Theory and Computations*,
Developments in Mathematics 53, https://doi.org/10.1007/978-981-13-0146-9_10

where

$$P_n = c_0 A^{n-1} + c_1 A^{n-2} + \cdots c_{n-2} A + c_{n-1} I,$$

which can be efficiently computed using the iterative Horner's Rule:

$$P_0 = O; \ c_0 = 1; \quad P_i = A P_{i-1} + c_{i-1} I, \ i = 1, \dots, n.$$

How can the coefficients $c_i, i = 1, \dots, n$ be obtained efficiently? It turns out that

$$c_i = -\frac{\mathrm{tr}(A P_i)}{i}, \quad i = 1, \dots n.$$

Putting all things together, the following algorithm, known as the Faddeev-LeVerrier algorithm [1, 2] and presented in Sect. 5.5, efficiently computes the coefficients c_i, $i = 1, \dots, n$, of the characteristic polynomial of A.

Algorithm 10.1.1 Given a nonsingular matrix $A \in \mathbb{C}^{n \times n}$, this algorithm computes the coefficients $c_i, i = 1, \dots, n$, of its characteristic polynomial (10.1.1) and its inverse A^{-1}.

1. $P_0 = O; c_0 = 1;$
2. for $i = 1$ to n

   $\quad P_i = A P_{i-1} + c_{i-1} I;$
   $\quad c_i = -\mathrm{tr}(A P_i)/i;$

3. $A^{-1} = -c_n^{-1} P_n = (-1)^{n-1} P_n / \det(A);$

As pointed out in Sect. 5.5, Decell generalized the above algorithm to general scalar matrices and the Moore-Penrose inverse [3]. Let $B = AA^*$ and

$$f(\lambda) = \det(\lambda I - B) = a_0 \lambda^n + a_1 \lambda^{n-1} + \cdots + a_{n-1} \lambda + a_n, \quad a_0 = 1, \quad (10.1.2)$$

be the characteristic polynomial of B. If $k > 0$ is the largest integer such that $a_k \neq 0$, then the Moore-Penrose inverse of A is given by

$$A^\dagger = -a_k^{-1} A^* (B^{k-1} + a_1 B^{k-2} + \cdots + a_{k-1} I).$$

If $k = 0$ is the largest integer such that $a_k \neq 0$, that is, a_0 is the only nonzero coefficient, then $A^\dagger = O$.

Analogous to Algorithm 10.1.1, the following Decell's algorithm computes $A^\dagger$.

Algorithm 10.1.2 [3] Given $A \in \mathbb{C}^{m \times n}$ and k the largest integer such that a_k in (10.1.2) is nonzero, this algorithm computes the Moore-Penrose inverse of A.

1. if $k = 0$ return $A^\dagger = O;$
2. $B = AA^*;$
3. $P_0 = O; A_0 = O; a_0 = 1;$

4. for $i = 1$ to k

$$P_i = A_{i-1} + a_{i-1}I;$$
$$A_i = BP_i;$$
$$a_i = -\text{tr}(A_i)/i;$$

5. $A^\dagger = -a_k^{-1}A^*P_k;$

In the following sections, we generalize the above algorithm to polynomial matrices and their generalized inverses.

10.2 Moore-Penrose Inverse of a Polynomial Matrix

The definition of the Moore-Penrose inverse of a polynomial matrix is the same as the scalar case, that is, it is the polynomial matrix satisfying the four Penrose conditions.

The Decell's Algorithm 10.1.2 for computing the Moore-Penrose inverse is generalized to polynomial matrices [4]. Consider the polynomial matrix

$$A(x) = A_0x^n + A_1x^{n-1} + \cdots + A_{n-1}x + A_n,$$

where $A_i \in \mathbb{R}^{m \times n}$, $i = 0, 1, ..., n$. Let $B(x) = A(x)A(x)^T$ and

$$
\begin{aligned}
p(\lambda, x) &= \det(\lambda I - B(x)) \\
&= a_0(x)\lambda^n + a_1\lambda^{n-1} + \cdots + a_{n-1}(x)\lambda + a_n(x), \quad (10.2.1)
\end{aligned}
$$

where $a_0(x) = 1$, be the characteristic polynomial of $B(x)$. It is shown in [4] that if k is the largest integer such that $a_k(x) \neq 0$ and $\mathbb{Z}$ is the set containing the zeros of $a_k(x)$, then the Moore-Penrose inverse $A(x)^\dagger$ of $A(x)$ for $x \in \mathbb{R} \backslash \mathbb{Z}$ is given by

$$A(x)^\dagger = -a_k(x)^{-1}A(x)^T \left(B(x)^{k-1} + a_1(x)B(x)^{k-2} + \cdots + a_{k-1}(x)I \right).$$

If $k = 0$ is the largest integer such that $a_k(x) \neq 0$, then $A(x)^\dagger = O$. Moreover, for each $x_i \in \mathbb{Z}$, if $k_i < k$ is the largest integer such that $a_{k_i}(x_i) \neq 0$, then the Moore-Penrose inverse $A(x_i)^\dagger$ of $A(x_i)$ is given by

$$
\begin{aligned}
&A(x_i)^\dagger \\
&= -a_{k_i}(x_i)^{-1}A(x_i)^T \left(B(x_i)^{k_i-1} + a_1(x_i)B(x_i)^{k_i-2} + \cdots + a_{k_i-1}(x_i)I \right).
\end{aligned}
$$

The algorithm for computing the polynomial matrix $A(x)^\dagger$ is completely analogous to Algorithm 10.1.2, replacing the scalar matrices A, A_i, B, and P_i with the polynomial matrices $A(x)$, $A_i(x)$, $B(x)$, and $P(x)$ respectively and the scalars a_i with the polynomials $a_i(x)$. Obviously, the algorithm involves symbolic computation. Also in [4], a two-dimensional algorithm that avoids symbolic computation is presented.

From the definition $B(x) = A(x)A(x)^T$, the degree of $B(x)$ can be as high as $2n$. Consequently, $A_i(x)$, $a_i(x)$, and $P_{i+1}(x)$ are of degrees up to $2in$. Let

$$a_i(x) = \sum_{j=0}^{2in} a_{i,j} x^j, \quad i = 1, ..., k,$$

where $a_{i,j}$ are scalars, and

$$P_i(x) = \sum_{j=0}^{2(i-1)n} P_{i,j} x^j, \quad i = 1, ..., k,$$

where $P_{i,j}$ are scalar matrices, then $A(x)^\dagger$ can be written as

$$A(x)^\dagger = -\left(\sum_{j=0}^{2kn} a_{k,j} x^j \right)^{-1} \left(\sum_{j=0}^{n} A_j^T x_j \right) \left(\sum_{j=0}^{2(k-1)n} P_{k,j} x^j \right)$$

$$= -\left(\sum_{j=0}^{2kn} a_{k,j} x^j \right)^{-1} \left(\sum_{j=0}^{(2k-1)n} \sum_{l=0}^{j} (A_{j-l}^T P_{k,l}) x^j \right). \qquad (10.2.2)$$

Now we derive $a_{i,j}$ and $P_{i,j}$. Following Algorithm 10.1.2, first we have

$$A_i(x) = B(x) P_i(x)$$

$$= \left(\sum_{j=0}^{n} A_j x_j \right) \left(\sum_{j=0}^{n} A_j^T x_j \right) \left(\sum_{j=0}^{2(i-1)n} P_{i,j} x^j \right)$$

$$= \left(\sum_{j=0}^{2n} \left(\sum_{p=0}^{j} A_{j-p} A_p^T \right) x^j \right) \left(\sum_{j=0}^{2(i-1)n} P_{i,j} x^j \right)$$

$$= \sum_{j=0}^{2in} \left(\sum_{p=0}^{j} \left(\sum_{l=0}^{j-p} (A_{j-p-l} A_l^T) \right) P_{i,p} \right) x^j.$$

It then follows that

$$a_i(x) = -i^{-1} \mathrm{tr}(A_i(x))$$

$$= -i^{-1} \sum_{j=0}^{2in} \mathrm{tr} \left(\sum_{p=0}^{j} \left(\sum_{l=0}^{j-p} (A_{j-p-l} A_l^T) \right) P_{i,p} \right) x^j,$$

which gives

$$a_{i,j} = -\frac{1}{i}\text{tr}\left(\sum_{p=0}^{j}\left(\sum_{l=0}^{j-p}(A_{j-p-l}A_l^T)\right)P_{i,p}\right), \quad j = 0, ..., 2in. \tag{10.2.3}$$

Moreover,

$$P_i(x) = A_{i-1}(x) - a_{i-1}(x)I$$

$$= \sum_{j=0}^{2(i-1)n}\left(\sum_{p=0}^{j}\left(\sum_{l=0}^{j-p}(A_{j-p-l}A_l^T)\right)P_{i-1,p}\right)x^j + \sum_{j=0}^{2(i-1)n}a_{i-1,j}x^j$$

$$= \sum_{j=0}^{2(i-1)n}\left(\left(\sum_{p=0}^{j}\left(\sum_{l=0}^{j-p}(A_{j-p-l}A_l^T)P_{i-1,p}\right)\right) + a_{i-1,j}I\right)x^j,$$

which gives

$$P_{i,j} = \left(\sum_{p=0}^{j}\left(\sum_{l=0}^{j-p}(A_{j-p-l}A_l^T)P_{i-1,p}\right)\right) + a_{i-1,j}I, \tag{10.2.4}$$

for $j = 0, ..., 2(i-1)n$.

Finally, we have the following two-dimensional algorithm for computing $A(x)^\dagger$.

Algorithm 10.2.1 [4] Given a polynomial matrix $A(x) \in \mathbb{R}^{m \times n}[x]$ and k, the largest integer such that $a_k(x)$ in (10.2.1) is nonzero, this algorithm computes the Moore-Penrose inverse of $A(x)$.

1. if $k = 0$ return $A(x)^\dagger = O$;
2. $P_{0,0} = O$; $a_{0,0} = 1$;
3. for $i = 1$ to k

   compute $P_{i,j}$, $\quad j = 0, ..., 2(i-1)n$, by (10.2.4);
   compute $a_{i,j}$, $\quad j = 0, ..., 2in$, by (10.2.3);

4. compute $A(x)^\dagger$ by (10.2.2).

Note that in the above algorithm it is assumed that $P_{i,j} = O$ when $j > 2(i-1)n$. This algorithm is called two-dimensional, since it involves the computation of two-dimensional variables $a_{i,j}$ and $P_{i,j}$.

10.3 Drazin Inverse of a Polynomial Matrix

The definition of the Drazin inverse $A(x)_d$ of a polynomial matrix $A(x) \in \mathbb{R}^{n \times n}[x]$ is defined as the same as the scalar case, that is, $A(x)_d$ is the matrix satisfying the three conditions:

$$A(x)^{k+1} A(x)_d = A(x)^k,$$
$$A(x)_d A(x) A(x)_d = A(x)_d,$$
$$A(x) A(x)_d = A(x)_d A(x),$$

where $k = \mathrm{ind}(A(x))$, the index of $A(x)$, defined as the smallest integer such that $\mathrm{rank}(A(x)^k) = \mathrm{rank}(A(x)^{k+1})$.

Let

$$p(\lambda, x) = \det(\lambda I - A(x))$$
$$= a_0(x)\lambda^n + a_1(x)\lambda^{n-1} + \cdots + a_{n-1}(x)\lambda + a_n(x),$$

where $a_0(x) = 1$, be the characteristic polynomial of $A(x)^{k+1}$, then

$$a_{r+1}(x) = \ldots = a_n(x) = 0, \quad \text{and} \quad a_r(x) \neq 0,$$

where $r = \mathrm{rank}(A(x)^{k+1})$. Let $\mathbb{Z}$ be the set containing the zeros of $a_r(x)$, then the Drazin inverse $A(x)_d$ of $A(x)$ for $x \in \mathbb{R} \backslash \mathbb{Z}$ is given by

$$A(x)_d = -a_r(x)^{-1} A(x)^k ((A(x)^{k+1})^{r-1} + a_1(x)(A(x)^{k+1})^{r-2} + \cdots$$
$$+ a_{r-2}(x)A(x)^{k+1} + a_{r-1}(x)I).$$

If $r = 0$, then $A(x)_d = O$.

Following the Decell's Algorithm 10.1.2, we have the following finite algorithm for computing the Drazin inverse $A(x)_d$ of $A(x)$ [5, 6].

Algorithm 10.3.1 [5]　　Given　　$A(x) \in \mathbb{R}^{n \times n}[x]$,　　$k = \mathrm{Ind}(A(x))$　　and $r = \mathrm{rank}(A(x)^{k+1})$, this algorithm computes the Drazin inverse $A(x)_d$ of $A(x)$.

1. if $r = 0$ return $A(x)_d = O$;
2. $B(x) = A(x)^{k+1}$;
3. $P_0(x) = O$; $a_0(x) = 1$;
4. for $i = 1$ to r

$$P_i(x) = B(x)P_{i-1}(x) + a_{i-1}(x)I;$$
$$a_i = -\mathrm{tr}(B(x)P_i(x))/i;$$

5. $A(x)_d = -a_r(x)^{-1} A(x)^k P_r(x)$.

Following derivation of the two-dimensional Algorithm 10.2.1, we can obtain a two-dimensional algorithm for computing the Drazin inverse of a polynomial matrix [5, 6].

Notice that the degrees of $B(x) = A(x)^{k+1}$, $A(x)^k$, $P_i(x)$, and $a_i(x)$ are respectively $(k+1)n, kn, (i-1)(k+1)n$, and $i(k+1)n$. Let

$$B(x) = \sum_{j=0}^{(k+1)n} B_j x^j, \quad A(x)^k = \sum_{j=0}^{kn} \widehat{A}_j x^j, \quad P_i(x) = \sum_{j=0}^{(i-1)(k+1)n} P_{i,j} x^j,$$

where B_j, $\widehat{A}_j$, and $P_{i,j}$ are scalar matrices, and

$$a_i(x) = \sum_{j=0}^{i(k+1)n} a_{i,j}x^j,$$

where $a_{i,j}$ are scalars, then

$$A(x)_d$$

$$= -\left(\sum_{j=0}^{r(k+1)n} a_{r,j}x^j\right)^{-1} \left(\sum_{j=0}^{kn} \widehat{A}_j x^j\right)\left(\sum_{j=0}^{(r-1)(k+1)n} P_{r,j}x^j\right)$$

$$= -\left(\sum_{j=0}^{r(k+1)n} a_{r,j}x^j\right)^{-1} \left(\sum_{j=0}^{(kr+r-1)n} \left(\sum_{l=0}^{j}(\widehat{A_{j-l}P_{r,l}})\right) x^j\right). \quad (10.3.1)$$

Now we derive $a_{i,j}$ and $P_{i,j}$. Firstly,

$$a_i(x) = -i^{-1}\mathrm{tr}(B(x)P_i(x))$$

$$= -i^{-1}\mathrm{tr}\left(\left(\sum_{j=0}^{(k+1)n} B_j x^j\right)\left(\sum_{j=0}^{(i-1)(k+1)n} P_{i,j}(x)x^j\right)\right)$$

$$= -i^{-1}\mathrm{tr}\left(\sum_{j=0}^{i(k+1)n} \left(\sum_{l=0}^{j} B_{j-l}P_{i,l}\right) x^j\right),$$

implying that

$$a_{i,j} = -\frac{1}{i}\mathrm{tr}\left(\sum_{l=0}^{j} B_{j-l}P_{i,l}\right). \quad j = 0, ..., i(k+1)n. \quad (10.3.2)$$

Secondly,

$$P_i(x) = B(x)P_{i-1}(x) - a_{i-1}(x)I$$

$$= \left(\sum_{j=0}^{(k+1)n} B_j x^j\right)\left(\sum_{j=0}^{(i-2)(k+1)n} P_{i-1,j}x^j\right) + \sum_{j=0}^{(i-1)(k+1)n} a_{i-1,j}Ix^j$$

$$= \sum_{j=0}^{(i-1)(k+1)n} \left(\sum_{l=0}^{j} B_{j-l}P_{i-1,l}\right) x^j + \sum_{j=0}^{(i-1)(k+1)n} a_{i-1,j}x^j$$

$$= \sum_{j=0}^{(i-1)(k+1)n} \left(\left(\sum_{l=0}^{j} B_{j-l}P_{i-1,l}\right) a_{i-1,j}I\right) x^j,$$

implying that

$$P_{i,j} = \left(\sum_{l=0}^{j} B_{j-l} P_{i-1,l} \right) + a_{i-1,j} I, \quad j = 0, ..., (i-1)(k+1)n. \qquad (10.3.3)$$

Finally, we have the following two-dimensional algorithm for computing the Drazin inverse $A(x)_d$ of $A(x)$. Comparing with Algorithm 10.3.1, this algorithm avoids symbolic computation.

Algorithm 10.3.2 [5]   Given $A(x) \in \mathbb{R}^{n \times n}[x]$, $k = \mathrm{Ind}(A(x))$ and $r = \mathrm{rank}(A(x)^{k+1})$, this algorithm computes the Drazin inverse $A(x)_d$ of $A(x)$.

1. if $k = 0$ return $A(x)_d = O$;
2. $P_{0,0} = O$; $a_{0,0} = 1$;
3. for $i = 1$ to k

    compute $P_{i,j}$, $j = 0, ..., (i-1)(k+1)n$, by (10.3.3);
    compute $a_{i,j}$, $j = 0, ..., i(k+1)n$, by (10.3.2);

4. compute $A(x)_d$ by (10.3.1).

In the algorithm, it is assumed that $B_j = O$, when $j > (k+1)n$, and $P_{i,j} = O$, when $j > (i-1)(k+1)n$.

Example 10.3.1 [5] Let

$$A(x) = \begin{bmatrix} x-1 & 1 & 0 & 0 \\ 0 & 1 & x & 0 \\ 0 & 0 & 0 & x \\ 0 & 0 & 0 & 0 \end{bmatrix},$$

then $n = 1$.

It can be determined

$$\mathrm{rank}(A(x)^2) = \mathrm{rank}(A(x)^3) = 2, \quad \text{when } x \neq 1.$$

Thus $r = k = 2$. Initially, we have

$$P_{0,0} = O \quad \text{and} \quad a_{0,0} = 1.$$

When $i = 1$, $P_{1,0} = I$ and

$$a_{1,0} = -\mathrm{tr}(B_0 P_{1,0}) = 0,$$
$$a_{1,1} = -\mathrm{tr}(B_1 P_{1,0}) = -3,$$
$$a_{1,2} = -\mathrm{tr}(B_2 P_{1,0}) = 3,$$
$$a_{1,3} = -\mathrm{tr}(B_3 P_{1,0}) = -1.$$

When $i = 2$,

$$P_{2,0} = B_0 P_{1,0} + a_{1,0}I = \begin{bmatrix} -1 & 1 & 0 & 0 \\ 0 & 1 & 0 & 0 \\ 0 & 0 & 0 & 0 \\ 0 & 0 & 0 & 0 \end{bmatrix}$$

$$P_{2,1} = B_1 P_{1,0} + a_{1,1}I = \begin{bmatrix} 0 & -1 & 0 & 0 \\ 0 & -3 & 1 & 0 \\ 0 & 0 & -3 & 0 \\ 0 & 0 & 0 & -3 \end{bmatrix}$$

$$P_{2,2} = B_2 P_{1,0} + a_{1,2}I = \begin{bmatrix} 0 & 1 & 1 & 1 \\ 0 & 3 & 0 & 1 \\ 0 & 0 & 3 & 0 \\ 0 & 0 & 0 & 3 \end{bmatrix}$$

$$P_{2,3} = B_3 P_{1,0} + a_{1,3}I = \begin{bmatrix} 0 & 0 & 0 & 0 \\ 0 & -1 & 0 & 0 \\ 0 & 0 & -1 & 0 \\ 0 & 0 & 0 & -1 \end{bmatrix}$$

and

$$a_{2,0} = -\frac{1}{2}\text{tr}(B_0 P_{2,0}) = -1$$

$$a_{2,1} = -\frac{1}{2}\text{tr}(B_1 P_{2,0} + B_0 P_{2,1}) = 3$$

$$a_{2,2} = -\frac{1}{2}\text{tr}(B_2 P_{2,0} + B_1 P_{2,1} + B_0 P_{2,2}) = -3$$

$$a_{2,3} = -\frac{1}{2}\text{tr}(B_3 P_{2,0} + B_2 P_{2,1} + B_1 P_{2,2} + B_0 P_{2,3}) = 1$$

$$a_{2,4} = -\frac{1}{2}\text{tr}(B_3 P_{2,1} + B_2 P_{2,2} + B_1 P_{2,3}) = 0$$

$$a_{2,5} = -\frac{1}{2}\text{tr}(B_3 P_{2,2} + B_2 P_{2,3}) = 0$$

$$a_{2,6} = -\frac{1}{2}\text{tr}(B_3 P_{2,3}) = 0$$

Finally, we obtain

$$A(x)_d = -\frac{1}{(x-1)^3} \begin{bmatrix} -(x-1)^2 & (x-1)^2 & x^2(x-1) & x^2(x^2-x+1) \\ 0 & -(x-1)^3 & -x(x-1)^3 & -x^2(x-1)^3 \\ 0 & 0 & 0 & 0 \\ 0 & 0 & 0 & 0 \end{bmatrix},$$

for $x \neq 1$. The case when $x = 1$ can be dealt with as a special case.   □

Remarks

An algorithm for computing the Moore-Penrose inverse of a polynomial matrix with two variables is presented in [4]. In [7], Karampetakis and Tzekis improved Algorithm 10.2.1 for the case when there are big gaps between the powers of x, for example, $A(x) = A_0 x^{80} + A_{79} x + A_{80}$. The above algorithms can be generalized to rational matrices by using the least common denominator of $P_i(x)$ [6].

References

1. D.K. Fadeev, V.N. Fadeeva, *Computational Methods of Linear Algebra* (W.H. Freeman & Co., Ltd., San Francisco, 1963)
2. F.R. Gantmacher, *The Theory of Matrices*, vol. 1–2 (Chelsea Publishing Co., New York, 1960)
3. H.P. Decell Jr., An application of the Cayley-Hamilton theorem to generalized matrix inversion. SIAM Rev. **7**(4), 526–528 (1965)
4. N.P. Karampetakis, Computation of the generalized inverse of a polynomial matrix and applications. Linear Algebra Appl. **252**(1), 35–60 (1997)
5. J. Gao, G. Wang, Two algorithms for computing the Drazin inverse of a polynomial matrix. J. Shanghai Teach. Univ. (Nat. Sci.) **31**(2), 31–38 (2002). In Chinese
6. J. Ji, A finite algorithm for the Drazin inverse of a polynomial matrix. Appl. Math. Comput. **130**(2–3), 243–251 (2002)
7. N.P. Karampetakis, P. Tzekis, On the computation of the generalized inverse of a polynomial matrix. IMA J. Math. Control Inform. **18**(1), 83–97 (2001)

Chapter 11
Moore-Penrose Inverse of Linear Operators

Before Moore introduced the generalized inverse of matrices by algebraic methods, Fredholm, Hilbert, Schmidt, Bounitzky, Hurwitz and other mathematicians had studied the generalized inverses of integral operators and differential operators. Recently, due to the development of science and technology and the need for practical problems, researchers are very interested in the study of the generalized inverses of linear operators in abstract spaces.

In this and the following chapter, we will introduce the concepts, properties, representation theorems and computational methods for the generalized inverses of bounded linear operators in Hilbert spaces. This chapter is based on [1], Chap. 12 contains our recent research results [2, 3].

We introduce the following notations used in these two chapters: X_1 and X_2 are Hilbert spaces over the same field; $B(X_1, X_2)$ denotes the set of bounded linear operators from X_1 to X_2; $\mathcal{R}(T)$ and $\mathcal{N}(T)$ represent the range and null space of the operator T, respectively; $\sigma(T)$ and $\sigma_r(T)$ stand for the spectrum and spectral radius of the operator T; T^* is the conjugate operator of the operator T; $T\mid_S$ is the restriction of T on the subspace S; $M^\perp$ represents the orthogonal complement of M.

11.1 Definition and Basic Properties

Suppose that X_1 and X_2 are Hilbert spaces over the same field of scalars. We consider the fundamental problem of solving a general linear equation of the type

$$T\mathbf{x} = \mathbf{b}, \tag{11.1.1}$$

where $\mathbf{b} \in X_2$ and $T \in B(X_1, X_2)$.

© Springer Nature Singapore Pte Ltd. and Science Press 2018
G. Wang et al., *Generalized Inverses: Theory and Computations*,
Developments in Mathematics 53, https://doi.org/10.1007/978-981-13-0146-9_11

The most prevalent example of an equation of the type (11.1.1) is obtained when $X_1 = \mathbb{R}^n$, $X_2 = \mathbb{R}^m$ and T is an m-by-n matrix. If $X_1 = X_2 = L^2[0, 1]$, then the integral operator defined by

$$(T\mathbf{x})(s) = \int_0^1 k(s, t)\mathbf{x}(t)dt, \quad s \in [0, 1],$$

where $k(s, t) \in L^2([0, 1] \times [0, 1])$, provides another important example.

If the inverse T^{-1} of the operator T exists, then the Eq. (11.1.1) always has the unique solution $\mathbf{x} = T^{-1}\mathbf{b}$. But in general such a linear equation may have more than one solution when $\mathcal{N}(T) \neq \{\mathbf{0}\}$, or may have no solution at all when $\mathbf{b} \notin \mathcal{R}(T)$. Even if the equation has no solution in the traditional meaning, it is still possible to assign what is in a sense of a "best possible" solution to the problem. In fact, if we let P denote the projection of X_2 onto $\mathcal{R}(T)$, then $P\mathbf{b}$ is the vector in $\mathcal{R}(T)$ which is closest to $\mathbf{b}$ and it is reasonable to consider a solution $\mathbf{u} \in X_1$ of the equation

$$T\mathbf{x} = P\mathbf{b}, \tag{11.1.2}$$

as a generalized solution of (11.1.1).

Another natural approach to assigning generalized solutions to the Eq. (11.1.1) is to find a $\mathbf{u} \in X_1$ which "comes closest" to solving (11.1.1) in the sense that

$$\|T\mathbf{u} - \mathbf{b}\| \leq \|T\mathbf{x} - \mathbf{b}\|,$$

for any $\mathbf{x} \in X_1$.

The next theorem shows the equivalence between (11.1.2) and the above problem.

Theorem 11.1.1 *Suppose $T \in B(X_1, X_2)$ has closed range $\mathcal{R}(T)$ and $\mathbf{b} \in X_2$, then the following conditions on $\mathbf{u} \in X_1$ are equivalent.*
(1) $T\mathbf{u} = P\mathbf{b}$;
(2) $\|T\mathbf{u} - \mathbf{b}\| \leq \|T\mathbf{x} - \mathbf{b}\|$ *for any $\mathbf{x} \in X_1$;*
(3) $T^*T\mathbf{u} = T^*\mathbf{b}$.

Proof (1) $\Rightarrow$ (2): Suppose $T\mathbf{u} = P\mathbf{b}$. Then by applying the Pythagorean theorem and the fact that $P\mathbf{b} - \mathbf{b} \in \mathcal{R}(T)^\perp$, we have

$$\|T\mathbf{x} - \mathbf{b}\|^2 = \|T\mathbf{x} - P\mathbf{b}\|^2 + \|P\mathbf{b} - \mathbf{b}\|^2$$
$$= \|T\mathbf{x} - P\mathbf{b}\|^2 + \|T\mathbf{u} - \mathbf{b}\|^2$$
$$\geq \|T\mathbf{u} - \mathbf{b}\|^2,$$

for any $\mathbf{x} \in X_1$.

$(2) \Rightarrow (3)$: If $\|T\mathbf{u} - \mathbf{b}\| \leq \|T\mathbf{x} - \mathbf{b}\|$ for all $\mathbf{x} \in X_1$, then again by applying the Pythagorean theorem and the equality $P\mathbf{b} = T\mathbf{x}$ for some $\mathbf{x} \in X_1$, we have

$$\|T\mathbf{u} - \mathbf{b}\|^2$$
$$= \|T\mathbf{u} - P\mathbf{b}\|^2 + \|P\mathbf{b} - \mathbf{b}\|^2$$
$$= \|T\mathbf{u} - P\mathbf{b}\|^2 + \|T\mathbf{x} - \mathbf{b}\|^2$$
$$\geq \|T\mathbf{u} - P\mathbf{b}\|^2 + \|T\mathbf{u} - \mathbf{b}\|^2.$$

Therefore $T\mathbf{u} - P\mathbf{b} = \mathbf{0}$ and

$$T\mathbf{u} - \mathbf{b} = P\mathbf{b} - \mathbf{b} \in \mathcal{R}(T)^\perp = \mathcal{N}(T^*)$$

implying that $T^*(T\mathbf{u} - \mathbf{b}) = \mathbf{0}$.
$(3) \Rightarrow (1)$: If $T^*T\mathbf{u} = T^*\mathbf{b}$, then $T\mathbf{u} - \mathbf{b} \in \mathcal{R}(T)^\perp$, therefore

$$\mathbf{0} = P(T\mathbf{u} - \mathbf{b}) = T\mathbf{u} - P\mathbf{b}.$$

This completes the proof. □

Definition 11.1.1 A vector $\mathbf{u} \in X_1$ satisfying one of the three equivalent conditions (1) to (3) of Theorem 11.1.1 is called a least squares solution of the equation $T\mathbf{x} = \mathbf{b}$.

Remark: Since $\mathcal{R}(T)$ is closed, a least squares solution of (11.1.1) exists for each $\mathbf{b} \in X_2$. Also, if $\mathcal{N}(T) \neq \{\mathbf{0}\}$, then there are infinitely many least squares solutions of (11.1.1), since if $\mathbf{u}$ is a least squares solution, then so is $\mathbf{u} + \mathbf{v}$ for any $\mathbf{v} \in \mathcal{N}(T)$.

It follows from Theorem 11.1.1 that the set of least squares solutions of (11.1.1) can be written as

$$\{\mathbf{u} \in X_1 : T^*T\mathbf{u} = T^*\mathbf{b}\},$$

which, by the continuity and linearity of T and T^*, is a closed convex set. This set contains a unique vector of minimal norm and we choose this vector to be the least squares solution uniquely associated with $\mathbf{b}$ by way of the generalized inversion process.

Definition 11.1.2 Let $T \in B(X_1, X_2)$ have closed range $\mathcal{R}(T)$. The mapping $T^\dagger : X_2 \to X_1$ defined by $T^\dagger\mathbf{b} = \mathbf{u}$, where $\mathbf{u}$ is the minimal norm least squares solution of the equation $T\mathbf{x} = \mathbf{b}$, is called the generalized inverse of T.

We will refer to Definition 11.1.2 above as the variational definition, and denote it by definition (V). Note that if the operator T is invertible, then we certainly have $T^\dagger = T^{-1}$.

The generalized inverse $T^\dagger$ has the following basic properties.

Theorem 11.1.2 *If $T \in B(X_1, X_2)$ has closed range $\mathcal{R}(T)$, then*

$$\mathcal{R}(T^\dagger) = \mathcal{R}(T^*) = \mathcal{R}(T^\dagger T).$$

Proof Let $\mathbf{b} \in X_2$. First, we show that $T^\dagger \mathbf{b} \in \mathcal{N}(T)^\perp = \mathcal{R}(T^*)$. Suppose

$$T^\dagger \mathbf{b} = \mathbf{u}_1 + \mathbf{u}_2 \in \mathcal{N}(T)^\perp \oplus \mathcal{N}(T),$$

then $\mathbf{u}_1$ is a least squares solution of $T\mathbf{x} = \mathbf{b}$, since

$$T\mathbf{u}_1 = T(\mathbf{u}_1 + \mathbf{u}_2) = TT^\dagger \mathbf{b} = P\mathbf{b}.$$

Also, if $\mathbf{u}_2 \neq \mathbf{0}$, By the Pythagorean theorem, we have

$$\|\mathbf{u}_1\|^2 < \|\mathbf{u}_1 + \mathbf{u}_2\|^2 = \|T^\dagger \mathbf{b}\|^2,$$

contradicting the fact that $T^\dagger \mathbf{b}$ is the least squares solution of minimal norm. Therefore $T^\dagger \mathbf{b} = \mathbf{u}_1 \in \mathcal{N}(T)^\perp = \mathcal{R}(T^*)$.

Second, suppose that $\mathbf{u} \in \mathcal{N}(T)^\perp$. Let $\mathbf{b} = T\mathbf{u}$. We claim that $T\mathbf{u} = PT\mathbf{u} = P\mathbf{b}$. Thus $\mathbf{u}$ is the least squares solution. Indeed, if $\mathbf{x}$ is another least squares solution, then

$$T\mathbf{x} = P\mathbf{b} = T\mathbf{u}$$

and hence $\mathbf{x} - \mathbf{u} = \widehat{\mathbf{u}} \in \mathcal{N}(T)$. It then follows that

$$\|\mathbf{x}\|^2 = \|\mathbf{x} - \mathbf{u} + \mathbf{u}\|^2 = \|\widehat{\mathbf{u}}\|^2 + \|\mathbf{u}\|^2 \geq \|\mathbf{u}\|^2.$$

Hence $\mathbf{u}$ is the least squares solution of minimal norm, that is, $\mathbf{u} = T^\dagger \mathbf{b} \in \mathcal{R}(T^\dagger)$ Thus we see that $\mathcal{R}(T^\dagger) = \mathcal{R}(T^*)$.

Note that for any $\mathbf{b} \in X_2$, $T^\dagger \mathbf{b} = T^\dagger P\mathbf{b} \in \mathcal{R}(T^\dagger T)$, thus $\mathcal{R}(T^\dagger) \subset \mathcal{R}(T^\dagger T)$. It is obvious that $\mathcal{R}(T^\dagger T) \subset \mathcal{R}(T^\dagger)$ and hence $\mathcal{R}(T^\dagger) = \mathcal{R}(T^\dagger T)$. $\quad\square$

Corollary 11.1.1 *If $T \in B(X_1, X_2)$ has closed range $\mathcal{R}(T)$, then $T^\dagger \in B(X_2, X_1)$.*

Proof First we prove that $T^\dagger$ is linear. Let $\mathbf{b}, \bar{\mathbf{b}} \in X_2$, then

$$TT^\dagger \mathbf{b} = P\mathbf{b} \quad \text{and} \quad TT^\dagger \bar{\mathbf{b}} = P\bar{\mathbf{b}}.$$

Therefore

$$TT^\dagger \mathbf{b} + TT^\dagger \bar{\mathbf{b}} = P(\mathbf{b} + \bar{\mathbf{b}}) = TT^\dagger(\mathbf{b} + \bar{\mathbf{b}})$$

and hence, by Theorem 11.1.2,

$$T^\dagger \mathbf{b} + T^\dagger \bar{\mathbf{b}} - T^\dagger(\mathbf{b} + \bar{\mathbf{b}}) \in \mathcal{N}(T)^\perp \cap \mathcal{N}(T) = \{\mathbf{0}\}.$$

Similarly, it can be shown that for any scalar α, $T^\dagger(\alpha \mathbf{b}) = \alpha T^\dagger(\mathbf{b})$.

Next we show that $T^\dagger$ is bounded. Since $\mathcal{R}(T^\dagger) = \mathcal{R}(T^*) = \mathcal{N}(T)^\perp$, there exists a positive number m such that

$$\|TT^\dagger \mathbf{b}\| \geq m\|T^\dagger \mathbf{b}\|,$$

for all $\mathbf{b} \in X_2$. Since $TT^\dagger \mathbf{b} = P\mathbf{b}$, it follows that

$$\|\mathbf{b}\| \geq \|P\mathbf{b}\| \geq m\|T^\dagger \mathbf{b}\|, \quad \text{that is,} \quad \frac{\|T^\dagger \mathbf{b}\|}{\|\mathbf{b}\|} \leq \frac{1}{m}$$

and hence $T^\dagger$ is bounded. □

Next we give some alternative definitions of the generalized inverse $T^\dagger$ and prove their equivalence.

Definition 11.1.3 If $T \in B(X_1, X_2)$ has closed range $\mathcal{R}(T)$, then $T^\dagger$ is the unique operator in $B(X_2, X_1)$ satisfying

$$(1)\ TT^\dagger = P_{\mathcal{R}(T)}; \quad (2)\ T^\dagger T = P_{\mathcal{R}(T^\dagger)}.$$

Definition 11.1.4 If $T \in B(X_1, X_2)$ has closed range $\mathcal{R}(T)$, then $T^\dagger$ is the unique operator in $B(X_2, X_1)$ satisfying

$$(1)\ TT^\dagger T = T; \quad (2)\ T^\dagger TT^\dagger = T^\dagger;$$
$$(3)\ (TT^\dagger)^* = TT^\dagger; \quad (4)\ (T^\dagger T)^* = T^\dagger T.$$

Definition 11.1.5 If $T \in B(X_1, X_2)$ has closed range $\mathcal{R}(T)$, then $T^\dagger$ is the unique operator in $B(X_2, X_1)$ satisfying

$$(1)\ T^\dagger T\mathbf{x} = \mathbf{x}, \mathbf{x} \in \mathcal{N}(T)^\perp;$$
$$(2)\ T^\dagger \mathbf{y} = \mathbf{0}, \quad \mathbf{y} \in \mathcal{R}(T)^\perp.$$

We call Definitions 11.1.3, 11.1.4, and 11.1.5 as the Moore definition, Penrose definition, and Desoer-Whalen definition, and denote them by definitions (M), (P) and (D-W), respectively. Next we will prove the uniqueness of $T^\dagger$.

Theorem 11.1.3 *There can be at most one operator $T^\dagger \in B(X_2, X_1)$ satisfying the definition* (P).

Proof From (2) and (3) of the definition (P), we have

$$T^\dagger = T^\dagger (TT^\dagger)^* = T^\dagger T^{\dagger*} T^*. \tag{11.1.3}$$

It follows from (2) and (4) of the definition (P) that

$$T^\dagger = (T^\dagger T)^* T^\dagger = T^* T^{\dagger *} T^\dagger. \tag{11.1.4}$$

By (1) and (3) of the definition (P), we have

$$T = (TT^\dagger)^* T = T^{\dagger *} T^* T \tag{11.1.5}$$

and therefore,

$$T^* = T^* TT^\dagger. \tag{11.1.6}$$

Finally, by (1) and (4) of the definition (P), we see that

$$T = TT^\dagger T = TT^* T^{\dagger *}.$$

Therefore,

$$T^* = T^\dagger TT^*. \tag{11.1.7}$$

Suppose now that X and Y are two operators in $B(X_2, X_1)$ satisfying the definition (P), then

$$
\begin{aligned}
X &= XX^* T^* & \text{by (11.1.3)} \\
&= XX^* T^* TY & \text{by (11.1.6)} \\
&= XTY & \text{by (11.1.5)} \\
&= XTT^* Y^* Y & \text{by (11.1.4)} \\
&= T^* Y^* Y & \text{by (11.1.7)} \\
&= Y & \text{by (11.1.4).}
\end{aligned}
$$

The proof is completed.   □

Theorem 11.1.4 *The definitions* (M) *and* (P) *are equivalent.*

Proof If $T^\dagger$ satisfies the definition (M), then we see that $TT^\dagger$ and $T^\dagger T$ are self-adjoint. Also,

$$TT^\dagger T = P_{\mathcal{R}(T)} T = T$$

and

$$T^\dagger TT^\dagger = P_{\mathcal{R}(T^\dagger)} T^\dagger = T^\dagger.$$

Hence $T^\dagger$ satisfies the definition (P).

Conversely, if $T^\dagger$ satisfies the definition (P), then

$$(TT^\dagger)(TT^\dagger) = T(T^\dagger TT^\dagger) = TT^\dagger.$$

Therefore $TT^\dagger$ is a self-adjoint idempotent operator and hence is a projection operator onto the subspace $S = \{\mathbf{y} : TT^\dagger\mathbf{y} = \mathbf{y}\}$. Since $TT^\dagger T = T$, it follows that $\mathcal{R}(T) \subset S$. Also, if $\mathbf{y} \in S$, then for any $\mathbf{z} \in \mathcal{N}(T^*)$,

$$(\mathbf{y}, \mathbf{z}) = (TT^\dagger\mathbf{y}, \mathbf{z}) = (T^\dagger\mathbf{y}, T^*\mathbf{z}) = \mathbf{0},$$

hence $\mathbf{y} \in \mathcal{N}(T^*)^\perp = \mathcal{R}(T)$ and $S \subset \mathcal{R}(T)$. It follows that

$$TT^\dagger = P_{\mathcal{R}(T)}.$$

It remains to show that if $T^\dagger$ satisfies the definition (P), then $T^\dagger T = P_{\mathcal{R}(T^\dagger)}$. Its proof is left as an exercise. $\qquad\qquad\square$

It follows from Theorem 11.1.4 that the definitions (M) and (P) are equivalent and we refer to these definitions as the Moore-Penrose inverse definition, denoted by definition (M-P).

Theorem 11.1.5 *The definitions* (V), (M-P) *and* (D-W) *are equivalent.*

Proof (D-W) $\Rightarrow$ (V): Suppose that $T^\dagger$ satisfies the definition (D-W). Let

$$\mathbf{b} = \mathbf{b}_1 + \mathbf{b}_2 \in \mathcal{R}(T) \oplus \mathcal{R}(T)^\perp = X_2,$$

then $TT^\dagger\mathbf{b} = TT^\dagger\mathbf{b}_1$. Since $\mathbf{b}_1 \in \mathcal{R}(T)$, we have $T\mathbf{x} = \mathbf{b}_1$ for some $\mathbf{x} \in \mathcal{N}(T)^\perp$, therefore

$$TT^\dagger\mathbf{b} = TT^\dagger T\mathbf{x} = T\mathbf{x} = \mathbf{b}_1 = P_{\mathcal{R}(T)}\mathbf{b}.$$

That is, $T^\dagger\mathbf{b}$ is a least squares solution. Suppose that $\mathbf{u}$ is another least squares solution, then $\mathbf{u} - T^\dagger\mathbf{b} \in \mathcal{N}(T)$. Also, since $\mathbf{b} = \mathbf{b}_1 + \mathbf{b}_2 \in \mathcal{R}(T) \oplus \mathcal{R}(T)^\perp$ and $T^\dagger\mathbf{b}_2 = \mathbf{0}$, from (2) of the definition (D-W), we have

$$T^\dagger\mathbf{b} = T^\dagger\mathbf{b}_1 = T^\dagger T\mathbf{x} = \mathbf{x},$$

for some $\mathbf{x} \in \mathcal{N}(T)^\perp$. Therefore, $T^\dagger\mathbf{b} \in \mathcal{N}(T)^\perp$, implying that $T^\dagger\mathbf{b} \perp (\mathbf{u} - T^\dagger\mathbf{b})$. By the Pythagorean theorem, we have $\|T^\dagger\mathbf{b}\|^2 \le \|\mathbf{u}\|^2$ and hence $T^\dagger\mathbf{b}$ is the minimal norm least squares solution, that is, $T^\dagger$ satisfies the definition (V).

(V) $\Rightarrow$ (M-P): If $T^\dagger$ satisfies the definition (V), then clearly $TT^\dagger = P_{\mathcal{R}(T)}$. Also for any $\mathbf{x} \in X_1$, by Theorem 11.1.2, we have

$$\mathbf{x} = \mathbf{x}_1 + \mathbf{x}_2 \in \mathcal{N}(T)^\perp \oplus \mathcal{N}(T) = \mathcal{R}(T^\dagger) \oplus \mathcal{N}(T).$$

Therefore
$$TT^\dagger T\mathbf{x} = TT^\dagger T\mathbf{x}_1 = P_{\mathcal{R}(T)}T\mathbf{x}_1 = T\mathbf{x}_1,$$

and hence
$$T^\dagger T\mathbf{x} - \mathbf{x}_1 \in \mathcal{N}(T) \cap \mathcal{N}(T)^\perp = \{\mathbf{0}\},$$

that is,
$$T^\dagger T\mathbf{x} = \mathbf{x}_1 = P_{\mathcal{R}(T^\dagger)}\mathbf{x},$$

which implies that $T^\dagger T = P_{\mathcal{R}(T^\dagger)}$. Thus $T^\dagger$ satisfies the definition (M-P).

(M-P) $\Rightarrow$ (D-W): Suppose that $T^\dagger$ satisfies the definition (M-P) and $\mathbf{y} \in \mathcal{R}(T)^\perp$, then
$$T^\dagger \mathbf{y} = T^\dagger TT^\dagger \mathbf{y} = T^\dagger P_{\mathcal{R}(T)}\mathbf{y} = \mathbf{0}.$$

It remains to show that $T^\dagger T\mathbf{x} = \mathbf{x}$ for $\mathbf{x} \in \mathcal{N}(T)^\perp$. Indeed, since $TT^\dagger T = T$, it follows that $T^\dagger T\mathbf{x} - \mathbf{x} \in \mathcal{N}(T)$ for any $\mathbf{x} \in X_1$. Now if $\mathbf{x} \in \mathcal{N}(T)^\perp$, by the Pythagorean theorem, we have

$$\|\mathbf{x}\|^2 \geq \|P_{\mathcal{R}(T^\dagger)}\mathbf{x}\|^2 = \|T^\dagger T\mathbf{x}\|^2 = \|T^\dagger T\mathbf{x} - \mathbf{x} + \mathbf{x}\|^2$$
$$= \|T^\dagger T\mathbf{x} - \mathbf{x}\| + \|\mathbf{x}\|^2.$$

Therefore $T^\dagger T\mathbf{x} - \mathbf{x} = \mathbf{0}$ for $\mathbf{x} \in \mathcal{N}(T)^\perp$, that is, $T^\dagger T\mathbf{x} = \mathbf{x}$. Hence $T^\dagger$ satisfies the definition (D-W). $\qquad\square$

Since $\mathcal{R}(T^\dagger) = \mathcal{N}(T)^\perp$, if $\mathbf{u}$ is a least squares solution and $\mathbf{v} \in \mathcal{N}(T)$, then $\mathbf{u} + \mathbf{v}$ is also a least squares solution. Thus the set of the least squares solution of (11.1.1) is $T^\dagger \mathbf{b} + \mathcal{N}(T)$.

If $\mathcal{R}(T)$ is closed, it is well known that $\mathcal{R}(T^*)$ is also closed. Suppose $\widetilde{T} = T^*T\big|_{\mathcal{R}(T^*)}$, we have

$$(\widetilde{T}\mathbf{x}, \ \mathbf{x}) = \|T\mathbf{x}\|^2 \geq m^2\|\mathbf{x}\|^2 \quad (m > 0),$$

for $\mathbf{x} \in \mathcal{R}(T^*)$. Thus we can define $\widetilde{T}^{-1}$ on $\mathcal{R}(\widetilde{T})$, and $\mathcal{R}(\widetilde{T}) = \mathcal{R}(T^*T) = \mathcal{R}(T^*)$. Thus $\widetilde{T}^{-1} \in B(\mathcal{R}(T^*), \mathcal{R}(T^*))$, from which it is easy to prove the following theorem.

Theorem 11.1.6 *Suppose that* $T \in B(X_1, X_2)$ *has closed range* $\mathcal{R}(T)$ *and* $\widetilde{T} = T^*T\big|_{\mathcal{R}(T^*)}$, *then*

$$T^\dagger = \widetilde{T}^{-1}T^*. \tag{11.1.8}$$

Proof From $(\widetilde{T}^{-1}T^*)T\mathbf{x} = \mathbf{x}$, for $\mathbf{x} \in \mathcal{N}(T)^\perp = \mathcal{R}(T^*)$, and $(\widetilde{T}^{-1}T^*)\mathbf{y} = \mathbf{0}$, for $\mathbf{y} \in \mathcal{R}(T)^\perp = \mathcal{N}(T^*)$, we have $\widetilde{T}^{-1}T^* = T^\dagger$ by the definition (D-W). $\qquad\square$

11.2 Representation Theorem

In this section, we will present the representation theorem of the generalized inverse $T^\dagger$ of a bounded linear operator. It states that $T^\dagger$ can be represented as the limit of a sequence of operators.

Before the representation theorem, we provide some necessary background [4]. Suppose $A \in B(X, X)$ and self-adjoint, $A^* = A$, let

$$m = \inf_{\|x\|=1} (A\mathbf{x}, \mathbf{x}) \quad \text{and} \quad M = \sup_{\|x\|=1} (A\mathbf{x}, \mathbf{x}).$$

If $S_n(x)$ is a continuous real valued function on $[m, M]$, then

$$\|S_n(x)\| = \max_{x \in \sigma(A)} |S_n(x)| = \sigma_r(S_n(A)),$$

where $\sigma(A) \subset \Omega \subset (-\infty, +\infty)$. Let $\{S_n(x)\}$ be a sequence of continuous real valued functions on Ω with $\lim_{n \to \infty} S_n(x) = S(x)$ uniformly on $\sigma(A)$, then $\lim_{n \to \infty} S_n(A) = S(A)$ uniformly on $B(X, X)$.

Suppose $T \in B(X_1, X_2)$ and $\mathcal{R}(T)$ is closed. Let $H = \mathcal{R}(T^*)$, then H is a Hilbert space. Define the operator $\widetilde{T} = T^*T \mid_H$. The spectrum of the operator $\widetilde{T} \in B(H, H)$ satisfies $\sigma(\widetilde{T}) \subset (0, +\infty)$. Indeed, $H = \mathcal{R}(T^*) = \mathcal{N}(T)^\perp$, for every $\mathbf{x} \in H$, we have

$$((T^*T \mid_H)\mathbf{x}, \mathbf{x}) = \|T\mathbf{x}\|^2 \geq m^2 \|\mathbf{x}\|^2, \quad m > 0.$$

Thus $T^*T \mid_H$ is a self-adjoint positive operator, and $\sigma(\widetilde{T}) \subset (0, +\infty)$ holds.

Theorem 11.2.1 (Representation theorem) *Suppose that the range $\mathcal{R}(T)$ of $T \in B(X_1, X_2)$ is closed and let $\widetilde{T} = T^*T \mid_H$, where $H = \mathcal{R}(T^*)$. If Ω is an open set such that $\sigma(\widetilde{T}) \subset \Omega \subset (0, +\infty)$ and $\{S_n(x)\}$ is a sequence of continuous real valued functions on Ω with $\lim_{n \to \infty} S_n(x) = x^{-1}$ uniformly on $\sigma(\widetilde{T})$, then*

$$T^\dagger = \lim_{n \to \infty} S_n(\widetilde{T})T^*,$$

where the convergence is in the uniform topology for $B(X_2, X_1)$. Furthermore,

$$\|S_n(\widetilde{T})T^* - T^\dagger\| \leq \sup_{x \in \sigma(\widetilde{T})} |x S_n(x) - 1| \|T^\dagger\|.$$

Proof Using the spectral theorem for self-adjoint linear operators, we have

$$\lim_{n \to \infty} S_n(\widetilde{T}) = \widetilde{T}^{-1}$$

uniformly in $B(H, H)$. It follows from (11.1.8) that

$$\lim_{n \to \infty} S_n(\widetilde{T})T^* = \widetilde{T}^{-1}T^* = T^\dagger.$$

To obtain the error bound, we note that $T^* = \widetilde{T}T^\dagger$ and therefore

$$S_n(\widetilde{T})T^* - T^\dagger = (S_n(\widetilde{T})\widetilde{T} - I)T^\dagger.$$

Since $\widetilde{T}$ is self-adjoint and $S_n(x)x$ is a real-valued continuous function on $\sigma(\widetilde{T})$, $S_n(\widetilde{T})\widetilde{T}$ is also self-adjoint. Using the spectral radius formula for self-adjoint operators and the spectral mapping theorem, we have

$$\begin{aligned}
\|S_n(\widetilde{T})\widetilde{T} - I\| &= |\sigma_r(S_n(\widetilde{T})\widetilde{T} - I)| \\
&= \sup_{x \in \sigma(\widetilde{T})} |S_n(x)x - 1|
\end{aligned}$$

and hence

$$\|S_n(\widetilde{T})T^* - T^\dagger\| \leq \sup_{x \in \sigma(\widetilde{T})} |S_n(x)x - 1| \, \|T^\dagger\|.$$

The proof is completed.   □

This theorem suggests that we construct an iterative process of computing a sequence that converges to the generalized inverse. This is the basis for the computational methods in the next section. Before that, we present the following bound for $\sigma(\widetilde{T})$ to be used in the computational formulas in the next section.

Theorem 11.2.2 *Suppose $T \in B(X_1, X_2)$ has closed range $\mathcal{R}(T)$ and let $\widetilde{T} = T^*T \mid_H$, where $H = \mathcal{R}(T^*)$, then for each $\lambda \in \sigma(\widetilde{T})$, we have*

$$\|T^\dagger\|^{-2} \leq \lambda \leq \|T\|^2.$$

Proof Since in the Hilbert space $H = \mathcal{R}(T^*)$,

$$\|T^*T \mid_H \| \leq \|T^*T\|,$$

we have

$$\|\widetilde{T}\| \leq \|T^*T\| \leq \|T^*\| \, \|T\| = \|T\|^2.$$

Thus

$$\lambda \leq \|\widetilde{T}\| \leq \|T\|^2.$$

On the other hand, if $\mathbf{x} \in H = \mathcal{R}(T^*)$, then

$$\begin{aligned}
\|\mathbf{x}\|^2 &= \|P_{\mathcal{R}(T^*)}\mathbf{x}\|^2 = \|T^\dagger T\mathbf{x}\|^2 \\
&\leq \|T^\dagger\|^2 \|T\mathbf{x}\|^2 = \|T^\dagger\|^2 (\widetilde{T}\mathbf{x}, \mathbf{x}).
\end{aligned}$$

Therefore

$$(\tilde{T}\mathbf{x},\ \mathbf{x}) - \|T^\dagger\|^{-2}(\mathbf{x},\ \mathbf{x}) \geq 0,$$

that is,

$$\tilde{T} \geq \|T^\dagger\|^{-2}I,$$

from which the result follows.   □

11.3 Computational Methods

In this section, we will describe Euler-Knopp methods, Newton methods, hyperpower methods and the methods based on interpolating function theory for computing the generalized inverse $T^\dagger$. These methods are based on Theorem 11.2.1 (the Representation theorem).

11.3.1 Euler-Knopp Methods

Let

$$S_n(x) = \alpha \sum_{k=0}^{n}(1 - \alpha x)^k \qquad (11.3.1)$$

be the Euler-Knopp transformation of the series $\sum_{k=0}^{\infty}(1 - x)^k$ [5], where α is a fixed parameter. Clearly,

$$\lim_{n\to\infty} S_n(x) = \frac{1}{x}$$

uniformly on compact subsets of the set

$$E_\alpha = \{x :\ |1 - \alpha x| < 1\} = \{x :\ 0 < x < 2/\alpha\}.$$

It follows from Theorem 11.2.2 that $\sigma(\tilde{T}) \subset [\,\|T^\dagger\|^{-2},\ \|T\|^2\,] \subset (0,\ \|T\|^2\,]$, if we choose the parameter α such that $0 < \alpha < 2\|T\|^{-2}$, then $\sigma(\tilde{T}) \subset (0,\ \|T\|^2\,] \subset E_\alpha$. For such a parameter, applying the Representation theorem, we obtain

$$T^\dagger = \lim_{n\to\infty} S_n(\tilde{T})T^* = \alpha \sum_{k=0}^{\infty}(I - \alpha T^*T)^k T^*.$$

Note that if we set $T_n = \alpha \sum_{k=0}^{n} (I - \alpha T^*T)^k T^*$, then we get the iterative process:

$$\begin{cases} T_0 = \alpha T^*, \\ T_{n+1} = (I - \alpha T^*T)T_n + \alpha T^*. \end{cases} \tag{11.3.2}$$

Therefore $\lim_{n \to \infty} T_n = T^\dagger$. In order to apply Theorem 11.2.1 to estimate the error between T_n and $T^\dagger$, we first estimate $|x S_n(\mathbf{x}) - 1|$. From (11.3.1),

$$\begin{cases} S_0(x) = \alpha, \\ S_{n+1}(x) = (1 - \alpha x)S_n(x) + \alpha. \end{cases} \tag{11.3.3}$$

Thus we have

$$S_{n+1}(x)x - 1 = (1 - \alpha x)(S_n(x)x - 1).$$

Therefore

$$|S_n(x)x - 1| = |1 - \alpha x|^{n+1}.$$

It follows from Theorem 11.2.2 that $\|T^\dagger\|^{-2} \leq x \leq \|T\|^2$ for $x \in \sigma(\tilde{T})$ and $0 < \alpha < 2\|T\|^{-2}$, thus

$$|1 - \alpha x| \leq \beta,$$

where

$$\beta = \max \left\{ |1 - \alpha\|T\|^2|, \ |1 - \alpha\|T^\dagger\|^{-2}| \right\}. \tag{11.3.4}$$

Therefore

$$|S_n(x)x - 1| \leq \beta^{n+1}. \tag{11.3.5}$$

Since $\|T\| \|T^\dagger\| \geq \|T T^\dagger\| = \|P\| = 1$, we have $2 > \alpha\|T\|^2 \geq \alpha\|T^\dagger\|^{-2} > 0$, which implies $0 < \beta < 1$. By Theorem 11.2.1 and (11.3.5), we get the estimate of the error

$$\|T_n - T^\dagger\| \leq \|T^\dagger\|\beta^{n+1},$$

which implies that the Euler-Knopp method defined by (11.3.2) is a first order iterative method. Next we will present a faster convergent iterative sequence.

11.3.2   Newton Methods

Suppose that for $\alpha > 0$, we define a sequence of functions $\{S_n(x)\}$ by

$$\begin{cases} S_0(x) = \alpha, \\ S_{n+1}(x) = S_n(x)(2 - x S_n(x)). \end{cases} \tag{11.3.6}$$

We will show that (11.3.6) converges to x^{-1} uniformly on the compact subset $\sigma(\widetilde{T})$ of the set $E_\alpha = \{x : 0 < x < 2/\alpha\}$. Specifically,

$$x S_{n+1}(x) - 1 = x S_n(x)(2 - x S_n(x)) - 1$$
$$= -(x S_n(x) - 1)^2.$$

Furthermore

$$|x S_n(x) - 1| = |x S_{n-1}(x) - 1|^2 = \cdots = |\alpha x - 1|^{2^n}.$$

It follows from Theorem 11.2.2 that $\|T^\dagger\|^{-2} \le x \le \|T\|^2$ for $x \in \sigma(\widetilde{T})$. Choose α such that $0 < \alpha < 2\|T\|^{-2}$, then

$$|\alpha x - 1| \le \beta < 1,$$

where β is given by (11.3.4). Thus

$$|x S_n(x) - 1| = |\alpha x - 1|^{2^n} \le \beta^{2^n} \to 0 \quad \text{as } n \to \infty. \tag{11.3.7}$$

Therefore $\lim_{n \to \infty} S_n(x) = x^{-1}$ uniformly on $\sigma(\widetilde{T})$.

Let the sequence of operators $\{S_n(\widetilde{T})\}$ defined by (11.3.6) be

$$\begin{cases} S_0(\widetilde{T}) = \alpha I, \\ S_{n+1}(\widetilde{T}) = S_n(\widetilde{T})(2I - T^* T S_n(\widetilde{T})), \end{cases}$$

then $\{S_n(\widetilde{T})\} \subset B(H, H)$, $H = \mathcal{R}(T^*)$. Applying Theorem 11.2.1, we have

$$T^\dagger = \lim_{n \to \infty} S_n(\widetilde{T}) T^*$$

uniformly on $B(X_2, X_1)$.

Setting $T_n = S_n(\widetilde{T}) T^*$, we obtain the iterative process

$$\begin{cases} T_0 = \alpha T^*, \\ T_{n+1} = T_n(2I - T T_n). \end{cases} \tag{11.3.8}$$

Therefore $\lim_{n \to \infty} T_n = T^\dagger$. Applying Theorem 11.2.1 and (11.3.7), we get the following bound for the difference between T_n and $T^\dagger$:

$$\|T_n - T^\dagger\| = \|T^\dagger\| \sup_{x \in \sigma(\widetilde{T})} |x S_n(x) - 1| \le \|T^\dagger\| \beta^{2^n}.$$

Thus the Newton method determined by (11.3.8) is a second order iterative method.

11.3.3   Hyperpower Methods

The so-called hyperpower method is a technique for extrapolating on the desirable quadratic convergence property of the Newton methods. It has the pth order convergence rate. But each iteration requires more computation than the Newton method when $p > 2$.

Given an integer $p \geq 2$, define a sequence of functions $\{S_n^p(x)\}$ by

$$
\begin{cases}
S_0^p(x) = \alpha > 0, \\
S_{n+1}^p(x) = S_n^p(x) \sum_{k=0}^{p-1}(1 - x S_n^p(x))^k.
\end{cases}
\tag{11.3.9}
$$

If $p = 2$, (11.3.9) coincides (11.3.6). Noting that

$$
|x S_{n+1}^p(x) - 1| = |x S_n^p(x) - 1|^p,
$$

we get

$$
|x S_n^p(x) - 1| = |\alpha x - 1|^{p^n}.
$$

It follows from Theorem 11.2.2 that $\|T^\dagger\|^{-2} \leq x \leq \|T\|^2$ for $x \in \sigma(\widetilde{T})$. Choose α such that $0 < \alpha < 2\|T\|^{-2}$, then

$$
|\alpha x - 1| \leq \beta < 1,
$$

where β is the same as in (11.3.4). Thus

$$
|x S_n^p(x) - 1| = |\alpha x - 1|^{p^n} \leq \beta^{p^n} \to 0 \quad \text{as } n \to \infty.
\tag{11.3.10}
$$

Therefore $\lim_{n \to \infty} S_n^p(x) = x^{-1}$ uniformly on $\sigma(\widetilde{T})$.

Let the sequence of operators $\{S_n^p(\widetilde{T})\}$ defined by (11.3.9) be

$$
\begin{cases}
S_0^p(\widetilde{T}) = \alpha I, \\
S_{n+1}^p(\widetilde{T}) = S_n^p(\widetilde{T}) \sum_{k=0}^{p-1}(I - T^*T S_n^p(\widetilde{T}))^k.
\end{cases}
$$

Then $\{S_n^p(\widetilde{T})\} \subset B(H, H)$, where $H = \mathcal{R}(T^*)$. Applying Theorem 11.2.1, we have

$$
\lim_{n \to \infty} S_n^p(\widetilde{T})T^* = T^\dagger
$$

uniformly on $B(X_2, X_1)$. Note that

$$S_{n+1}^p(\widetilde{T})T^* = S_n^p(\widetilde{T}) \sum_{k=0}^{p-1}(I - T^*TS_n^p(\widetilde{T}))^k T^*$$

$$= S_n^p(\widetilde{T})T^* \sum_{k=0}^{p-1}(I - TS_n^p(\widetilde{T})T^*)^k.$$

Thus, defining $T_n^p = S_n^p(\widetilde{T})T^*$, we get the iterative process

$$\begin{cases} T_0^p = \alpha T^*, \\ T_{n+1}^p = T_n^p \sum_{k=0}^{p-1}(I - TT_n^p)^k. \end{cases}$$

Therefore $\lim_{n\to\infty} T_n^p = T^\dagger$. Applying Theorem 11.2.1 and (11.3.10), we obtain the following bound for the difference between T_n^p and $T^\dagger$:

$$\|T_n^p - T^\dagger\| \le \|T^\dagger\| \sup_{x \in \sigma(\widetilde{T})} |xS_n^p(x) - 1| \le \|T^\dagger\|\beta^{p^n}.$$

It is easy to see that the hyperpower method is a pth order iterative method.

11.3.4 Methods Based on Interpolating Function Theory

In this subsection, we shall use the Representation theorem and the Newton interpolation and the Hermite interpolation for the function $f(x) = 1/x$ to approximate the generalized inverse $T^\dagger$ and present its error bound.

First we introduce the Newton interpolation method. If $p_n(x)$ denotes the unique polynomial of degree n which interpolates the function $f(x) = 1/x$ at the points $x = 1, 2, \ldots, n+1$, then the Newton interpolation formula [6] gives the interpolating polynomial

$$p_n(x) = \sum_{j=0}^n \binom{x-1}{j} \Delta^j f(1),$$

where Δ is the forward difference operator defined by

$$\Delta f(x) = f(x+1) - f(x), \quad \Delta^j f(x) = \Delta(\Delta^{j-1}f)(x),$$

and

$$\binom{x-1}{j} = \frac{(x-1)(x-2)\cdots(x-j)}{j!}.$$

It is easy to verify that $\Delta^j f(1) = (-1)^j (j+1)^{-1}$, which gives

$$p_n(x) = \sum_{j=0}^{n} \frac{1}{j+1} \prod_{l=0}^{j-1} \left(1 - \frac{x}{l+1}\right). \tag{11.3.11}$$

Here the product from 0 to -1 by convention is defined to be 1. A simple induction argument shows that

$$1 - x\, p_n(x) = \prod_{l=0}^{n} \left(1 - \frac{x}{l+1}\right), \quad n = 0, 1, 2, \cdots. \tag{11.3.12}$$

This statement is clearly true for $n = 0$. Assuming that it is true for n, we have

$$1 - xp_{n+1}(x) = 1 - xp_n(x) - \frac{x}{n+2} \prod_{l=0}^{n} \left(1 - \frac{x}{l+1}\right)$$

$$= \left(1 - \frac{x}{n+2}\right) \prod_{l=0}^{n} \left(1 - \frac{x}{l+1}\right)$$

$$= \prod_{l=0}^{n+1} \left(1 - \frac{x}{l+1}\right).$$

Thus (11.3.12) holds for $n + 1$. Therefore we can prove that the sequence of polynomials $\{p_n(x)\}$ satisfies $\lim_{n\to\infty} p_n(x) = 1/x$ uniformly on compact subsets of $(0, +\infty)$.

If $x > 0$, then by (11.3.12)

$$1 - x\, p_n(x) = \prod_{l=0}^{n} \left(1 - \frac{x}{l+1}\right).$$

Hence it suffices to show that

$$\lim_{n\to\infty} \prod_{l=0}^{n} \left(1 - \frac{x}{l+1}\right) = 0 \tag{11.3.13}$$

uniformly on compact subsets of $(0, +\infty)$.

If x lies in a fixed compact subset of $(0, +\infty)$, then there is a constant $K > 0$ such that

$$0 < 1 - \frac{x}{l+1} < 1 - \frac{K}{l+1},$$

therefore to establish (11.3.13) it suffices to show that

$$\prod_{l=0}^{\infty}\left(1 - \frac{x}{l+1}\right) = 0.$$

Indeed this is a consequence of a well-known fact about infinite products. Namely, if $\{a_l\}$ is a sequence of numbers with $0 < a_l < 1$ and $\sum_{j=0}^{\infty} a_j = \infty$, then the sequence $\prod_{j=0}^{n}(1 - a_j)$ is non-increasing and positive and therefore has a limit $a \geq 0$. It is easy to show by induction that

$$a_0 + \sum_{i=0}^{n} a_{i+1} \prod_{l=0}^{i}(1 - a_l) = 1 - \prod_{l=0}^{n+1}(1 - a_l)$$

and hence

$$1 \geq 1 - \prod_{l=0}^{n+1}(1 - a_l) = a_0 + \sum_{i=0}^{n} a_{i+1} \prod_{l=0}^{i}(1 - a_l) \geq a \sum_{i=0}^{n} a_i.$$

But it follows from $\sum_{j=0}^{\infty} a_j = \infty$ that $a = 0$, and so (11.3.13) holds.

Applying the Representation theorem, we conclude that

$$\lim_{n \to \infty} p_n(\widetilde{T})T^* = T^\dagger,$$

where $\widetilde{T} = T^*T|_{\mathcal{R}(T^*)}$. To phrase this result in a form suitable for computation, from (11.3.11) and (11.3.12), we note that

$$p_0(x) = 1,$$
$$p_{n+1}(x) = p_n(x) + \frac{1}{n+2} \prod_{l=0}^{n}\left(1 - \frac{x}{l+1}\right)$$
$$= p_n(x) + \frac{1}{n+2}(1 - xp_n(x)).$$

Therefore, setting $T_n = p_n(\widetilde{T})T^*$, we have the following Newton interpolation method

$$T_0 = T^*,$$
$$T_{n+1} = p_{n+1}(\widetilde{T})T^*$$
$$= p_n(\widetilde{T})T^* + \frac{1}{n+2}(T^* - \widetilde{T}p_n(\widetilde{T})T^*)$$
$$= T_n + \frac{1}{n+2}(T^* - T^*TT_n).$$

Thus $\lim_{n\to\infty} T_n = T^\dagger$. To obtain an asymptotic error bound for this method, we estimate $|1 - xp_n(x)|$. Note that for

$$x \in \sigma(\widetilde{T}) \subset [\,\|T^\dagger\|^{-2},\ \|T\|^2\,]$$

and $l \geq L = [\,\|T\|^2\,]$, we have

$$1 - \frac{x}{l+1} \leq \exp\left(-\frac{x}{l+1}\right).$$

Therefore

$$\prod_{l=L}^{n}\left(1 - \frac{x}{l+1}\right) \leq \exp\left(-x\sum_{l=L}^{n}\frac{1}{l+1}\right), \quad n \geq L.$$

Also,

$$\sum_{l=L}^{n}\frac{1}{l+1} \geq \int_{L+1}^{n+2}\frac{dt}{t} = \ln(n+2) - \ln(L+1),$$

and hence

$$\exp\left(-x\sum_{l=L}^{n}\frac{1}{l+1}\right) \leq (L+1)^x(n+2)^{-x}$$

$$= (\|T\|^2 + 1)^x(n+2)^{-x}.$$

Therefore, if we set

$$c = \max_{x\in\sigma(\widetilde{T})}\left|(1 + \|T\|^2)^x\prod_{l=0}^{L-1}\left(1 - \frac{x}{l+1}\right)\right|,$$

then it follows from (11.3.12) that

$$|1 - xp_n(x)| \leq c(n+2)^{-x}.$$

Applying the Representation theorem, we get the following bound for the difference between T_n and $T^\dagger$:

$$\|T_n - T^\dagger\| \leq c\|T^\dagger\|(n+2)^{-\|T^\dagger\|^{-2}}.$$

We now take the next natural step of investigating the use of the Hermite interpolation of the function $f(x) = 1/x$ [7]. We seek the unique polynomial $q_n(x)$ of degree $2n+1$ which satisfies $q_n(x) = 1/x$ and $q_n'(x) = -1/x^2$ at the points $x = 1, 2, \cdots, n+1$. By the Hermite interpolation formula [6]

$$q_n(x) = \sum_{i=0}^{n} (2(i+1) - x) \prod_{l=1}^{i} \left(\frac{l-x}{l+1}\right)^2. \qquad (11.3.14)$$

Here the product from 1 to 0, by convention, is defined to be 1. We first show that

$$1 - xq_n(x) = \prod_{l=0}^{n} \left(1 - \frac{x}{l+1}\right)^2, \quad n = 0, 1, 2, \cdots. \qquad (11.3.15)$$

This statement is clearly true for $n = 0$. The left-hand side of (11.3.15) equals $1 - xq_0(x) = 1 - x(2 - x) = (1 - x)^2$, which is the right-hand side of (11.3.15). Assuming the conclusion holds for n, we have

$$1 - xq_{n+1}(x)$$

$$= 1 - x\,q_n(x) - x(2(n+2) - x)\prod_{l=1}^{n+1} \left(\frac{l-x}{l+1}\right)^2$$

$$= \prod_{l=0}^{n} \left(\frac{l+1-x}{l+1}\right)^2 - ((2n+4)x - x^2)\prod_{l=0}^{n} \left(\frac{((l+1-x)/(l+1))^2}{(n+2)^2}\right)$$

$$= \prod_{l=0}^{n} \left(\frac{l+1-x}{l+1}\right)^2 \left(1 - \frac{(2n+4)x - x^2}{(n+2)^2}\right)$$

$$= \prod_{l=0}^{n+1} \left(1 - \frac{x}{l+1}\right)^2.$$

Next, similar to the sequence of polynomials $\{p_n(x)\}$, we can show that the sequence of polynomials $\{q_n(x)\}$ satisfies

$$\lim_{n \to \infty} q_n(x) = \frac{1}{x}$$

uniformly on the compact subsets of $(0, +\infty)$.

Applying the Representation theorem, we have

$$\lim_{n \to \infty} q_n(\widetilde{T})T^* = T^{\dagger},$$

where $\widetilde{T} = T^*T \,|_{\mathcal{R}(T^*)}$. To get the recurrence formula, by (11.3.14) and (11.3.15), we get

$$q_0(x) = 2 - x,$$

$$q_{n+1}(x) = q_n(x) + (2(n+2) - x) \prod_{l=1}^{n+1} \left(\frac{l-x}{l+1} \right)^2$$

$$= q_n(x) + (2(n+2) - x) \prod_{l=0}^{n} \left(\frac{((l+1-x)/(l+1))^2}{(n+2)^2} \right)$$

$$= q_n(x) + \frac{1}{n+2} \left(2 - \frac{x}{n+2} \right) (1 - x q_n(x)).$$

Setting $T_n = q_n(\widetilde{T})T^*$, we have the following Hermite interpolation method:

$$\begin{cases} T_0 = 2T^* - T^*TT^*, \\ T_{n+1} = T_n + \dfrac{1}{n+2} \left(2I - \dfrac{1}{n+2} T^*T \right) T^*(I - TT_n). \end{cases}$$

Thus $\lim_{n \to \infty} T_n = T^\dagger$. Similar to the Newton interpolation we can show that

$$\prod_{l=L}^{n} \left(1 - \frac{x}{l+1} \right)^2 \leq (1 + \|T\|^2)^{2x} (n+2)^{-2x}.$$

If we set

$$d = \max_{x \in \sigma(\widetilde{T})} \left| (1 + \|T\|^2)^{2x} \prod_{l=0}^{L-1} \left(1 - \frac{x}{l+1} \right)^2 \right|,$$

then

$$|1 - x q_n(x)| \leq d(n+2)^{-2x}.$$

Applying the Representation theorem, we get the following bound for the difference between T_n and $T^\dagger$:

$$\|T_n - T^\dagger\| \leq d \|T^\dagger\| (n+2)^{-\|T^\dagger\|^{-2}}.$$

Remarks

The steepest descent method, conjugate gradient method and Tikhonov regularization method for computing the generalized inverse $T^\dagger$ of a bounded linear operator with closed range can be found in [1]. A method for approximating infinite-dimensional Moore-Penrose inverses by finite-dimensional settings is presented in [8] and generally [9] gives various approximation methods for the generalized inverses of operators.

For a general perturbation theory of linear operators, there is an excellent book [10]. In particular, the books [11, 12] are about the generalized inverses of linear operators in Banach spaces. Perturbation analysis of the generalized inverse $T^\dagger$ and least squares solution in Hilbert space and Banach space can be found in [13–20] and perturbation analysis of oblique projections of operators can be found in [21].

Stable perturbations of operators are studied in [22]. In particular, the Moore-Penrose inverses of stable perturbation of Hilbert C^*-module operators are presented in [23].

A semi-continuity of generalized inverses in Banach algebras is presented in [24]. In [25], generalized condition numbers of bounded linear operators in Banach spaces are proposed.

Inner, outer and generalized inverses in Banach and Hilbert spaces are given in [26]. A more recent study of the outer inverse in Banach spaces can be found in [27]. The metric generalized inverses of linear operators in Banach spaces and their perturbation analysis are presented in [28, 29].

References

1. C.W. Groetsch, *Generalized Inverses of Linear Operators: Representation and Approximation* (Dekker, New York, 1977)
2. Y. Wei, S. Qiao, The representation and approximation of the Drazin inverse of a linear operator in Hilbert space. Appl. Math. Comput. **138**, 77–89 (2003)
3. V. Rakočević, Y. Wei, The representation and approximation of the W-weighted Drazin inverse of linear operators in Hilbert spaces. Appl. Math. Comput. **141**, 455–470 (2003)
4. A.E. Taylor, D.C. Lay, *Introduction to Functional Analysis*, 2nd edn. (Wiley, New York-Chichester-Brisbane, 1980)
5. G.H. Hardy, *Divergent Series* (Oxford University Press, London, 1949)
6. K.E. Atkinson, *An Introduction to Numerical Analysis*, 2nd edn. (Wiley, New York, 1989)
7. S.L. Campbell (ed.), *Recent Application of Generalized Inverses* (Pitman, London, 1982)
8. N. Du, Finite-dimensional approximation settings for infinite-dimensional Moore-Penrose inverses. SIAM J. Numer. Anal. **46**, 1454–1482 (2008)
9. J. Kuang, Approximate methods for generalized inverses of operators in Banach spaces. J. Comput. Math. **11**, 323–328 (1993)
10. T. Kato, *Perturbation Theory for Linear Operators*, 2nd edn. (Springer, New York, 1976)
11. M.Z. Nashed (ed.), *Generalized Inverses and Applications* (Academic Press, New York, 1976)
12. Y. Wang, *Generalized Inverses of Linear Operators in Banach Space* (Science Press, Beijing, 2005)
13. G. Chen, M. Wei, Y. Xue, Perturbation analysis of the least squares solution in Hilbert spaces. Linear Algebra Appl. **244**, 69–80 (1996)
14. G. Chen, Y. Xue, The expression of the generalized inverse of the perturbed operator under type I perturbation in Hilbert spaces. Linear Algebra Appl. **285**, 1–6 (1998)
15. G. Chen, Y. Xue, Perturbation analysis for the operator equation $Tx = b$ in Banach spaces. J. Math. Anal. Appl. **212**, 107–125 (1997)
16. J. Ding, J.L. Huang. Perturbation of generalized inverses in Hilbert spaces. J. Math. Anal. Appl. **198**, 506–515 (1996)
17. J. Ding, J.L. Huang, On the continuity of generalized inverses of linear operators in Hilbert spaces. Linear Algebra Appl. **262**, 229–242 (1997)
18. Y. Wei, G. Chen, Perturbation of least squares problem in Hilbert space. Appl. Math. Comput. **121**, 171–177 (2001)
19. Y. Wei, J. Ding, Representations for Moore-Penrose inverse in Hilbert space. Appl. Math. Lett. **14**, 599–604 (2001)

20. Q. Huang, J. Ma, Perturbation analysis of generalized inverses of linear operators in Banach spaces. Linear Algebra Appl. **389**, 355–364 (2004)
21. Q. Huang, On perturbations for oblique projection generalized inverses of closed linear operators in Banach spaces. Linear Algebra Appl. **434**(12), 2468–2474 (2011)
22. Y. Xue, *Stable Perturbations of Operators and Related Topics* (World Scientific Publishing Co Pte Ltd, Hackensack, NJ, 2012)
23. Q. Xu, Y. Wei, Y. Gu, Sharp norm estimations for Moore-Penrose inverses of stable perturbations of Hilbert C^*-module operators. SIAM J. Numer. Anal. **47**, 4735–4758 (2010)
24. Q. Huang, J. Ma, On the semi-continuity of generalized inverses in Banach algebras. Linear Algebra Appl. **419**, 172–179 (2006)
25. G. Chen, Y. Wei, Y. Xue, The generalized condition numbers of bounded linear operators in Banach spaces. J. Aust. Math. Soc. **76**, 281–290 (2004)
26. M.Z. Nashed, Inner, outer and generalized inverses in Banach and Hilbert spaces. Numer. Funct. Anal. Optim. **9**, 261–325 (1987)
27. X. Liu, Y. Yu, J. Zhong, Y. Wei, Integral and limit representations of the outer inverse in Banach space. Linear Multilinear Algebra **60**(3), 333–347 (2012)
28. H. Ma, H. Hudzik, Y. Wang, Continuous homogeneous selections of set-valued metric generalized inverses of linear operators in Banach spaces. Acta Math. Sin. (Engl. Ser.), **28**(1), 45–56 (2012)
29. H. Ma, S. Sun, Y. Wang, W. Zheng, Perturbations of Moore-Penrose metric generalized inverses of linear operators in Banach spaces. Acta Math. Sin. (Engl. Ser.), **30**(7), 1109–1124 (2014)

Chapter 12
Operator Drazin Inverse

Let X be a Hilbert space and $L(X)$ be the vector space of the linear operators from X into X. We denote the set of bounded linear operators from X into X by $B(X)$. In this chapter, we will investigate the definition, basic properties, representation theorem and computational methods for the Drazin inverse of an operator $T \in B(X)$, $\mathcal{R}(T^k)$ is closed, where $k = \mathrm{Ind}(T)$ is the index of T.

12.1 Definition and Basic Properties

This section introduces the definition of the Drazin inverse of an operator, its uniqueness, existence, and some basic properties.

Definition 12.1.1 Let $T \in L(X)$. If for some nonnegative integer $k \geq 0$, there exists $S \in L(X)$ such that

$$T S T^k = T^k, \tag{12.1.1}$$
$$S T S = S, \tag{12.1.2}$$
$$S T = T S, \tag{12.1.3}$$

then S is called the Drazin inverse of T and denoted by T^D. If $k = 1$, then S is called the group inverse of T and denoted by T_g.

It is easy to prove the uniqueness of the Drazin inverse of T.

Theorem 12.1.1 *If there exists the Drazin inverse of T, then it is unique.*

© Springer Nature Singapore Pte Ltd. and Science Press 2018
G. Wang et al., *Generalized Inverses: Theory and Computations*,
Developments in Mathematics 53, https://doi.org/10.1007/978-981-13-0146-9_12

Proof Suppose both S and S' satisfy (12.1.1)–(12.1.3), then

$$S = TS^2 = T^k S^{k+1} = S'T^{k+1}S^{k+1}$$
$$= (S')^{k+1}T^{2k+1}S^{k+1} = (S')^{k+1}T^{k+1}S = (S')^{k+1}T^k$$
$$= S',$$

which proves the uniqueness.   □

In order to show the existence of the Drazin inverse of an operator, we need the concepts of the ascent and descent of an operator [1].

Definition 12.1.2 Let $T \in L(X)$. If there exists the smallest nonnegative integer n such that $\mathcal{N}(T^n) = \mathcal{N}(T^{n+1})$, then this n is called the ascending index of T, denoted by $\alpha(T)$. If no such integer exists, then $\alpha(T) = \infty$.

If $\alpha(T) = n$, it is obvious that

$$\mathcal{N}(T) \subset \mathcal{N}(T^2) \subset \cdots \subset \mathcal{N}(T^n) = \mathcal{N}(T^{n+1}) = \mathcal{N}(T^{n+2}) = \cdots ,$$

where $\subset$ denotes strict inclusion.

If $\mathbf{x} \in \mathcal{N}(T^{n+2})$ then $T\mathbf{x} \in \mathcal{N}(T^{n+1}) = \mathcal{N}(T^n)$, thus $\mathbf{x} \in \mathcal{N}(T^{n+1})$ and $\mathcal{N}(T^{n+2}) \subset \mathcal{N}(T^{n+1})$, while $\mathcal{N}(T^{n+1}) \subset \mathcal{N}(T^{n+2})$ is obvious, therefore $\mathcal{N}(T^{n+1}) = \mathcal{N}(T^{n+2})$.

Definition 12.1.3 Let $T \in L(X)$. If there exists the smallest nonnegative integer n such that $\mathcal{R}(T^n) = \mathcal{R}(T^{n+1})$, then this n is called the descending index of T, denoted by $\delta(T)$. If no such integer exists, then $\delta(T) = \infty$.

When $\delta(T) = n$, it is obvious that

$$\mathcal{R}(T) \supset \mathcal{R}(T^2) \supset \cdots \supset \mathcal{R}(T^n) = \mathcal{R}(T^{n+1}) = \mathcal{R}(T^{n+2}) = \cdots .$$

Some properties of $\alpha(T)$ and $\delta(T)$ are listed in the following theorem.

Theorem 12.1.2 *Let $T \in L(X)$, then*

(1) $\alpha(T) = 0$ *if and only if T^{-1} exists;*
(2) $\delta(T) = 0$ *if and only if $\mathcal{R}(T) = X$;*
(3) *If $\alpha(T) < \infty$ and $\delta(T) < \infty$, then $\alpha(T) = \delta(T) = p$, and X has the direct sum decomposition $X = \mathcal{R}(T^p) \oplus \mathcal{N}(T^p)$;*
(4) *If X is finite dimensional, then $\alpha(T) = \delta(T) = p$ and $X = \mathcal{R}(T^p) \oplus \mathcal{N}(T^p)$.*

Proof See [1].   □

Definition 12.1.4 Let $T \in L(X)$, $\alpha(T) < \infty$, and $\delta(T) < \infty$, then the nonnegative integer $k = \alpha(T) = \delta(T)$ is called the index of the operator T, denoted by $\mathrm{Ind}(T) = k$. In particular, if T is invertible, then $\mathrm{Ind}(T) = 0$. For the zero operator 0, we adopt $\mathrm{Ind}(0) = 1$ by convention.

Now that we have defined the index of an operator, we are ready for the existence theorem of the Drazin inverse of an operator.

Theorem 12.1.3 *Let $T \in B(X)$, if $\mathrm{Ind}(T) = k < \infty$, then there exists the Drazin inverse $T^D \in L(X)$. Moreover, if $\mathcal{R}(T^k)$ is closed, then $T^D \in B(X)$.*

Proof Since $\mathrm{Ind}(T) = k$, X has the algebraic direct sum decomposition $X = \mathcal{R}(T^k) \oplus \mathcal{N}(T^k)$. Set $\tilde{T} = T|_{\mathcal{R}(T^k)}$, then $\tilde{T}$ is a one-to-one mapping from $\mathcal{R}(T^k)$ onto $\mathcal{R}(T^k)$. If $\tilde{T}\mathbf{x} = \mathbf{0}$, where $\mathbf{x} \in \mathcal{R}(T^k)$, then $\mathbf{x} \in \mathcal{N}(T^{k+1}) \cap \mathcal{R}(T^k)$, however $\mathcal{N}(T^{k+1}) = \mathcal{N}(T^k)$, thus $\mathbf{x} = \mathbf{0}$. On the other hand, for any $\mathbf{y} \in \mathcal{R}(T^k)$, since $\mathcal{R}(T^{k+1}) = \mathcal{R}(T^k)$, there exists some $\mathbf{x} \in X$ such that

$$\mathbf{y} = TT^k \mathbf{x} = \tilde{T}(T^k \mathbf{x}) \in \tilde{T}\mathcal{R}(T^k),$$

which proves the existence of $\tilde{T}^{-1}$ and $\tilde{T}^{-1} \in L(\mathcal{R}(T^k))$. Set

$$T^D = \tilde{T}^{-1}Q, \tag{12.1.4}$$

where Q is the projector along $\mathcal{N}(T^k)$ onto $\mathcal{R}(T^k)$. Next we will show that T^D is the Drazin inverse of T. Since any $\mathbf{x} \in X$ can be uniquely decomposed as

$$\mathbf{x} = \mathbf{x}_1 + \mathbf{x}_2 \in \mathcal{R}(T^k) \oplus \mathcal{N}(T^k),$$

by (12.1.4), we have

$$T^D\mathbf{x}_1 = \tilde{T}^{-1}Q\mathbf{x}_1 = \tilde{T}^{-1}\mathbf{x}_1 \quad \text{and} \quad T^D\mathbf{x}_2 = \tilde{T}^{-1}Q\mathbf{x}_2 = \mathbf{0}.$$

Consequently,

$$T^D\mathbf{x} = \tilde{T}^{-1}\mathbf{x}_1.$$

It then follows that

$$TT^DT^k\mathbf{x} = T\tilde{T}^{-1}T^k\mathbf{x}_1 = T^k\mathbf{x}_1 = T^k\mathbf{x},$$
$$T^DTT^D\mathbf{x} = T^DT\tilde{T}^{-1}\mathbf{x}_1 = T^D\mathbf{x}_1 = T^D\mathbf{x},$$
$$TT^D\mathbf{x} = T\tilde{T}^{-1}\mathbf{x}_1 = \mathbf{x}_1 = \tilde{T}^{-1}T\mathbf{x}_1 = \tilde{T}^{-1}QT\mathbf{x}_1 = T^DT\mathbf{x}.$$

Thus $T^D = \tilde{T}^{-1}Q$ is the Drazin inverse of T. If $\mathcal{R}(T^k)$ is closed, then $\tilde{T}^{-1}$ is bounded and $T^D \in B(X)$. $\qquad\square$

From the above proof, we can deduce that

$$\mathcal{R}(T^D) = \mathcal{R}(T^k) \quad \text{and} \quad \mathcal{N}(T^D) = \mathcal{N}(T^k).$$

The proof is left as an exercise.

Corollary 12.1.1 *Let $T \in B(X)$. If there exists the Drazin inverse T^D, for some nonnegative integer k, satisfying (12.1.1)–(12.1.3), then the smallest k is the index of T.*

Proof For $\mathbf{x} \in \mathcal{R}(T^k)$, there exists some $\mathbf{y} \in X$ such that $\mathbf{x} = T^k\mathbf{y}$. Due to the existence of the Drazin inverse of T, which satisfies (12.1.1)–(12.1.3), we know $\mathbf{x} = T^k\mathbf{y} = T^{k+1}T^D\mathbf{y} \in \mathcal{R}(T^{k+1})$, therefore $\mathcal{R}(T^k) = \mathcal{R}(T^{k+1})$. On the other hand, for $\mathbf{x} \in \mathcal{N}(T^{k+1})$, $T^k\mathbf{x} = T^D T^{k+1}\mathbf{x} = \mathbf{0}$, that is, $\mathbf{x} \in \mathcal{N}(T^k)$, thus $\mathcal{N}(T^k) = \mathcal{N}(T^{k+1})$.

It follows from Definition 12.1.4 that $\mathrm{Ind}(T) \leq k$, but $\mathrm{Ind}(T)$ cannot be less than k. Indeed, if $\mathrm{Ind}(T) = l < k$, by Theorem 12.1.3, there exists a Drazin inverse of T satisfying (12.1.1)–(12.1.3), which contradicts the assumption that k is the smallest, so we have $\mathrm{Ind}(T) = k$. $\square$

Corollary 12.1.2 *Let $T \in B(X)$ and $\mathrm{Ind}(T) = k$, then the Drazin inverse of T is the unique linear operator satisfying*

$$T^D T\mathbf{x} = TT^D\mathbf{x} = \mathbf{x}, \quad \mathbf{x} \in \mathcal{R}(T^k), \tag{12.1.5}$$

$$T^D\mathbf{y} = \mathbf{0}, \quad \mathbf{y} \in \mathcal{N}(T^k). \tag{12.1.6}$$

Proof In fact, (12.1.5) is equivalent to $T^D = (T|_{\mathcal{R}(T^k)})^{-1}$ on $\mathcal{R}(T^k)$ and (12.1.6) means that T^D maps $\mathcal{N}(T^k)$ into the zero element. Since $X = \mathcal{R}(T^k) \oplus \mathcal{N}(T^k)$, such T^D is consistent with the T^D in Theorem 12.1.3. $\square$

Now, we present some basic properties of the Drazin inverse of an operator.

Theorem 12.1.4 *Let $T \in B(X)$, $\mathrm{Ind}(T) = k$, and $\mathcal{R}(T^k)$ be closed, then $T^D T = TT^D$ is the projector along $\mathcal{N}(T^k) = \mathcal{N}(T^D)$ onto $\mathcal{R}(T^k) = \mathcal{R}(T^D)$, that is,*

$$T^D T = TT^D = P_{\mathcal{R}(T^k),\mathcal{N}(T^k)} = P_{\mathcal{R}(T^D),\mathcal{N}(T^D)}.$$

Proof Since T^D satisfies (12.1.2), $T^D T = TT^D$ is idempotent and

$$T^D T = TT^D = P_{\mathcal{R}(TT^D),\mathcal{N}(TT^D)}.$$

Since $T^D = (T^D T)T^D$, we have $\mathcal{R}(T^D) \subset \mathcal{R}(T^D T)$, while $\mathcal{R}(T^D T) \subset \mathcal{R}(T^D)$ is obvious, thus $\mathcal{R}(T^D) = \mathcal{R}(T^D T) = \mathcal{R}(TT^D)$. Similarly, we can prove $\mathcal{N}(T^D) = \mathcal{N}(TT^D)$. $\square$

Definition 12.1.5 Let $T \in B(X)$, $\mathrm{Ind}(T) = k$, and $\mathcal{R}(T^k)$ be closed, we call the product $TT^D T$ the core part of T, denoted by C_T. Let $N_T = T - C_T$, then

$$T = C_T + N_T$$

is the core-nilpotent decomposition of T.

It follows from Definition 12.1.5 that N_T is the nilpotent operator with index k, since

$$(N_T)^k = (T - TT^D T)^k = T^k(I - T^D T) = O,$$

and

$$(N_T)^l = T^l(I - TT^D) \neq O, \quad \text{for} \quad l < k.$$

Theorem 12.1.5 *Let $T \in B(X)$, $\text{Ind}(T) = k$, and $\mathcal{R}(T^k)$ be closed, then*

(1) $\text{Ind}(T^D) = \text{Ind}(C_T) = 1$, *when* $\text{Ind}(T) \geq 1$; *0, when* $\text{Ind}(T) = 0$;
(2) $N_T C_T = C_T N_T = O$;
(3) $N_T T^D = T^D N_T = O$;
(4) $C_T TT^D = TT^D C_T = C_T$;
(5) $(T^D)^D = C_T$;
(6) $T = C_T$ *if and only if* $\text{Ind}(T) \leq 1$;
(7) $((T^D)^D)^D = T^D$;
(8) $T^D = (C_T)^D$;
(9) $(T^D)^p = (T^p)^D$, *where p is an arbitrary positive integer;*
(10) $(T^D)^* = (T^*)^D$.

Proof It is left as an exercise. See [2]. □

The above content is presented in [2], which introduces the concept of the Drazin inverse of a linear operator, discusses the existence and uniqueness and some basic properties, also studies its relationship with other generalized inverses, however, it does not include the corresponding representation theorem and computational methods presented in the following sections. In [3], these problems are partially addressed.

12.2 Representation Theorem

In [4], a unified representation theorem of the Drazin inverse of a linear operator in Hilbert space is given. First, we give an expression for T^D, which is different from (12.1.4).

Theorem 12.2.1 *Let $T \in B(X)$ with $\text{Ind}(T) = k$ and $\mathcal{R}(T^k)$ be closed, then*

$$T^D = \widetilde{T}^{-1} T^k T^{*2k+1} T^k,$$

where $\widetilde{T} = (T^k T^{*2k+1} T^{k+1})|_{\mathcal{R}(T^k)}$ is the restriction of $T^k T^{*2k+1} T^{k+1}$ on $\mathcal{R}(T^k)$.

Proof It follows from [5, p. 247] that

$$T^D = T^k (T^{2k+1})^\dagger T^k,$$

where $(T^{2k+1})^\dagger$ is the Moore-Penrose inverse of T^{2k+1}. Also, it is easy to prove that

$$\mathcal{R}(T^k T^{*2k+1} T^k) = \mathcal{R}(T^D) \quad \text{and} \quad \mathcal{N}(T^k T^{*2k+1} T^k) = \mathcal{N}(T^D).$$

The conclusion then follows from [6, Theorem 2] and [7, Lemma 3.1]. $\square$

Remark The above theorem is a generalization of a result in [3] in that the conditions $\mathcal{N}(T^k) \subset \mathcal{N}(T^{k*})$ and $\mathcal{R}(T^k) \subset \mathcal{R}(T^{k*})$ required in [3] are removed. Also, this theorem generalizes Corollary 2.1 in [8] from matrices to linear operators.

Now we are ready to give the representation theorem.

Theorem 12.2.2 *Let $T \in B(X)$ with $\mathrm{Ind}(T) = k$ and $\mathcal{R}(T^k)$ be closed, and define $\widetilde{T} = (T^k T^{*2k+1} T^{k+1})|_{\mathcal{R}(T^k)}$. If Ω is an open set such that $\sigma(\widetilde{T}) \subset \Omega \subset (0, \infty)$ and $\{S_n(x)\}$ is a sequence of continuous real valued functions on Ω with $\lim_{n\to\infty} S_n(x) = 1/x$ uniformly on $\sigma(\widetilde{T})$, then*

$$T^D = \lim_{n\to\infty} S_n(\widetilde{T}) T^k T^{*2k+1} T^k.$$

Furthermore, for any $\epsilon > 0$, there is an operator norm $\| \cdot \|_*$ on X such that

$$\frac{\| S_n(\widetilde{T}) T^k T^{*2k+1} T^k - T^D \|_*}{\| T^D \|_*} \leq \max_{x \in \sigma(\widetilde{T})} |S_n(x)x - 1| + O(\epsilon). \qquad (12.2.1)$$

Proof It follows from [9] that

$$\sigma(T^k T^{*2k+1} T^{k+1}) = \sigma((T^{2k+1})^* (T^{2k+1}))$$

is nonnegative. Thus the spectrum of $\widetilde{T}$ is positive since $\widetilde{T}$ is nonsingular. Using [10, Theorem 10.27], we have

$$\lim_{n\to\infty} S_n(\widetilde{T}) = \widetilde{T}^{-1}$$

uniformly in $B(\mathcal{R}(T^k))$. It then follows from Theorem 12.2.1 that

$$\lim_{n\to\infty} S_n(\widetilde{T}) T^k T^{*2k+1} T^k = \widetilde{T}^{-1} T^k T^{*2k+1} T^k = T^D.$$

To obtain the error bound (12.2.1), we note that $T^k T^{*2k+1} T^k = \widetilde{T} T^D$. Therefore,

$$S_n(\widetilde{T}) T^k T^{*2k+1} T^k - T^D = (S_n(\widetilde{T})\widetilde{T} - I) T^D.$$

Also, for any $\epsilon > 0$, there is an operator norm $\| \cdot \|_*$ such that $\|T\|_* \leq \rho(T) + \epsilon$, see [11, p. 77]. Thus

$$\|S_n(\widetilde{T})T^k T^{*2k+1} T^k - T^D\|_*$$
$$\leq \|S_n(\widetilde{T})\widetilde{T} - I\|_* \|T^D\|_*$$
$$\leq (\max_{x \in \sigma(\widetilde{T})} |S_n(x)x - 1| + O(\epsilon))\|T^D\|_*,$$

which completes the proof. $\square$

To derive the specific error bounds, we need lower and upper bounds for $\sigma(\widetilde{T})$ given by the following theorem.

Theorem 12.2.3 *Let $T \in B(X)$ with $\mathrm{Ind}(T) = k$ and $\mathcal{R}(T^k)$ be closed. Define $\widetilde{T} = (T^k T^{*2k+1} T^{k+1})|_{\mathcal{R}(T^k)}$, then, for any $\lambda \in \sigma(\widetilde{T})$,*

$$\|(T^{2k+1})^\dagger\|^{-2} \leq \lambda \leq \|T\|^{4k+2}.$$

Proof For any $\lambda \in \sigma(\widetilde{T})$,

$$0 < \lambda \in \sigma(\widetilde{T}) \subset \sigma(T^k T^{*2k+1} T^{k+1}) = \sigma((T^{2k+1})^*(T^{2k+1})).$$

It is obvious that

$$\mathrm{Ind}((T^{2k+1})^* T^{2k+1}) = 1$$

and

$$\lambda^{-1} \in \sigma(((T^{2k+1})^* T^{2k+1})_g) = \sigma(((T^{2k+1})^* T^{2k+1})^\dagger)$$
$$= \sigma((T^{2k+1})^\dagger (T^{2k+1})^{\dagger*}).$$

It then follows that

$$\lambda^{-1} \leq \|(T^{2k+1})^\dagger (T^{2k+1})^{\dagger*}\| = \|(T^{2k+1})^\dagger\|^2,$$

that is,

$$\lambda \geq \|(T^{2k+1})^\dagger\|^{-2}.$$

On the other hand, since

$$\|T^k T^{*2k+1} T^{k+1}\| \geq \|(T^k T^{*2k+1} T^{k+1})|_{\mathcal{R}(T^k)}\|,$$

we get $\|\widetilde{T}\| \leq \|T\|^{4k+2}$. Thus $\lambda \leq \|T\|^{4k+2}$, for all $\lambda \in \sigma(\widetilde{T})$. $\square$

12.3   Computational Procedures

In this section, we apply Theorem 12.2.2 to five specific cases to derive five specific representations and five computational procedures for the Drazin inverse of a linear operator in Hilbert space and their corresponding error bounds.

12.3.1   Euler-Knopp Method

Consider the following sequence:

$$S_n(x) = \alpha \sum_{j=0}^{n} (1 - \alpha x)^j,$$

which can be viewed as the Euler-Knopp transform of the series $\sum_{n=0}^{\infty}(1 - x)^n$. Clearly $\lim_{n \to \infty} S_n(x) = 1/x$ uniformly on any compact subset of the set

$$E_\alpha = \{x : |1 - \alpha x| < 1\} = \{x : 0 < x < 2/\alpha\}.$$

By Theorem 12.2.3, we get

$$\sigma(\widetilde{T}) \subset \left[\|(T^{2k+1})^\dagger\|^{-2}, \ \|T\|^{4k+2} \right] \subset (0, \ \|T\|^{4k+2}].$$

If we choose the parameter α, $0 < \alpha < 2\|T\|^{-(4k+2)}$, such that $\sigma(\widetilde{T}) \subset (0, \|T\|^{4k+2}] \subset E_\alpha$, then we have the following representation of the Drazin inverse:

$$T^D = \alpha \sum_{n=0}^{\infty} (I - \alpha T^k T^{*2k+1} T^{k+1})^n T^k T^{*2k+1} T^k.$$

Setting

$$T_n = \alpha \sum_{j=0}^{n} (I - \alpha T^k T^{*2k+1} T^{k+1})^j T^k T^{*2k+1} T^k,$$

we have the following iterative procedure for the Drazin inverse:

$$\begin{cases} T_0 = \alpha T^k T^{*2k+1} T^k, \\ T_{n+1} = (I - \alpha T^k T^{*2k+1} T^{k+1}) T_n + \alpha T^k T^{*2k+1} T^k. \end{cases}$$

Therefore $\lim_{n \to \infty} T_n = T^D$. For the error bound, we note that the sequence $\{S_n(x)\}$ satisfies

$$S_{n+1}(x)x - 1 = (1 - \alpha x)(S_n(x)x - 1).$$

Thus

$$|S_n(x)x - 1| = |1 - \alpha x|^n \, |S_0(x)x - 1| = |1 - \alpha x|^{n+1}.$$

If $x \in \sigma(\widetilde{T})$ and $0 < \alpha < 2/\|T\|^{4k+2}$, then we see that $|1 - \alpha x| \le \beta < 1$, where

$$\beta = \max \left\{ |1 - \alpha\|T\|^{4k+2}|, \ |1 - \alpha\|(T^{2k+1})^\dagger\|^{-2}| \right\}. \qquad (12.3.1)$$

Therefore,

$$|S_n(x)x - 1| \le \beta^{n+1} \to 0, \quad \text{as } n \to \infty.$$

It follows from the above inequality and Theorem 12.2.2 that the error bound is

$$\frac{\|T_n - T^D\|_*}{\|T^D\|_*} \le \beta^{n+1} + O(\epsilon).$$

12.3.2 Newton Method

Suppose that for $\alpha > 0$, we define a sequence $\{S_n(x)\}$ of functions by

$$\begin{cases} S_0(x) = \alpha, \\ S_{n+1}(x) = S_n(x)(2 - x S_n(x)). \end{cases} \qquad (12.3.2)$$

Clearly, the sequence (12.3.2) satisfies

$$x S_{n+1}(x) - 1 = -(x S_n(x) - 1)^2.$$

Iterating on the above equality, we have

$$|x S_n(x) - 1| = |\alpha x - 1|^{2^n} \le \beta^{2^n} \to 0, \quad \text{as } n \to \infty,$$

for $0 < \alpha < 2/\|T\|^{4k+2}$, where β is given by (12.3.1).

One attractive feature of the Newton method is its quadratic rate of convergence in general. Using the above argument combined with Theorem 12.2.2, we see that the sequence $\{S_n(\widetilde{T})\}$ defined by

$$\begin{cases} S_0(\widetilde{T}) = \alpha I, \\ S_{n+1}(\widetilde{T}) = S_n(\widetilde{T})(2I - \widetilde{T} S_n(\widetilde{T})) \end{cases}$$

has the property that

$$\lim_{n \to \infty} S_n(\widetilde{T}) T^k T^{*2k+1} T^k = T^D.$$

Setting $T_n = S_n(\widetilde{T})T^k T^{*2k+1} T^k$, we have the following iterative procedure for the Drazin inverse:

$$\begin{cases} T_0 = \alpha T^k T^{*2k+1} T^k, \\ T_{n+1} = T_n(2I - TT_n). \end{cases}$$

For the error bound, we have

$$\frac{\|T_n - T^D\|_*}{\|T^D\|_*} \leq \beta^{2^n} + O(\epsilon).$$

12.3.3   Limit Expression

We give another limit expression of the Drazin inverse given by Meyer [12]. Specifically, for $k = \mathrm{Ind}(T)$,

$$T^D = \lim_{t \to 0^+} (tI + T^{k+1})^{-1} T^k.$$

It can be rewritten as

$$T^D = \lim_{t \to 0^+} (tI + T^k T^{*2k+1} T^{k+1})^{-1} T^k T^{*2k+1} T^k.$$

Setting $S_t(x) = (t + x)^{-1}$ $(t > 0)$, for $x \in \sigma(\widetilde{T})$, we can derive the following error bound for this method:

$$|x S_t(x) - 1| = \frac{t}{x + t} \leq \frac{t}{\|(T^{2k+1})^\dagger\|^{-2} + t} = \frac{\|(T^{2k+1})^\dagger\|^2 t}{1 + \|(T^{2k+1})^\dagger\|^2 t}.$$

Therefore, from Theorem 12.2.2, we have the error bound for the limit expression of the Drazin inverse

$$\frac{\|(tI + T^k T^{*2k+1} T^{k+1})^{-1} T^k T^{*2k+1} T^k - T^D\|_*}{\|T^D\|_*}$$

$$\leq \frac{\|(T^{2k+1})^\dagger\|^2 t}{1 + \|(T^{2k+1})^\dagger\|^2 t} + O(\epsilon).$$

The methods we have considered so far are based on approximating the function $f(x) = 1/x$. Next, we will apply Theorem 12.2.2 to polynomial interpolations of the function $f(x) = 1/x$ to derive iterative methods for computing T^D and their corresponding asymptotic error bounds.

12.3.4 Newton Interpolation

Let $P_n(x)$ denote the unique polynomial of degree n which interpolates the function $f(x) = 1/x$ at the points $x = 1, 2, \ldots, n+1$.

Just as discussed in Sect. 11.3.4, we have

$$P_n(x) = \sum_{j=0}^{n} \frac{1}{j+1} \prod_{l=0}^{j-1} \left(1 - \frac{x}{l+1}\right). \qquad (12.3.3)$$

It is easy to verify that

$$1 - x P_n(x) = \prod_{l=0}^{n} \left(1 - \frac{x}{l+1}\right), \quad n = 0, 1, 2, \ldots, \qquad (12.3.4)$$

and the polynomials $\{P_n(x)\}$ in (12.3.3) satisfy $\lim_{n \to \infty} P_n(x) = 1/x$ uniformly on any compact subset of $(0, \infty)$. It follows from Theorem 12.2.2 that

$$\lim_{n \to \infty} P_n(\widetilde{T}) T^k T^{*2k+1} T^k = T^D,$$

where $\widetilde{T} = (T^k T^{*2k+1} T^{k+1})|_{\mathcal{R}(T^k)}$.

In order to phrase this result in a form suitable for computation, we derive

$$\begin{cases} P_0(x) = 1, \\ P_{n+1}(x) = P_n(x) + \dfrac{1}{n+2}(1 - x P_n(x)). \end{cases}$$

Therefore, setting $T_n = P_n(\widetilde{T}) T^k T^{*2k+1} T^k$, we have the following iterative method for computing the Drazin inverse T^D:

$$\begin{cases} T_0 = T^k T^{*2k+1} T^k, \\ T_{n+1} = P_{n+1}(\widetilde{T}) T^k T^{*2k+1} T^k = T_n + \dfrac{T_0}{n+2}(I - T T_n). \end{cases}$$

So, $\lim_{n \to \infty} T_n = T^D$.

To derive an asymptotic error bound for this method, note that for

$$x \in \sigma(\widetilde{T}) \subset \left[\|(T^{2k+1})^\dagger\|^{-2}, \ \|T\|^{4k+2} \right]$$

and for $l \geq L = \left[\|T\|^{4k+2} \right]$, we have

$$1 - \frac{x}{l+1} \leq \exp\left(-\frac{x}{l+1}\right) \quad \text{for all } x \in \sigma(\widetilde{T}).$$

Therefore,

$$\prod_{l=L}^{n}\left(1-\frac{x}{l+1}\right) \leq \exp\left(-x\sum_{l=L}^{n}\frac{1}{l+1}\right), \quad n \geq L.$$

Also,

$$\sum_{l=L}^{n}\frac{1}{l+1} \geq \int_{L+1}^{n+2}\frac{dt}{t}$$

$$= \ln(n+2) - \ln(L+1)$$

and hence

$$\exp\left(-x\sum_{l=L}^{n}\frac{1}{l+1}\right) \leq (L+1)^x(n+2)^{-x}$$

$$= (1+\|T\|^{4k+2})^x(n+2)^{-x}.$$

If we set the constant

$$C = \max_{x\in\sigma(\widetilde{T})}\left|(1+\|T\|^{4k+2})^x\prod_{l=0}^{L-1}\left(1-\frac{x}{l+1}\right)\right|,$$

then from (12.3.3),

$$|1 - xP_n(x)| \leq C(n+2)^{-x}.$$

Finally, it follows from Theorem 12.2.2 that

$$\frac{\|T_n - T^D\|_*}{\|T^D\|_*} \leq C(n+2)^{-\|(T^{2k+1})^\dagger\|^{-2}} + O(\epsilon)$$

for sufficiently large n.

12.3.5 Hermite Interpolation

We consider approximating the Drazin inverse T^D by the Hermite interpolation of the function $f(x) = 1/x$ and deriving its asymptotic error bound.

Consider the unique polynomial $q_n(x)$ of degree $2n+1$ satisfying

$$q_n(i) = \frac{1}{i} \quad \text{and} \quad q'_n(i) = -\frac{1}{i^2}, \quad i = 1, 2, \ldots, n+1,$$

then the Hermite interpolation formula yields the representation

$$q_n(x) = \sum_{i=0}^{n} (2(i+1) - x) \prod_{l=1}^{i} \left(\frac{l-x}{l+1} \right)^2, \qquad (12.3.5)$$

where, by convention, the product term equals 1 when $l = 0$.

From the definition of $q_n(x)$ in (12.3.5), an inductive argument gives

$$1 - xq_n(x) = \prod_{i=0}^{n} \left(1 - \frac{x}{l+1} \right)^2.$$

The polynomials $q_n(x)$ in (12.3.5) satisfy $\lim_{n\to\infty} q_n(x) = 1/x$ uniformly on any compact subset of $(0, +\infty)$. It follows from Theorem 12.2.2 that

$$\lim_{n\to\infty} q_n(\widetilde{T}) T^k T^{*2k+1} T^k = T^D,$$

where $\widetilde{T} = (T^k T^{*2k+1} T^{k+1})|_{\mathcal{R}(T^k)}$.

Let

$$\begin{cases} q_0(x) = 2 - x, \\ q_{n+1}(x) = q_n(x) + \dfrac{1}{n+2} \left(2 - \dfrac{x}{n+2} \right) (1 - xq_n(x)), \end{cases}$$

and

$$T_n = q_n(\widetilde{T}) T^k T^{*2k+1} T^k.$$

We obtain the following iterative method for computing the Drazin inverse T^D:

$$\begin{cases} T_0 = (2I - MT)M, \\ T_{n+1} = T_n + \dfrac{1}{n+2} \left(2I - \dfrac{1}{n+2} MT \right) M(I - TT_n), \end{cases}$$

where $M = T^k T^{*2k+1} T^k$ and $\widetilde{T} = MT$.

Similar to the Newton interpolation method, we can establish the error bound as follows. For

$$l \geq L = \left[\|T\|^{4k+2} \right] \quad \text{and} \quad x \in \sigma(\widetilde{T}) \subset \left[\|(T^{2k+1})^\dagger\|^{-2}, \|T\|^{4k+2} \right],$$

we have

$$\prod_{l=L}^{n} \left(1 - \frac{x}{l+1} \right)^2 \leq (1 + \|T\|^{4k+2})^{2x} (n+2)^{-2x}.$$

Define the constant

$$d = \max_{x \in \sigma(\tilde{T})} (1 + \|T\|^{4k+2})^{2x} \prod_{l=0}^{L-1} \left(1 - \frac{x}{l+1}\right)^2,$$

then

$$|1 - xq_n(x)| \le d(n+2)^{-2x}.$$

By Theorem 12.2.2, we arrive at the error bound

$$\frac{\|T_n - T^D\|_*}{\|T^D\|_*} \le d(n+2)^{-2\|(T^{2k+1})^\dagger\|^{-2}} + O(\epsilon),$$

for sufficiently large n.

12.4 Perturbation Bound

The perturbation properties of the Drazin inverse of a matrix were investigated by Wei [13] and Wei and Wang [14] (see also Sects. 9.1 and 9.5 of Chap. 9).

In this section we study the perturbation of the generalized Drazin inverse introduced by Koliha [15]. We start with the Banach algebra setting, then move to bounded linear operators.

We denote by $\mathcal{A}$ a complex Banach algebra with identity 1. For an element $a \in \mathcal{A}$ we denote by $\sigma(a)$ the spectrum of a. We write $acc\ \sigma(a)$ for the set of all accumulation points of $\sigma(a)$. By $qNil(\mathcal{A})$ we denote the set of all quasi-nilpotent elements of $\mathcal{A}$. An element x is called a quasinilpotent element of $\mathcal{A}$ if x commutes with any $a \in \mathcal{A}$ and $1 - xa \in \text{Inv}(\mathcal{A})$, the set of all invertible elements in $\mathcal{A}$.

Definition 12.4.1 ([15]) Let $a \in \mathcal{A}$, we say that a is Drazin invertible if there exists an element $b \in \mathcal{A}$ such that

$$ab = ba, \quad ab^2 = b, \quad \text{and} \quad a^2 b - a \in qNil(\mathcal{A}).$$

If such b exists, it is unique [15], it is called the generalized Drazin inverse of a, and denoted by a^D. If $a^2 b - a$ is in fact nilpotent, then a^D is the standard Drazin inverse of a. The Drazin index $\text{ind}(a)$ of a is equal to k if $a^2 b - a$ is nilpotent of index k, otherwise, $\text{ind}(a) = \infty$. If $\text{ind}(a) = 1$, then a^D is denoted by a_g and called the group inverse of a. From this point on we use the term "Drazin inverse" for "generalized Drazin inverse". Recall [15] that a has a Drazin inverse if and only if $0 \notin acc\ \sigma(a)$.

Let $a \in \mathcal{A}$ be Drazin invertible. Following [14], we say that $b \in \mathcal{A}$ obeys Condition $(\mathcal{W})$ at a if

$$b - a = aa^D(b-a)aa^D \quad \text{and} \quad \|a^D(b-a)\| < 1. \tag{12.4.1}$$

We remark that the condition

$$b - a = aa^D(b - a)aa^D$$

is equivalent to the condition

$$b - a = aa^D(b - a) = (b - a)aa^D. \tag{12.4.2}$$

Basic auxiliary results are summarized in the following lemma (see also [14, Theorem 3.1 and 3.2]). For the sake of completeness we include a proof.

Lemma 12.4.1 *Let $a \in \mathcal{A}$ be Drazin invertible and $b \in \mathcal{A}$ obey Condition $(\mathcal{W})$ at a, then*

(1) $b = a(1 + a^D(b - a))$;
(2) $b = (1 + (b - a)a^D)a$;
(3) $1 + a^D(b - a)$ *and* $1 + (b - a)a^D$ *are invertible, and*

$$(1 + a^D(b - a))^{-1}a^D = a^D(1 + (b - a)a^D)^{-1}. \tag{12.4.3}$$

Proof To prove (1) and (2) let us remark that by (12.4.2) we have

$$b = a + (b - a) = a + aa^D(b - a) = a(1 + a^D(b - a))$$

and

$$b = a + (b - a) = a + (b - a)a^Da = (1 + (b - a)a^D)a.$$

Clearly, the condition $\|a^D(b - a)\| < 1$ implies that $1 + a^D(b - a)$ and $1 + (b - a)a^D$ are invertible. Finally, (12.4.3) follows by direct verification. $\square$

Now we show the main result of this section.

Theorem 12.4.1 *Let $a \in \mathcal{A}$ be Drazin invertible and $b \in \mathcal{A}$ obey Condition $(\mathcal{W})$ at a, then b is Drazin invertible and*

$$bb^D = aa^D, \quad b^D = (1 + a^D(b - a))^{-1}a^D = a^D(1 + (b - a)a^D)^{-1},$$

and

$$\mathrm{ind}(a) = \mathrm{ind}(b).$$

Proof By Lemma 12.4.1 (3), we know that $1 + a^D(b - a)$ and $1 + (b - a)a^D$ are invertible and

$$(1 + a^D(b - a))^{-1}a^D = a^D(1 + (b - a)a^D)^{-1}.$$

Setting $\widetilde{b} = (1 + a^D(b - a))^{-1}a^D = a^D(1 + (b - a)a^D)^{-1}$, we prove that b is Drazin invertible and $b^D = \widetilde{b}$. First we prove that b and $\widetilde{b}$ are commutable. By Lemma 12.4.1 (1) we have

$$b\widetilde{b} = a(1 + a^D(b - a))(1 + a^D(b - a))^{-1}a^D = aa^D \qquad (12.4.4)$$

and

$$\widetilde{b}b = a^D(1 + (b - a)a^D)^{-1}(1 + (b - a)a^D)a = a^Da. \qquad (12.4.5)$$

Hence, we get

$$b\widetilde{b} = \widetilde{b}b.$$

Therefore

$$\begin{aligned}
\widetilde{b} - b\widetilde{b}^2 &= \widetilde{b}(1 - b\widetilde{b}) \\
&= (1 + a^D(b - a))^{-1}a^D(1 - a^Da) \\
&= 0
\end{aligned}$$

and $b\widetilde{b}^2 = \widetilde{b}$. Finally, using (12.4.2) and (12.4.4), we get

$$\begin{aligned}
b - b^2\widetilde{b} &= b(1 - b\widetilde{b}) \\
&= b(1 - aa^D) \\
&= a(1 - aa^D) + (b - a)(1 - aa^D) \\
&= a - a^2a^D, \qquad (12.4.6)
\end{aligned}$$

which is quasinilpotent. We conclude that b is Drazin invertible with $b^D = \widetilde{b}$. The Eqs. (12.4.4) and (12.4.5) show that $bb^D = aa^D$. From (12.4.6) we conclude that $\mathrm{ind}(a) = \mathrm{ind}(b)$.   □

We remark that the known result for matrices [14, Theorem 3.2 and Corollaries 3.1 and 3.2] is a direct corollary of Theorem 12.4.1.

Corollary 12.4.1 *Let $a \in \mathcal{A}$ be Drazin invertible and $b \in \mathcal{A}$ obey* Condition ($\mathcal{W}$) *at a, then b is Drazin invertible and*

$$\frac{\|b^D - a^D\|}{\|a^D\|} \leq \frac{\|a^D(b - a)\|}{1 - \|a^D(b - a)\|}.$$

Corollary 12.4.2 *Let $a \in \mathcal{A}$ be Drazin invertible and $b \in \mathcal{A}$ obey* Condition ($\mathcal{W}$) *at a, then b is Drazin invertible and*

$$\frac{\|a^D\|}{1 + \|a^D(b - a)\|} \leq \|b^D\| \leq \frac{\|a^D\|}{1 - \|a^D(b - a)\|}.$$

Corollary 12.4.3 *Let $a \in \mathcal{A}$ be Drazin invertible, $b \in \mathcal{A}$ obey Condition $(\mathcal{W})$ at a, and $\|a^D(b - a)\| \leq 1/2$, then b is Drazin invertible and a obeys Condition $(\mathcal{W})$ at b.*

Corollary 12.4.4 *Let $a \in \mathcal{A}$ be Drazin invertible, $b \in \mathcal{A}$ obey Condition $(\mathcal{W})$ at a, and $\|a^D\| \|b - a\| < 1$, then b is Drazin invertible and*

$$\frac{\|b^D - a^D\|}{\|a^D\|} < \frac{\kappa_D(a)\|b - a\|/\|a\|}{1 - \kappa_D(a)\|b - a\|/\|a\|},$$

where $\kappa_D(a) = \|a\| \|a^D\|$ is defined as the condition number with respect to the Drazin inverse.

This section is based on [16].

12.5 Weighted Drazin Inverse of an Operator

The operator Drazin inverse discussed in the previous sections can be generalized by introducing a weight operator, as in the matrix case. Cline and Greville [17] introduced the concept of the W-weighted Drazin inverse of a rectangular matrix. Qiao [18] proposed the concept of the W-weighted Drazin inverse of a bounded linear operator and proved its existence, uniqueness and gave some basic properties.

In this section, we first introduce the definition, basic properties, and representations of the weighted operator Drazin inverse, then we present computational methods and perturbation analysis.

Let X_1 and X_2 be Hilbert spaces and $W \in B(X_2, X_1)$ a weight operator, then the W-weighted Drazin inverse of a bounded linear operator A is defined as follows.

Definition 12.5.1 Let $A \in B(X_1, X_2)$ and $W \in B(X_2, X_1)$. If for some nonnegative integer k, there exists $S \in L(X_1, X_2)$ satisfying

$$(AW)^{k+1}SW = (AW)^k, \tag{12.5.1}$$

$$SWAWS = S, \tag{12.5.2}$$

$$AWS = SWA, \tag{12.5.3}$$

then S is called the W-weighted Drazin inverse of A and denoted by $S = A_{d,W}$.

By comparing with Definition 12.1.1, we can see that the regular operator Drazin inverse is a special case of the W-weighted operator Drazin inverse, where $X_1 = X_2 = X$, $A \in B(X)$, and $W = I$.

The following theorem shows the uniqueness of the W-weighted Drazin inverse. Its proof is more involved than the proof of Theorem 12.1.1 of the uniqueness of the regular operator Drazin inverse, as we have to deal with two spaces X_1 and X_2.

Theorem 12.5.1 *Let $A \in B(X_1, X_2)$. If for some $W \in B(X_2, X_1)$, there exists an $S \in L(X_1, X_2)$ satisfying the Eqs. (12.5.1)–(12.5.3), then it must be unique.*

Proof Let $S_1, S_2 \in L(X_1, X_2)$ satisfy the Eqs. (12.5.1)–(12.5.3) for k_1 and k_2, respectively. Set $k = \max\{k_1, k_2\}$, it follows from Definition 12.5.1 that

$$
\begin{aligned}
S_1 &= (AW)^2 S_1 (WS_1)^2 \\
&= \cdots \\
&= (AW)^k S_1 (WS_1)^k \\
&= (AW)^{k+1} S_2 W S_1 (WS_1)^k \\
&= S_2 (WA)^{k+1} W S_1 (WS_1)^k \\
&= S_2 (WA)^k W (AWS_1 WS_1)(WS_1)^{k-1} \\
&= S_2 (WA)^k W S_1 (WS_1)^{k-1} \\
&= \cdots \\
&= S_2 WAWS_1.
\end{aligned}
$$

Repeating the first part of the above deduction, we have

$$
S_2 = (AW)^{k+1} S_2 (WS_2)^{k+1}.
$$

Using $AWS_2 W = S_2 WAW$, we obtain $S_2 W = (S_2 W)^{k+2} (AW)^{k+1}$. Thus

$$
\begin{aligned}
S_1 &= S_2 WAWS_1 \\
&= (S_2 W)^{k+2} (AW)^{k+1} AWS_1 \\
&= (S_2 W)^{k+2} (AW)^{k+1} S_1 WA \\
&= (S_2 W)^{k+2} (AW)^k A \\
&= S_2 (WS_2)^{k+1} (WA)^{k+1} \\
&= (S_2 W)^{k+1} S_2 (WA)^{k+1} \\
&= (S_2 W)^k S_2 W S_2 WA(WA)^k \\
&= (S_2 W)^k S_2 WAWS_2 (WA)^k \\
&= (S_2 W)^k S_2 (WA)^k \\
&= \cdots \\
&= (S_2 W) S_2 (WA) \\
&= S_2 WAWS_2 \\
&= S_2,
\end{aligned}
$$

which shows the uniqueness. $\qquad\qquad\qquad\qquad\qquad\qquad\qquad\qquad\qquad\qquad\qquad\square$

Next, we will establish a relation between the W-weighted Drazin inverse $A_{d,W}$ and the regular Drazin inverse $(AW)^D$. First, we derive some properties of $(AW)^D$.

Theorem 12.5.2 *Let $A \in B(X_1, X_2)$ and $W \in B(X_2, X_1)$. If there exists the Drazin inverse of WA, then there exists the Drazin inverse of AW and*

$$(AW)^D = A((WA)^D)^2 W \tag{12.5.4}$$

and their indices have the relation:

$$\text{Ind}(AW) \le \text{Ind}(WA) + 1.$$

Proof Since $(WA)^D$ exists, supposing $\text{Ind}(WA) = k$, we have

$$(WA)^D (WA)^{k+1} = (WA)^k, \tag{12.5.5}$$
$$((WA)^D)^2 (WA) = (WA)^D, \tag{12.5.6}$$
$$(WA)^D (WA) = (WA)(WA)^D. \tag{12.5.7}$$

From the above Eqs. (12.5.5)–(12.5.7), it is easy to verify that

$$A((WA)^D)^2 W(AW)^{k+2} = (AW)^{k+1}, \tag{12.5.8}$$
$$A((WA)^D)^2 W(AW)A((WA)^D)^2 W = A((WA)^D)^2 W, \tag{12.5.9}$$
$$A((WA)^D)^2 W(AW) = (AW)A((WA)^D)^2 W. \tag{12.5.10}$$

It then follows from Definition 12.1.1 and Theorem 12.1.1 that

$$A((WA)^D)^2 W = (AW)^D.$$

From Corollary 12.1.1, $\text{Ind}(AW)$ is the smallest nonnegative integer satisfying (12.5.1)–(12.5.3). Thus from (12.5.8)–(12.5.10), we know

$$\text{Ind}(AW) \le k + 1 = \text{Ind}(WA) + 1,$$

which completes the proof.   □

Theorem 12.5.3 *Under the assumptions in Theorem 12.5.2, for any positive integer p, we have*

$$W((AW)^D)^p = ((WA)^D)^p W \tag{12.5.11}$$

and

$$A((WA)^D)^p = ((AW)^D)^p A. \tag{12.5.12}$$

Proof When $p = 1$, it follows from (12.5.4), (12.5.6) and (12.5.7) that

$$
\begin{aligned}
W(AW)^D &= (WA)((WA)^D)^2 W \\
&= ((WA)^D)^2 WAW \\
&= (WA)^D W.
\end{aligned}
$$

By the induction, (12.5.11) holds. Similarly, we can prove (12.5.12).   □

The following theorem shows an explicit expression of the W-weighted operator Drazin inverse $A_{d,W}$ in terms of the regular operator Drazin inverse $(WA)^D$. Thus, it proves the existence of the W-weighted operator Drazin inverse.

Theorem 12.5.4 *Let $A \in B(X_1, X_2)$, $W \in B(X_2, X_1)$, and* $\text{Ind}(AW) = k$, *then*

$$
\begin{aligned}
A_{d,W} &= A((WA)^D)^2 \qquad\qquad\qquad (12.5.13) \\
&= ((AW)^D)^2 A \in L(X_1, X_2).
\end{aligned}
$$

Furthermore, if $\mathcal{R}((AW)^k)$ is closed, then $A_{d,W} \in B(X_1, X_2)$.

Proof Since $\text{Ind}(AW) = k$, it follows from Theorem 12.1.3 that there exists $(AW)^D \in L(X_2)$. By Theorems 12.5.2 and 12.5.3, it is easy to verify that $A((WA)^D)^2$ satisfies

$$
\begin{aligned}
(AW)^{k+1} A((WA)^D)^2 W &= (AW)^k, \\
A((WA)^D)^2 WAW A((WA)^D)^2 &= A((WA)^D)^2, \\
AWA((WA)^D)^2 &= A((WA)^D)^2 WA.
\end{aligned}
$$

It follows from the above equations and Theorem 12.5.1 of the uniqueness of the W-weighted Drazin inverse that

$$
A_{d,W} = A((WA)^D)^2 \in L(X_1, X_2).
$$

If $\mathcal{R}((AW)^k)$ is closed, by Theorem 12.1.3, we have $(AW)^D \in B(X_2)$. Similarly, we can show

$$
(WA)^D = W((AW)^D)^2 A \in B(X_1).
$$

Thus

$$
A_{d,W} = A((WA)^D)^2 \in B(X_1, X_2).
$$

The other expression of $A_{d,W}$ in (12.5.13) can be obtained similarly. The proof is omitted here.   □

Theorem 12.5.4 gives an expression of $A_{d,W}$ in terms of $(WA)^D$. In the following theorem, $A_{d,W}$ is expressed in terms of A and W [19]. The theorem itself is analogous to Theorem 12.2.1, however, its proof is more involved than the proof of the regular case, because we have to deal with two spaces X_1 and X_2. So, we give the theorem and its proof.

Theorem 12.5.5 *Suppose that* $A \in B(X_1, X_2)$, $W \in B(X_2, X_1)$ *with* $k = \max\{\text{Ind}(AW),\ \text{Ind}(WA)\}$ *and* $\mathcal{R}((AW)^k)$ *is closed, then*

$$A_{d,W} = \tilde{A}^{-1} A(WA)^k (A(WA)^{2k+2})^* A(WA)^k,$$

where

$$\tilde{A} = (A(WA)^k (A(WA)^{2k+2})^* (AW)^{k+2})|_{\mathcal{R}(A(WA)^k)}$$

is the restriction of $A(WA)^k (A(WA)^{2k+2})^* (AW)^{k+2}$ *on* $\mathcal{R}(A(WA)^k)$.

Proof Setting $G = A(WA)^k$, we know that $\mathcal{R}(G)$ is a closed subspace of X_2. It is obvious that $\mathcal{R}(GWAWG) \subset \mathcal{R}(G)$. Since

$$\mathcal{R}(G) = \mathcal{R}(GWAWA_{d,W}) = \mathcal{R}(GWAWGG^\dagger A_{d,W})$$
$$\subset \mathcal{R}(GWAWG),$$

we have $\mathcal{R}(GWAWG) = \mathcal{R}(G)$. Similarly, we get $\mathcal{N}(GWAWG) = \mathcal{N}(G)$. Clearly, $\mathcal{R}(G^\dagger GWAWGG^\dagger) \subset \mathcal{R}(G^\dagger)$. Now,

$$\mathcal{R}(G^\dagger) = \mathcal{R}(G^\dagger G) = \mathcal{R}(G^\dagger GWAWG) = \mathcal{R}(G^\dagger GWAWGG^\dagger G)$$
$$\subset \mathcal{R}(G^\dagger GWAWGG^\dagger)$$

implies $\mathcal{R}(G^\dagger GWAWGG^\dagger) = \mathcal{R}(G^\dagger)$. Now, it follows from [20, p. 70] that

$$A_{d,W} = (WAW)^{(2)}_{\mathcal{R}(G),\mathcal{N}(G)} = (P_{\mathcal{N}(G)^\perp} WAW P_{\mathcal{R}(G)})^\dagger = (G^\dagger GWAWGG^\dagger)^\dagger.$$

Next, we prove that $A_{d,W} = G(GWAWG)^\dagger G$. Set $X = G(GWAWG)^\dagger G$. By direct computation, we have

$$G^\dagger GWAWGG^\dagger X = G^\dagger (GWAWG(GWAWG)^\dagger)G$$
$$= G^\dagger P_{\mathcal{R}(GWAWG)} G$$
$$= G^\dagger G,$$

and

$$XG^\dagger GWAWGG^\dagger = G((GWAWG)^\dagger GWAWG)G^\dagger$$
$$= GP_{\mathcal{R}((GWAWG)^*)}G^\dagger$$
$$= GG^\dagger,$$

that is

$$G^\dagger GWAWGG^\dagger X = (G^\dagger GWAWGG^\dagger X)^*$$

and

$$XG^\dagger GWAWGG^\dagger = (XG^\dagger GWAWGG^\dagger)^*.$$

On the other hand,

$$(G^\dagger GWAWGG^\dagger)X(G^\dagger GWAWGG^\dagger) =$$

and

$$X(G^\dagger GWAWGG^\dagger)X = G(GWAWG)^\dagger G = X.$$

Thus we arrive at $A_{d,W} = G(GWAWG)^\dagger G$.

Also, it is easy to prove

$$\mathcal{R}(A(WA)^k(A(WA)^{2k+2})^*A(WA)^k) = \mathcal{R}(A_{d,W})$$

and

$$\mathcal{N}(A(WA)^k(A(WA)^{2k+2})^*A(WA)^k) = \mathcal{N}(A_{d,W}).$$

The conclusion then follows from [21, Theorem 2.2].   □

Remark The above theorem is a generalization of a result in [22] in that the conditions $\mathcal{N}((AW)^k) \subset \mathcal{N}((AW)^{k*})$ and $\mathcal{R}((AW)^k) \subset \mathcal{R}((AW)^{k*})$ are removed.

12.5.1 Computational Methods

The following theorem says that if we have a sequence of real valued functions that converges to x^{-1} then we can represent $A_{d,W}$ as a limit of a sequence of operator functions.

Theorem 12.5.6 *Suppose that* $A \in B(X_1, X_2)$, $W \in B(X_2, X_1)$ *with* $k = \max\{\text{Ind} (AW), \text{Ind}(WA)\}$ *and* $\mathcal{R}((AW)^k)$ *is closed. Define*

$$\widetilde{A} = (A(WA)^k(A(WA)^{2k+2})^*(AW)^{k+2})|_{\mathcal{R}(A(WA)^k)}.$$

If Ω *is an open set such that* $\sigma(\widetilde{A}) \subset \Omega \subset (0, +\infty)$ *and* $\{S_n(x)\}$ *is a sequence of continuous real valued functions on* Ω *with* $\lim_{n\to\infty} S_n(x) = 1/x$ *uniformly on* $\sigma(\widetilde{A})$, *then*

$$A_{d,W} = \lim_{n\to\infty} S_n(\widetilde{A})A(WA)^k(A(WA)^{2k+2})^*A(WA)^k.$$

Furthermore, for any $\epsilon > 0$, there is an operator norm $\| \cdot \|_*$ on X_1 such that

$$\frac{\| S_n(\widetilde{A})A(WA)^k(A(WA)^{2k+2})^* A(WA)^k - A_{d,w} \|_*}{\| A_{d,w} \|_*} \tag{12.5.14}$$
$$\leq \max_{x \in \sigma(\widetilde{A})} |S_n(x)x - 1| + O(\epsilon).$$

Proof Following the proof of Theorem 12.5.5 and replacing $\widetilde{T}$ there with $\widetilde{A}$, we first can show that the spectrum of $\widetilde{A}$ is positive, then using [10, Theorem 10.27], we have

$$\lim_{n \to \infty} S_n(\widetilde{A}) = \widetilde{A}^{-1}$$

uniformly in $B(\mathcal{R}(A(WA)^k))$. It then follows from Theorem 12.5.5 that

$$\lim_{n \to \infty} S_n(\widetilde{A})A(WA)^k(A(WA)^{2k+2})^* A(WA)^k$$
$$= \widetilde{A}^{-1}A(WA)^k(A(WA)^{2k+2})^* A(WA)^k$$
$$= A_{d,w}.$$

To obtain the error bound (12.5.14), we note that

$$A(WA)^k(A(WA)^{2k+2})^* A(WA)^k = \widetilde{A}A_{d,w}.$$

Therefore,

$$S_n(\widetilde{A})A(WA)^k(A(WA)^{2k+2})^* A(WA)^k - A_{d,w} = (S_n(\widetilde{A})\widetilde{A} - I)A_{d,w}.$$

Also, for any $\epsilon > 0$, there is an operator norm $\| \cdot \|_*$ such that $\| \widetilde{A} \|_* \leq \rho(\widetilde{A}) + \epsilon$, see [11, p. 77]. Thus

$$\| S_n(\widetilde{A})A(WA)^k(A(WA)^{2k+2})^* A(WA)^k - A_{d,w} \|_*$$
$$\leq \| S_n(\widetilde{A})\widetilde{A} - I \|_* \| A_{d,w} \|_*$$
$$\leq (\max_{x \in \sigma(\widetilde{A})} |S_n(x)x - 1| + O(\epsilon)) \| A_{d,w} \|_*,$$

which completes the proof. □

Similarly, replacing $\widetilde{T}$ in Theorem 12.2.3 with $\widetilde{A}$, we can derive lower and upper bounds for $\lambda \in \sigma(\widetilde{A})$:

$$\| (A(WA)^{2k+2})^\dagger \|^{-2} \leq \lambda \leq \| A \|^2 \| AW \|^{4k+4}. \tag{12.5.15}$$

Now, by using various sequences $\{S_n(x)\}$ that converge to x^{-1}, we can obtain various methods for computing $A_{d,w}$.

Euler-Knopp Sequence:

$$S_n(x) = \alpha \sum_{j=0}^{n} (1 - \alpha x)^j.$$

For $\alpha \in (0, \ 2/(\|A\|^{-2} \|AW\|^{-(4k+4)}))$, we have

$$A_{d,W} = \alpha \sum_{n=0}^{\infty} (I - \alpha A(WA)^k (A(WA)^{2k+2})^* (AW)^{k+2})^n$$

$$\cdot A(WA)^k (A(WA)^{2k+2})^* A(WA)^k.$$

Setting

$$A_n = \alpha \sum_{j=0}^{n} (I - \alpha A(WA)^k (A(WA)^{2k+2})^* (AW)^{k+2})^j$$

$$\cdot A(WA)^k (A(WA)^{2k+2})^* A(WA)^k,$$

we have the following iterative procedure for computing the W-weighted Drazin inverse:

$$\begin{cases} A_0 = \alpha A(WA)^k (A(WA)^{2k+2})^* A(WA)^k, \\ A_{n+1} = (I - \alpha A(WA)^k (A(WA)^{2k+2})^* (AW)^{k+2}) A_n \\ \qquad + \alpha A(WA)^k (A(WA)^{2k+2})^* A(WA)^k. \end{cases}$$

For the error bound, we note that the sequence $\{S_n(x)\}$ satisfies

$$S_{n+1}(x)x - 1 = (1 - \alpha x)(S_n(x)x - 1).$$

Thus

$$|S_n(x)x - 1| = |1 - \alpha x|^n |S_0(x)x - 1| = |1 - \alpha x|^{n+1}.$$

If $x \in \sigma(\widetilde{A})$ and

$$0 < \alpha < \frac{2}{\|A\|^2 \|AW\|^{4k+4}},$$

then $|1 - \alpha x| \leq \beta < 1$, where

$$\beta = \max\{|1 - \alpha \|A\|^2 \|AW\|^{4k+4}|, \ |1 - \alpha \|(A(WA)^{2k+2})^\dagger\|^{-2}|\}. \qquad (12.5.16)$$

Therefore,

$$|S_n(x)x - 1| \leq \beta^{n+1} \to 0, \quad \text{as } n \to \infty.$$

It follows from the above limit and Theorem 12.5.6 that the error bound is

$$\frac{\|A_n - A_{d,W}\|_*}{\|A_{d,W}\|_*} \leq \beta^{n+1} + O(\epsilon).$$

Newton's Iteration:

$$\begin{cases} S_0(x) = \alpha, \\ S_{n+1}(x) = S_n(x)(2 - x S_n(x)), \end{cases}$$

for $\alpha > 0$.

Applying Theorem 12.5.6, we get

$$\lim_{n \to \infty} S_n(\widetilde{A}) A(WA)^k (A(WA)^{2k+2})^* A(WA)^k = A_{d,W}.$$

Setting

$$A_n = S_n(\widetilde{A}) A(WA)^k (A(WA)^{2k+2})^* A(WA)^k,$$

we have the following Newton's iterative procedure for computing the W-weighted Drazin inverse:

$$\begin{cases} A_0 = \alpha A(WA)^k (A(WA)^{2k+2})^* A(WA)^k, \\ A_{n+1} = A_n(2I - WAWA_n), \end{cases}$$

for $\alpha \in (0, \ 2/(\|A\|^2 \|AW\|^{4k+4}))$, and an error bound

$$\frac{\|A_n - A_{d,W}\|_*}{\|A_{d,W}\|_*} \leq \beta^{2^n} + O(\epsilon),$$

where β is given by (12.5.16).

One attractive feature of the Newton method is its quadratic rate of convergence in general.

Considering an alternative real valued function $S_t(x) = (t + x)^{-1}, t > 0$, for $x \in \sigma(\widetilde{A})$, we have another limit representation of the W-weighted Drazin inverse [23]. Let $k = \max\{\mathrm{Ind}(AW), \ \mathrm{Ind}(WA)\}$, then

$$A_{d,W} = \lim_{t \to 0^+} (tI + (AW)^{k+2})^{-1}(AW)^k A,$$

which can be rewritten as

$$A_{d,W} = \lim_{t \to 0^+} (tI + X(AW)^{k+2})^{-1} X A(WA)^k,$$

where

$$X = A(WA)^k (A(WA)^{2k+2})^*.$$

Furthermore, from

$$|x S_t(x) - 1| = \frac{t}{x + t}$$

$$\leq \frac{t}{\|(A(WA)^{2k+2})^\dagger\|^{-2} + t}$$

$$= \frac{\|(A(WA)^{2k+2})^\dagger\|^2 t}{1 + \|(A(WA)^{2k+2})^\dagger\|^2 t}$$

and Theorem 12.5.6, we have the following error bound.

$$\frac{\|(tI + X(AW)^{k+2})^{-1} XA(WA)^k - A_{d,W}\|_*}{\|A_{d,W}\|_*} \leq \frac{\|(A(WA)^{2k+2})^\dagger\|^2 t}{1 + \|(A(WA)^{2k+2})^\dagger\|^2 t} + O(\epsilon).$$

Analogous to the regular Drazin inverse case, we can apply Theorem 12.5.6 to polynomial interpolations of the function $f(x) = 1/x$ to derive iterative methods for computing $A_{d,W}$ and their corresponding asymptotic error bounds.

Newton's Polynomial Interpolation

Similar to Sect. 12.3.4, considering the Newton's polynomial interpolation

$$P_n(x) = \sum_{j=0}^{n} \frac{1}{j+1} \prod_{l=0}^{j-1} \left(1 - \frac{x}{l+1}\right)$$

of $f(x) = x^{-1}$ at $x = 1, 2, ..., n + 1$. It can be verified that

$$1 - x P_n(x) = \prod_{l=0}^{n} \left(1 - \frac{x}{l+1}\right).$$

Applying Theorem 12.5.6, we get

$$\lim_{n \to \infty} P_n(\widetilde{A}) A(WA)^k (A(WA)^{2k+2})^* A(WA)^k = A_{d,W},$$

where

$$\widetilde{A} = (A(WA)^k (A(WA)^{2k+2})^* (AW)^{k+2})|_{R(A(WA)^k)}.$$

Setting

$$A_n = P_n(\widetilde{A}) A(WA)^k (A(WA)^{2k+2})^* A(WA)^k,$$

we have the following iterative method for computing the W-weighted Drazin inverse $A_{d,W}$:

$$\begin{cases} A_0 = A(WA)^k(A(WA)^{2k+2})^*A(WA)^k, \\ A_{n+1} = P_{n+1}(\tilde{A})A(WA)^k(A(WA)^{2k+2})^*A(WA)^k \\ \qquad = A_n + \dfrac{A_0}{n+2}(I - WAWA_n), \end{cases}$$

To derive an asymptotic error bound for this method, note that for

$$x \in \sigma(\tilde{A}) \subset [\,\|(A(WA)^{2k+2})^\dagger\|^{-2}, \ \|A\|^2\,\|AW\|^{4k+4}\,]$$

and for $l \geq L$, where $L = [\|A\|^2\,\|AW\|^{4k+4}]$ is the integer closest to $\|A\|^2\,\|AW\|^{4k+4}$, we have

$$1 - \frac{x}{l+1} \leq \exp\left(-\frac{x}{l+1}\right) \quad \text{for all } x \in \sigma(\tilde{A}).$$

Therefore

$$\prod_{l=L}^{n}\left(1 - \frac{x}{l+1}\right) \leq \exp\left(-x\sum_{l=L}^{n}\frac{1}{l+1}\right), \quad n \geq L.$$

Also,

$$\sum_{l=L}^{n}\frac{1}{l+1} \geq \int_{L+1}^{n+2}\frac{dt}{t} = \ln(n+2) - \ln(L+1),$$

hence

$$\exp\left(-x\sum_{l=L}^{n}\frac{1}{l+1}\right) \leq (L+1)^x(n+2)^{-x}$$

$$= (1 + \|A\|^2\,\|AW\|^{4k+4})^x(n+2)^{-x}.$$

If we set the constant

$$C = \max_{x\in\sigma(\tilde{A})}(1 + \|A\|^2\,\|AW\|^{4k+4})^x\prod_{l=0}^{L-1}\left(1 - \frac{x}{l+1}\right),$$

where $L = [\|A\|^2\,\|AW\|^{4k+4}]$ is the integer closest to $\|A\|^2\,\|AW\|^{4k+4}$, then

$$|1 - xP_n(x)| \leq C(n+2)^{-x}.$$

Finally, it follows from Theorem 12.5.6 that

$$\frac{\|A_n - A_{d,W}\|_*}{\|A_{d,W}\|_*} \le C(n+2)^{-\|(A(WA)^{2k+2})^\dagger\|^{-2}} + O(\epsilon),$$

for sufficiently large n.

Hermite Interpolation

Similar to Sect. 12.3.5, consider the unique polynomial $q_n(x)$ of degree $2n + 1$ which satisfies

$$q_n(i) = \frac{1}{i} \quad \text{and} \quad q_n'(i) = -\frac{1}{i^2}, \quad i = 1, 2, \ldots, n+1.$$

The Hermite interpolation formula yields the representation

$$q_n(x) = \sum_{i=0}^{n} (2(i+1) - x) \prod_{l=1}^{i} \left(\frac{l-x}{l+1}\right)^2, \tag{12.5.17}$$

where, by convention, the product term equals 1 when $l = 0$.

Applying Theorem 12.5.6 to the polynomials $q_n(x)$ in (12.5.17), we have

$$A_{d,W} = \lim_{n\to\infty} q_n(\tilde{A}) A(WA)^k (A(WA)^{2k+2})^* A(WA)^k.$$

Setting

$$A_n = q_n(\tilde{A}) A(WA)^k (A(WA)^{2k+2})^* A(WA)^k,$$

we get the following iterative method for computing the W-weighted Drazin inverse $A_{d,W}$:

$$\begin{cases} A_0 = (2I - MWAW)M, \\ A_{n+1} = A_n + \dfrac{1}{n+2}\left(2I - \dfrac{1}{n+2}MWAW\right)M(I - WAWA_n), \end{cases}$$

where

$$M = A(WA)^k (A(WA)^{2k+2})^* A(WA)^k.$$

To derive an asymptotic error bound, for $l \ge L = [\|A\|^2 \|AW\|^{4k+4}]$ and

$$x \in \sigma(\tilde{A}) \subset [(A(WA)^{2k+2})^\dagger\|^{-2}, \|A\|^2 \|AW\|^{4k+4}],$$

we have

$$\prod_{l=L}^{n} \left(1 - \frac{x}{l+1}\right)^2 \le (1 + \|A\|^2 \|AW\|^{4k+4})^{2x}(n+2)^{-2x}.$$

Let the constant

$$d = \max_{x \in \sigma(\tilde{A})} (1 + \|A\|^2 \|AW\|^{4k+4})^{2x} \prod_{l=0}^{L-1} \left(1 - \frac{x}{l+1}\right)^2,$$

then

$$|1 - xq_n(x)| \leq d(n+2)^{-2x}.$$

By Theorem 12.5.6, we arrive at the error bound:

$$\frac{\|A_n - A_{d,W}\|_*}{\|A_{d,W}\|_*} \leq d(n+2)^{-2\|(A(WA)^{2k+2})^\dagger\|^{-2}} + O(\epsilon),$$

for sufficiently large n.

12.5.2 Perturbation Analysis

In this section, we study the perturbation of the W-weighted Drazin inverse of a bounded linear operator between Banach spaces. Specifically, suppose $B = A + E$ is a perturbed A, we investigate the error in $B_{d,W}$ in terms of the perturbation E.

Fix $W \in B(X_2, X_1)$. For $A, B \in B(X_1, X_2)$, we define the W-product of A and B by

$$A * B \equiv AWB.$$

Also, for $A \in B(X_1, X_2)$, we denote the W-product of A with itself m times by A^{*m}. For $A \in B(X_1, X_2)$, define

$$\|\|A\|\| \equiv \|A\| \|W\|,$$

then $(B(X_1, X_2), *, \|\| \cdot \|\|)$ is a Banach algebra. If W is a one-to-one map of X_2 to X_1, then $W^{-1} \in B(X_1, X_2)$ is the unit of this algebra. If the inverse of W does not exist, then we adjoin a unit to the algebra. In either case, we may assume that we are working in a unital algebra. Now suppose that $A, B \in B(X_1, X_2)$ satisfy:

(1) $(AW)^{k+1} BW = (AW)^k$,
(2) $BWAWB = B$,
(3) $AWB = BWA$.

Postmultiplying (1) with A and then using (3), we have

(1') $(A)^{*k+2} * B = (A)^{*k+1}$,
(2') $B * A * B = B$,
(3') $A * B = B * A$.

Conditions (2') and (3') are (2) and (3) written in terms of the W-product. Thus, A has a Drazin inverse in the algebra constructed above. In this case, we write $B = A^D$ or more precisely $B = A_{d,W}$.

According to the matrix case [17], the unique solution, if it exists, of (1), (2) and (3) is called the W-weighted Drazin inverse of A. In this case we say that A is W-Drazin invertible.

In the following proposition, we give several equivalent conditions for the existence of the W-Drazin inverse.

Proposition 12.5.1 *Let X_1 and X_2 be Banach spaces, $A \in B(X_1, X_2)$ and $W \in B(X_2, X_1)$, then the following five conditions are equivalent:*

(1) *A is W-Drazin invertible, that is the three equations:*

   (a) $(AW)^{k+1}XW = (AW)^k$, *for some nonnegative integer k,*
   (b) $XWAWX = X$,
   (c) $AWX = XWA$.

   have a common solution $X \in B(X_2, X_1)$;
(2) *AW is Drazin invertible;*
(3) *WA is Drazin invertible;*
(4) $\alpha(AW) = p < +\infty$, $\mathcal{R}((AW)^{p+k})$ *is closed for some $k \geq 1$, and $\delta(WA) < +\infty$, recalling that $\alpha(T)$ is the ascending index of T defined in Definition 12.1.2 and $\delta(T)$ is the descending index of T defined in Definition 12.1.3;*
(5) $\alpha(WA) = q < +\infty$, $\mathcal{R}((WA)^{q+l})$ *is closed for some $l \geq 1$, and $\delta(AW) < +\infty$.*

If any of the five conditions is satisfied, then the above three Eqs. (a)–(c) have a unique solution

$$X = A_{d,W} = A((WA)^D)^2 = ((AW)^D)^2 A.$$

Open problem. In connection with Proposition 12.5.1 and the characteristics of the Drazin inverse of a bounded operator on Banach space, it would be interesting to prove or disprove that $\alpha(AW) < +\infty$ and $\delta(WA) < +\infty$ imply that AW is Drazin invertible.

Recalling (12.4.1), $b \in \mathcal{A}$ is said to obey Condition $(\mathcal{W})$ at a if

$$b - a = aa^D(b - a)aa^D \quad \text{and} \quad \|a^D(b - a)\| < 1.$$

For convenience, we state the main perturbation result from [16, Theorem 2.1].

Lemma 12.5.1 *Let $a \in \mathcal{A}$ be Drazin invertible and $b \in \mathcal{A}$ obey Condition $(\mathcal{W})$ at a, then b is Drazin invertible,*

$$bb^D = aa^D, \quad b^D = (1 + a^D(b - a))^{-1}a^D = a^D(1 + (b - a)a^D)^{-1},$$

and

$$\text{ind}(a) = \text{ind}(b).$$

Now, let A, $B \in B(X_1, X_2)$ and $W \in B(X_2, X_1)$. Suppose that A is W-Drazin invertible and B satisfies Condition $(\mathcal{W})$ at A, that is,

$$B - A = A * A^D * (B - A) * A * A^D \text{ and } \||A^D * (B - A)|\| < 1,$$

which can be rewritten as

$$B - A = AWA_{d,w}W(B - A)WAWA_{d,w} \text{ and } \|A_{d,w}W(B - A)\|\|W\| < 1.$$

Set the perturbation $E = B - A$. Now, based on the above lemma, we have the following result

Theorem 12.5.7 *Let A, $B \in B(X_1, X_2)$, $W \in B(X_2, X_1)$, A be W-Drazin invertible and B obey Condition $(\mathcal{W})$ at A, then B is W-Drazin invertible and*

$$(BW)(B_{d,w}W) = (AW)(A_{d,w}W), \quad \text{Ind}(BW) = \text{Ind}(AW),$$

and $B_{d,w}$ can be given by

$$B_{d,w} = (I + A_{d,w}WEW)^{-1}A_{d,w} = A_{d,w}(I + WEWA_{d,w})^{-1} \quad (12.5.18)$$

and

$$\mathcal{R}(B_{d,w}) = \mathcal{R}(A_{d,w}) \quad \text{and} \quad \mathcal{N}(B_{d,w}) = \mathcal{N}(A_{d,w}). \quad (12.5.19)$$

Proof Note that (12.5.18) implies (12.5.19). Because B obeys Condition $(\mathcal{W})$ at A, we know that $I + A_{d,w}WEW$ and $I + WEWA_{d,w}$ are invertible. Now, from Lemma 12.5.1, we complete the proof of the theorem. □

The next corollary gives absolute and relative perturbation errors and lower and upper bounds for $\|B_{d,w}\|$.

Corollary 12.5.1 *Under the assumptions in Theorem 12.5.7, B is W-Drazin invertible and the absolute error*

$$B_{d,w} - A_{d,w} = -B_{d,w}WEWA_{d,w} = -A_{d,w}WEWB_{d,w},$$

the relative error

$$\frac{\|B_{d,w} - A_{d,w}\|}{\|A_{d,w}\|} \leq \frac{\|A_{d,w}WEW\|}{1 - \|A_{d,w}WEW\|},$$

and

$$\frac{\|A_{d,w}\|}{1 + \|A_{d,w}WEW\|} \leq \|B_{d,w}\| \leq \frac{\|A_{d,w}\|}{1 - \|A_{d,w}WEW\|}.$$

Next, we present a condition number that measures the sensitivity of $B_{d,w}$ to the perturbation E.

Corollary 12.5.2 *Under the assumptions in Theorem 12.5.7, we have*

(1) *if* $\|A_{d,w}\| \, \|WEW\| < 1$, *then B is W-Drazin invertible and*

$$\frac{\|B_{d,w} - A_{d,w}\|}{\|A_{d,w}\|} \leq \frac{\kappa_{d,w}(A)\|WEW\|/\|WAW\|}{1 - \kappa_{d,w}(A)\|WEW\|/\|WAW\|},$$

where $\kappa_{d,w}(A) = \|WAW\| \, \|A_{d,w}\|$ *is the condition number of the W-weighted Drazin inverse of A;*

(2) *if* $\|A_{d,w}WEW\| < 1/2$, *then B is W-Drazin invertible and A obeys Condition* $(\mathcal{W})$ *at B.*

The following result is motivated by the index splitting of a matrix [24].

Theorem 12.5.8 *Let* $A, U, V \in B(X_1, X_2)$ *and* $W \in B(X_2, X_1)$. *Suppose that* $A = U - V$ *is W-Drazin invertible,* UW *is Drazin invertible,* $\mathrm{Ind}(AW) = k_1$, $\mathrm{Ind}(WA) = k_2$, $\mathrm{Ind}(UW) = \mathrm{Ind}(WU) = 1$, $U_{d,w}WVW$ *is a compact operator,* $\mathcal{R}((UW)^D) = \mathcal{R}((AW)^{k_1})$ *and* $\mathcal{N}((WU)^D) = \mathcal{N}((WA)^{k_2})$, *then* $I - U_{d,w}WVW$ *is invertible and*

$$A_{d,w} = (I - U_{d,w}WVW)^{-1}U_{d,w} = U_{d,w}(I - WVWU_{d,w})^{-1}. \qquad (12.5.20)$$

Proof To prove that $I - U_{d,w}WVW$ is invertible it suffices to show that $\mathcal{N}(I - U_{d,w}WVW) = \{0\}$. Suppose that $\mathbf{x} \in \mathcal{N}(I - U_{d,w}WVW)$, which means that $U_{d,w}WVW\mathbf{x} = \mathbf{x}$. Since

$$\begin{aligned}
(U_{d,w}WVW)&(U_{d,w}WVW)\mathbf{x} \\
&= U_{d,w}W(U - A)WU_{d,w}WVW\mathbf{x} \\
&= (U_{d,w}WUWU_{d,w})WVW\mathbf{x} - U_{d,w}WAW(U_{d,w}WVW\mathbf{x}) \\
&= U_{d,w}WVW\mathbf{x} - U_{d,w}WAW\mathbf{x},
\end{aligned}$$

we have

$$\begin{aligned}
U_{d,w}WAW\mathbf{x} &= U_{d,w}WVW\mathbf{x} - (U_{d,w}WVW)(U_{d,w}WVW)\mathbf{x} \\
&= U_{d,w}WVW(I - U_{d,w}WVW)\mathbf{x} \\
&= \mathbf{0}.
\end{aligned}$$

Thus we obtain

$$WAW\mathbf{x} \in \mathcal{N}(U_{d,w}) = \mathcal{N}((WU)^D) = \mathcal{N}((WA)^{k_2}) = \mathcal{N}(A_{d,w}),$$

and $A_{d,w}WAW\mathbf{x} = \mathbf{0}$, thus $\mathbf{x} \in \mathcal{N}(A_{d,w}WAW) = \mathcal{N}((AW)^{k_1})$. However,

$$\mathbf{x} \in \mathcal{R}((AW)^{k_1}) \cap \mathcal{N}((AW)^{k_1}) = \{\mathbf{0}\}.$$

Hence, $\mathbf{x} = \mathbf{0}$ and $I - U_{d,w} W V W$ is invertible. Notice that

$$
\begin{aligned}
&(I - U_{d,w} W V W) A_{d,w} \\
&= (I - U_{d,w}(WUW - WAW)) A_{d,w} \\
&= A_{d,w} - U_{d,w} WUW A_{d,w} + U_{d,w} WAW A_{d,w} \\
&= U_{d,w}.
\end{aligned}
$$

Thus, we get (12.5.20).   □

When $W = I$ and A is a square matrix, Theorem 12.5.8 reduces to the results in [24].

Corollary 12.5.3 *Let $A = U - V \in \mathbb{C}^{n \times n}$ and $\mathrm{Ind}(A) = k$. Suppose that $\mathcal{R}(U) = \mathcal{R}(A^k)$ and $\mathcal{N}(U) = \mathcal{N}(A^k)$, then $I - U^D V$ is invertible and*

$$
A^D = (I - U^D V)^{-1} U^D = U^D (I - V U^D)^{-1}.
$$

This section is based on [21].

Remarks

The representation theorem of the Drazin inverse of a linear operator in Banach space is given in [2–4, 7, 25, 26] and more recent results on representations, properties, and characterizations of the Drazin inverse of a linear operator are given in [26–28]. The necessary and sufficient condition for the existence of the Drazin inverse of a linear operator in Banach space and the applications in infinite-dimensional linear systems can be found in [29] and [30], respectively. The Drazin inverse of an element of a Banach algebra is given in [31, 32]. Additional results for the generalized Drazin inverse are presented in [33–35] and weighted g-Drazin in [36]. Wang [22, 37, 38] first studied the iterative methods, the representations and approximations of the operator W-weighted Drazin inverse in Banach space. Perturbation analysis of the weighted Drazin inverse of a linear operator can be found in [16, 19, 21, 39–41].

References

1. A.E. Taylor, D.C. Lay, *Introduction to Functional Analysis*, 2nd edn. (Wiley, New York-Chichester-Brisbane, 1980)
2. S. Qiao, The Drazin inverse of a linear operator on Banach spaces. J. Shanghai Normal Univ. **10**, 11–18 (1981). in Chinese
3. J. Kuang, The representation and approximation for Drazin inverses of linear operators. Numer. Math., J. Chinese Univ., **4**, 97–106 (1982). in Chinese
4. Y. Wei, S. Qiao, The representation and approximation of the Drazin inverse of a linear operator in Hilbert space. Appl. Math. Comput. **138**, 77–89 (2003)
5. S.L. Campbell (ed.), *Recent Application of Generalized Inverses* (Pitman, London, 1982)
6. Y. Wei, A characterization and representation of the Drazin inverse. SIAM J. Matrix Anal. Appl. **17**, 744–747 (1996)

7. Y. Wei, Representation and perturbation of the Drazin inverse in Banach space. Chinese J. Contemp. Math. **21**, 39–46 (2000)

8. Y. Chen, Representation and approximation for the Drazin inverse $A^{(d)}$. Appl. Math. Comput. **119**, 147–160 (2001)

9. B.A. Barnes, Common operator properties of the linear operators RS and SR. Proc. Amer. Math. Soc. **126**, 1055–1061 (1998)

10. R. Walter, *Functional Analysis*, 2nd edn. (McGraw-Hill, New York, 1991)

11. F. Chatelin, *Spectral Approximation of Linear Operators* (Academic Press, New York, 1983)

12. C.D. Meyer, Limits and the index of a square matrix. SIAM J. Appl. Math. **26**, 469–478 (1974)

13. Y. Wei, On the perturbation of the group inverse and oblique projection. Appl. Math. Comput. **98**, 29–42 (1999)

14. Y. Wei, G. Wang, The perturbation theory for the Drazin inverse and its applications. Linear Algebra Appl. **258**, 179–186 (1997)

15. J.J. Koliha, A generalized Drazin inverse. Glasgow Math. J. **38**, 367–381 (1996)

16. V. Rakočević, Y. Wei, The perturbation theory for the Drazin inverse and its applications II. J. Austral. Math. Soc. **70**, 189–197 (2001)

17. R.E. Cline, T.N.E. Greville, A Drazin inverse for rectangular matrices. Linear Algebra Appl. **29**, 54–62 (1980)

18. S. Qiao, The weighted Drazin inverse of a linear operator on Banach spaces and its approximation. Numer. Math., J. Chinese Univ., **3**, 1–8 (1981). in Chinese

19. V. Rakočević, Y. Wei, The representation and approximation of the W-weighted Drazin inverse of linear operators in Hilbert spaces. Appl. Math. Comput. **141**, 455–470 (2003)

20. A. Ben-Israel, T.N.E. Greville, *Generalized Inverses: Theory and Applications*, 2nd edn. (Springer, New York, 2003)

21. V. Rakočević, Y. Wei, A weighted Drazin inverse and applications. Linear Algebra Appl. **350**, 25–39 (2002)

22. G. Wang, Iterative methods for computing the Drazin inverse and the W-weighted Drazin inverse of linear operators based on functional interpolation. Numer. Math., J. Chinese Univ., **11**, 269–280 (1989)

23. Y. Wei, A characterization and representation of the generalized inverse $A_{T,S}^{(2)}$ and its applications. Linear Algebra Appl. **280**, 87–96 (1998)

24. Y. Wei, Index splitting for Drazin inverse and the singular linear system. Appl. Math. Comput. **95**, 115–124 (1998)

25. D. Cai, The Drazin generalized inverse of linear operators. J. Math. (Wuhan) **1**, 81–88 (1985). in Chinese

26. H. Du, C. Deng, The representation and characterization of Drazin inverses of operators on a Hilbert space. Linear Algebra Appl. **407**, 117–124 (2005)

27. P.S. Stanimirović, V.N. Katsikis, H. Ma, Representations and properties of the W-weighted Drazin inverse. Linear Multilinear Algebra **65**(6), 1080–1096 (2017)

28. C. Deng, H. Du, The reduced minimum modulus of Drazin inverses of linear operators on Hilbert spaces. Proc. Amer. Math. Soc. **134**, 3309–3317 (2006)

29. S.R. Caradus, *Generalized Inverses and Operator Theory*, vol. 50, Queen's Papers in Pure and Applied Mathematics (Queen's University, Kingston, Ontario, 1978)

30. S.L. Campbell, The Drazin inverse of an operator, in *Recent Applications of Generalized Inverses*, ed. by S.L. Cambbell (Pitman, 1982)

31. D. Huang, Group inverses and Drazin inverses over Banach algebras. Integral Equ. Oper. Theory **17**, 54–67 (1993)

32. N.C. González, J.J. Koliha, Y. Wei, On integral representations of the Drazin inverse in Banach algebras. Proc. Edinburgh Math. Soc. **45**, 327–331 (2002)

33. D.S. Djordjević, Y. Wei, Additive results for the generalized Drazin inverse. J. Austral. Math. Soc. **73**, 115–125 (2002)

34. J.J. Koliha, V. Rakčević, Differentiability of the g-Drazin inverse. Studia Math. **168**(3), 193–201 (2005)

35. C. Deng, Y. Wei, New additive results for the generalized Drazin inverse. J. Math. Anal. Appl. **370**, 313–321 (2010)
36. A. Dajić, J.J. Koliha, The weighted g-Drazin inverse for operators. J. Aust. Math. Soc. **82**(2), 163–181 (2007)
37. G. Wang, Y. Wei, Iterative methods for computing the W-weighted Drazin inverses of bounded linear operators in Banach spaces. J. Shanghai Normal Univ. **28**, 1–7 (1999). in Chinese
38. G. Wang, Several approximate methods for the W-weighted Drazin inverse of bounded linear operator in Banach spaces. Numer. Math., J. Chinese Univ., **10**, 76–81 (1988). in Chinese
39. N. Castro González, J.J. Koliha, Y. Wei, Error bounds for perturbation of the Drazin inverse of closed operators with equal spectral projections. Appl. Anal. **81**(4), 915–928 (2002)
40. X. Wang, H. Ma, R. Nikolov, A note on the perturbation bounds of W-weighted Drazin inverse. Linear Multilinear Algebra **64**(10), 1960–1971 (2016)
41. X. Wang, H. Ma, M. Cvetković-Ilić, A note on the perturbation bounds of W-weighted Drazin inverse of linear operator in Banach space. Filomat **31**(2), 505–511 (2017)

Index